NEUROMETHODS

Series Editor
Wolfgang Walz
University of Saskatchewan
Saskatoon, SK, Canada

For further volumes:
http://www.springer.com/series/7657

Neuromethods publishes cutting-edge methods and protocols in all areas of neuroscience as well as translational neurological and mental research. Each volume in the series offers tested laboratory protocols, step-by-step methods for reproducible lab experiments and addresses methodological controversies and pitfalls in order to aid neuroscientists in experimentation. *Neuromethods* focuses on traditional and emerging topics with wide-ranging implications to brain function, such as electrophysiology, neuroimaging, behavioral analysis, genomics, neurodegeneration, translational research and clinical trials. *Neuromethods* provides investigators and trainees with highly useful compendiums of key strategies and approaches for successful research in animal and human brain function including translational "bench to bedside" approaches to mental and neurological diseases.

Multiphoton Microscopy

Edited by

Espen Hartveit

Department of Biomedicine, University of Bergen, Bergen, Norway

 Humana Press

Editor
Espen Hartveit
Department of Biomedicine
University of Bergen
Bergen, Norway

ISSN 0893-2336   ISSN 1940-6045   (electronic)
Neuromethods
ISBN 978-1-4939-9704-6   ISBN 978-1-4939-9702-2   (eBook)
https://doi.org/10.1007/978-1-4939-9702-2

This Humana imprint is published by the registered company Springer Science+Business Media, LLC, part of Springer Nature.
The registered company address is: 233 Spring Street, New York, NY 10013, U.S.A.

Preface to the Series

Experimental life sciences have two basic foundations: concepts and tools. The *Neuromethods* series focuses on the tools and techniques unique to the investigation of the nervous system and excitable cells. It will not, however, shortchange the concept side of things as care has been taken to integrate these tools within the context of the concepts and questions under investigation. In this way, the series is unique in that it not only collects protocols but also includes theoretical background information and critiques which led to the methods and their development. Thus it gives the reader a better understanding of the origin of the techniques and their potential future development. The *Neuromethods* publishing program strikes a balance between recent and exciting developments like those concerning new animal models of disease, imaging, in vivo methods, and more established techniques, including, for example, immunocytochemistry and electrophysiological technologies. New trainees in neurosciences still need a sound footing in these older methods in order to apply a critical approach to their results.

Under the guidance of its founders, Alan Boulton and Glen Baker, the *Neuromethods* series has been a success since its first volume published through Humana Press in 1985. The series continues to flourish through many changes over the years. It is now published under the umbrella of Springer Protocols. While methods involving brain research have changed a lot since the series started, the publishing environment and technology have changed even more radically. Neuromethods has the distinct layout and style of the Springer Protocols program, designed specifically for readability and ease of reference in a laboratory setting.

The careful application of methods is potentially the most important step in the process of scientific inquiry. In the past, new methodologies led the way in developing new disciplines in the biological and medical sciences. For example, physiology emerged out of anatomy in the nineteenth century by harnessing new methods based on the newly discovered phenomenon of electricity. Nowadays, the relationships between disciplines and methods are more complex. Methods are now widely shared between disciplines and research areas. New developments in electronic publishing make it possible for scientists that encounter new methods to quickly find sources of information electronically. The design of individual volumes and chapters in this series takes this new access technology into account. Springer Protocols makes it possible to download single protocols separately. In addition, Springer makes its print-on-demand technology available globally. A print copy can therefore be acquired quickly and for a competitive price anywhere in the world.

Saskatoon, SK, Canada   *Wolfgang Walz*

Preface

The ability to visualize live and intact neural tissue at the cellular and subcellular level has seen incredible advances over the past two decades. One of the core techniques driving this continued progression and development is multiphoton excitation (MPE) microscopy. It is remarkable that a clever application of a fundamental principle in physics has provided this powerful tool to visualize and interrogate complex neural tissue in 3D. For those of us who stubbornly insist on working with intact neural tissue, either a slice preparation in vitro or a whole animal in vivo, the ability to look deep into the tissue with detailed subcellular resolution is nothing less than magic. For a young scientist entering the field today, where MPE microscopy can be found at almost every neuroscience research center, it must be difficult to imagine the momentous transition into the MPE world of today when fluorescence imaging of live neural tissue with supreme optical sectioning and minimal phototoxicity was enabled.

During the early years, establishing a laboratory for MPE microscopy was very costly and demanded specialized technical skills (or an in-house physicist) to manage the necessary lasers. But this has changed, and the technique, although still on the expensive side for individual laboratories, has matured and become much more accessible. Nevertheless, even with a number of turnkey systems offered by commercial companies, many scientists still decide to build their own microscopes. For readers who would like to try this, the first chapter offers some invaluable advice.

My primary goal in editing this book was to assemble a collection of chapters covering different applications of MPE microscopy that would highlight techniques and experimental strategies that expert researchers have found to be useful and successful in their own work and laboratories. I have encouraged the authors to include as much detail as possible when it comes to practical and effective "tips and tricks" that hopefully will benefit both established researchers and newcomers with little or no previous experience in MPE microscopy. For some applications, this has resulted in chapters with a series of step-by-step procedures. For other applications, it has resulted in detailed discussions and explanations of creative solutions to difficult problems. In both cases, I hope the chapters will inspire, guide, and ease the path from novice to master. Essentially, I wanted to produce the book I would have liked to read myself, more than 10 years ago when I eagerly unpacked and assembled the "Movable Objective Microscope" (also known as the "Denk scope") from Sutter Instrument. Whereas the research and applications enabled by MPE microscopy are not limited to neuroscience, the list of important discoveries in this field is already long and continues to increase.

In this book, the reader will find chapters that together cover the fundamentals of MPE microscopy as applied to both in vitro and in vivo experimental systems. Scientific advances often involve combining different techniques, and several chapters provide information on how to combine MPE microscopy with targeted electrophysiological recording and with synthetic and genetically encoded sensors for signaling molecules, calcium, and transmembrane voltage. Pulsed laser light is useful not only for imaging but also for uncaging neuroactive substances, and this is covered in several chapters that illustrate the application of this technique for investigating neural signaling and plasticity. Additional chapters provide valuable information for using MPE microscopy to investigate cellular and large-scale neural morphology, signaling in astrocytes, and multi-site signaling with subcellular resolution in neural networks of behaving animals. Finally, no book on MPE microscopy is complete without a chapter detailing how this imaging technique was first used, and continues to be used, to study the retina, the body's own light sensor and image processor.

I would like to thank the series editor, Wolfgang Walz, for his patience, guidance, and encouragement during this project. Last, but not least, I would also like to take this opportunity to express my sincere gratitude to each of the authors for their efforts in providing their valuable contributions to this volume which I hope will enable many more to enjoy the benefits of MPE microscopy.

Bergen, Norway   *Espen Hartveit*

Contents

Contributors

APOORVA D. AJAY • *Department of Neurology and Neurotherapeutics, University of Texas Southwestern Medical Center, Dallas, TX, USA; Peter O'Donnell, Jr. Brain Institute, University of Texas Southwestern Medical Center, Dallas, TX, USA*

TOM BADEN • *Institute for Ophthalmic Research, University of Tübingen, Tübingen, Germany; School of Life Sciences, University of Sussex, Brighton, UK*

BART G. BORGHUIS • *University of Louisville School of Medicine, Department of Anatomical Sciences and Neurobiology, Louisville, KY, USA*

QIANG CHEN • *Committee on Computational Neuroscience, The University of Chicago, Chicago, IL, USA*

DAVID A. DIGREGORIO • *Laboratory of Dynamic Neuronal Imaging, Department of Neuroscience, Institut Pasteur, Paris, France; CNRS UMR 3571, Paris, France*

RUNE ENGER • *Oslo University Hospital, Department of Neurology, Oslo, Norway; GliaLab and Letten Centre, Division of Physiology, Department of Molecular Medicine, Institute of Basic Medical Sciences, University of Oslo, Oslo, Norway*

THOMAS EULER • *Institute for Ophthalmic Research, University of Tübingen, Tübingen, Germany; Werner Reichardt Centre for Integrative Neuroscience (CIN), University of Tübingen, Tübingen, Germany*

PAUL R. EVANS • *Max Planck Florida Institute for Neuroscience, Jupiter, FL, USA*

KATRIN FRANKE • *Institute for Ophthalmic Research, University of Tübingen, Tübingen, Germany; Bernstein Centre for Computational Neuroscience Tübingen, University of Tübingen, Tübingen, Germany*

MARK P. GOLDBERG • *Department of Neurology and Neurotherapeutics, University of Texas Southwestern Medical Center, Dallas, TX, USA; Peter O'Donnell, Jr. Brain Institute, University of Texas Southwestern Medical Center, Dallas, TX, USA*

ESPEN HARTVEIT • *Department of Biomedicine, University of Bergen, Bergen, Norway*

TRAVIS C. HILL • *Department of Neurosurgery, New York University, New York, NY, USA*

ANDREAS HOEHNE • *Laboratory of Dynamic Neuronal Imaging, Department of Neuroscience, Institut Pasteur, Paris, France; CNRS UMR 3571, Paris, France*

GERGELY KATONA • *3D Functional Network and Dendritic Imaging, Institute of Experimental Medicine, Hungarian Academy of Sciences, Budapest, Hungary; Two-Photon Measurement Technology Research Group, The Faculty of Information Technology, Pázmány Péter Catholic University, Budapest, Hungary*

BERND KUHN • *Okinawa Institute of Science and Technology Graduate University, Okinawa, Japan*

MATE MAROSI • *3D Functional Network and Dendritic Imaging, Institute of Experimental Medicine, Hungarian Academy of Sciences, Budapest, Hungary*

JULIAN P. MEEKS • *Department of Neurology and Neurotherapeutics, University of Texas Southwestern Medical Center, Dallas, TX, USA; Peter O'Donnell, Jr. Brain Institute, University of Texas Southwestern Medical Center, Dallas, TX, USA; Department of Neuroscience, University of Texas Southwestern Medical Center, Dallas, TX, USA*

ERLEND A. NAGELHUS • *Oslo University Hospital, Department of Neurology, Oslo, Norway; GliaLab and Letten Centre, Division of Physiology, Department of Molecular Medicine, Institute of Basic Medical Sciences, University of Oslo, Oslo, Norway*

WON CHAN OH • *Department of Pharmacology, University of Colorado Anschutz Medical Campus, Denver, CO, USA*

LUCY M. PALMER • *Florey Institute of Neuroscience and Mental Health, University of Melbourne, Parkville, VIC, Australia*

LAXMI K. PARAJULI • *Department of Cell Biology and Neuroscience, Juntendo University, Tokyo, Japan*

DENISE M. O. RAMIREZ • *Department of Neurology and Neurotherapeutics, University of Texas Southwestern Medical Center, Dallas, TX, USA; Peter O'Donnell, Jr. Brain Institute, University of Texas Southwestern Medical Center, Dallas, TX, USA*

NELSON REBOLA • *Laboratory of Dynamic Neuronal Imaging, Department of Neuroscience, Institut Pasteur, Paris, France; CNRS UMR 3571, Paris, France*

CHRISTOPHER J. ROOME • *Okinawa Institute of Science and Technology Graduate University, Okinawa, Japan*

BALÁZS RÓZSA • *3D Functional Network and Dendritic Imaging, Institute of Experimental Medicine, Hungarian Academy of Sciences, Budapest, Hungary; Two-Photon Measurement Technology Research Group, The Faculty of Information Technology, Pázmány Péter Catholic University, Budapest, Hungary*

SPENCER LAVERE SMITH • *University of California Santa Barbara, Santa Barbara, CA, USA*

ROLF SPRENGEL • *Research Group of the Max Planck Institute for Medical Research, Institute for Anatomy and Cell Biology, Heidelberg University, Heidelberg, Germany*

IVAR S. STEIN • *Center for Neuroscience, University of California, Davis, CA, USA*

GERGELY SZALAY • *3D Functional Network and Dendritic Imaging, Institute of Experimental Medicine, Hungarian Academy of Sciences, Budapest, Hungary*

WANNAN TANG • *Oslo University Hospital, Department of Neurology, Oslo, Norway; GliaLab and Letten Centre, Division of Physiology, Department of Molecular Medicine, Institute of Basic Medical Sciences, University of Oslo, Oslo, Norway*

ALEXANDRA TRAN-VAN-MINH • *Laboratory of Dynamic Neuronal Imaging, Department of Neuroscience, Institut Pasteur, Paris, France; CNRS UMR 3571, Paris, France*

MARGARET LIN VERUKI • *Department of Biomedicine, University of Bergen, Bergen, Norway*

WEI WEI • *Department of Neurobiology, The University of Chicago, Chicago, IL, USA*

LONG YAN • *Max Planck Florida Institute for Neuroscience, Jupiter, FL, USA*

RYOHEI YASUDA • *Max Planck Florida Institute for Neuroscience, Jupiter, FL, USA*

BAS-JAN ZANDT • *Department of Biomedicine, University of Bergen, Bergen, Norway*

KAREN ZITO • *Center for Neuroscience, University of California, Davis, CA, USA*

Chapter 1

Building a Two-Photon Microscope Is Easy

Spencer LaVere Smith

Abstract

Building a two-photon microscope is easy because most of the work is done by the laser itself. All the microscope needs to do is to focus the laser light to a point, move it across the preparation, and measure the fluorescence photons emitted. These jobs are done by an objective, a scan engine, and a detector, respectively. That's all there is to it.

Key words Two-photon, Multiphoton, Optical design, Laser scanning microscope, Imaging, Microscopy, Design

1 Introduction

1.1 Fluorescence

When we look around the world, our eyes are usually detecting photons that originated with a familiar light source (e.g., the sun or a light bulb) and bounced off and through objects eventually reaching our eyes. However, when we view fluorescence (e.g., fluorescent paint in a blacklight poster), the fluorescence photons that reach our eyes did not originate with the light source. Instead, the fluorescent molecules absorbed photons from the light source and then emitted different photons.

We can illustrate the process in a Jablonski diagram, which illustrates the energy levels of a molecule. In the case of green fluorescent protein [1], a blue photon is absorbed, and the absorbed energy sends the molecule into an excited state. A small fraction of the energy is dissipated through heat, and then the molecule relaxes back down to the ground state, emitting a green photon in the process (Fig. 1a). The entire process of absorption and emission takes a few nanoseconds, and most of that is on the emission side.

1.2 Two-Photon Absorption

In two-photon excitation, instead of a single high photon being absorbed, two lower-energy infrared photons are absorbed. This is disallowed in classical physics, because each absorption is a single event. If the photon has insufficient energy to get the molecule to

Espen Hartveit (ed.), *Multiphoton Microscopy*, Neuromethods, vol. 148,
https://doi.org/10.1007/978-1-4939-9702-2_1, © Springer Science+Business Media, LLC, part of Springer Nature 2019

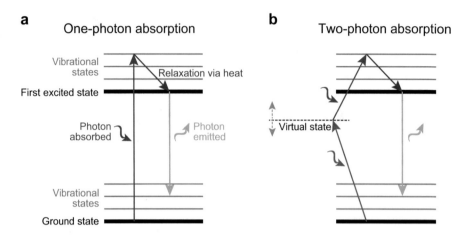

Fig. 1 Energy-level (Jablonski) diagram for conventional one-photon excitation and two-photon excitation. (**a**) In one-photon absorption, a high energy (short wavelength, toward the blue end of the visible spectrum) photon is absorbed by a fluorescent molecule which then has a change in electronic structure (is "excited"). The molecule moves from the ground state to the first excited state. The molecule relaxes, via emission of heat, to the lowest vibrational state of the first excited state. Then, a lower energy photon (toward the red end of the visible spectrum) is emitted as the molecule relaxes down to the ground state. (**b**) In two-photon absorption, two low-energy (near infrared) photons are absorbed nearly simultaneously. A virtual state is drawn on the diagram, but this is not a true energy state for the molecule. The energy level of the virtual state is undefined (unlike true energy states with precisely defined energies). Note that the two photons that are absorbed do not have to be of equal wavelength

the excited state, then the molecule simply does not make the transition. Multiple absorptions do not help in the classical case. For example, to jump a six-foot fence, a six-foot leap is required—two three-foot leaps are insufficient. However, in quantum mechanics, there is a level of uncertainty, formalized by Heisenberg's Uncertainty Principle [2]. This uncertainty permits two low-energy photons to excite a molecule, provided that they arrive at nearly the same place, at nearly the same time.

To illustrate two-photon absorption in a Jablonski diagram, scientists often draw a "virtual state" between the ground state and the first excited state, as if there is an intermediate energy level (Fig. 1b). This is useful for explaining the concept of two-photon excitation, but it is incorrect to consider the virtual state this way, because the virtual state is not a well-defined energy level. In fact, Maria Göppert-Mayer drew the virtual state *above* the first excited state in the Jablonski diagrams in her thesis [3]. The two photons absorbed do not have to be of the same wavelength. Nondegenerate (i.e., two photons of different wavelengths) two-photon excitation has been explored for microscopy applications [4].

2   Materials and Methods

2.1   The Laser

The laser is the workhorse of the system, and the first two-photon microscope [5] was developed following improvements in laser technology [6, 7]. Two-photon absorption is a highly unlikely event under bright sunlight (~ once every ten million years) [8]. For two-photon excitation, the incident light must have an intensity many orders of magnitude higher. There must be so many photons that the molecule absorbing them can't—quantum-mechanically speaking—tell where one photon ends and the next one begins [3]. Thus, extremely high light intensities are required to obtain sufficient rates of fluorescence for imaging. These high intensities can be generated in the brief pulses of light from ultra-fast lasers.

The likelihood of a two-photon excitation event for a given molecule is measured by its two-photon cross section, which is a property of the fluorescent molecule and the wavelengths of the exciting photons. The cross section is measured in units of cm^4s, which may seem like unusual units at first but are trivial to derive. Start with a more familiar, macroscopic cross-sectional area, which can be measured in cm^2. The two-photon cross section is the product of two of these conventional cross sections (each measured in cm^2) and the lifetime of the virtual state (measured in seconds). Typical one-photon cross sections for fluorescent molecules are approximately 10^{-17} cm^2 and virtual state lifetimes are approximately 10^{-15} s. Thus, two-photon cross sections are on the order of 10^{-49} cm^4s. For convenience, a special unit is used, and for respect, it is named after Maria Göppert-Mayer: 10 Göppert-Mayer units, or 10 GM = 10^{-49} cm^4s. Fluorescent dyes and proteins that are commonly used in two-photon neuroimaging have cross sections from about 10 to 300 GM.

Ultrafast lasers used for two-photon excitation emit incredibly intense pulses of light. Today, the most commonly used ultrafast laser used in two-photon imaging is the titanium-doped sapphire (Ti:Sapph) laser [9]. These systems generate average powers of about 1–10 watts. However, they emit few if any photons 99.999% of the time. All the energy they produce is squeezed into ~100 fs pulses, and these are emitted at the rate of ~80 million times per second. It is difficult to grasp how brief 100 fs is, so let's scale it up for comparison. If a 100 fs long pulse would be scaled up to 1 s, then there would be about a day and a half between adjacent one-second-long pulses. Thus, with these parameters, we have both high peak powers to ensure two-photon excitation events and also moderate average power at the sample (~0.01–0.1 watts) to minimize tissue heating and damage [10–12].

Ti:Sapph laser systems are convenient due to their tunability (~700–1000 nm), and these remain the most popular technology for two-photon microscopy. However, they are relatively complex and expensive, typically making up the bulk of the system cost. Three-photon imaging requires wavelengths (e.g., 1300 nm) out of the range of conventional Ti:Sapph systems, and thus optical parametric amplifiers are used [13]. Laser technology is a rapidly developing area, and new options are emerging, including fiber lasers [14].

The laser is the most critical component in a two-photon imaging system, and where our system diagram starts (Fig. 2). Next, a scan engine will move the beam around the specimen, and an objective will focus the beam to a small volume in the sample. Finally, a collection system will direct the emitted photons to a detector (usually a photomultiplier tube, PMT). Those are the parts we will discuss in the following sections. Note that most of these optics, in their basic configurations, are quite simple. Galileo Galilei would recognize many of these optical systems.

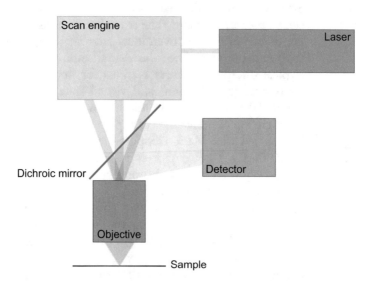

Fig. 2 The basic elements of a two-photon microscope. A block diagram of the key components of a two-photon laser scanning microscope shows that it is essentially simple. The laser beam is scanned such that its center is stationary on the back focal plane of the objective, but its angle of approach varies. The beam should also be expanded to use the full numerical aperture of the objective. Underfilling the objective back aperture will lead to a blurry excitation volume—which leads to both lower-resolution and less-efficient two-photon excitation (due to a lower concentration of photons). The objective performs two jobs: it focuses the excitation light into the sample, and it collects the emitted fluorescence photons. The emitted light is directed to a sensitive detector. The dichroic mirror separates the excitation light and emitted light. This is usually not perfect, and an additional infrared blocking filter in the detection pathway can help prevent excitation light from saturating the sensitive detector

2.2   The Scan Engine

The scan engine takes the laser beam and angles it into the objective so that it excites different regions of the sample. The beam width extends across the entire back aperture of the objective, and its center should not move. When the beam approaches the back aperture of the objective along the central optical axis, the center of the focal plane is excited. When the beam hits the back aperture at an angle, a lateral point in the focal plane is excited. Thus, the job of the scan engine is to rapidly alter the angle of approach of the beam (Fig. 2).

One might not even need a scan engine if the sample can be moved fast enough. Instead, one could leave the beam fixed in space and simply move the sample to image it. However, this is often not an option for biological samples, and even if it were, it is unlikely that the sample could be moved rapidly enough to provide fast, high-quality sampling that is free from movement artifacts. In particular, many in vivo preparations would not appreciate such jostling. Thus, we usually opt to scan the beam and leave the sample stationary.

A great deal of creativity and optimization can be exercised in the design of the scan engine, perhaps more so than any other component of a two-photon imaging system. Let us start with the basics. Usually, we are using an infinity-corrected objective, so we need to provide collimated light to the back focal plane of the objective and vary the angle of approach (the angle between the optical axes of the laser beam and the objective) to translate the excitation volume across the sample.

The most conventional way to do so is to use scanning mirrors. The deflection angle of a scan mirror can be relayed to the back aperture of the objective using a scan lens and tube lens in series [15]. Scan mirrors typically scan in one axis only (x or z), and thus two are placed in series to provide access to the full field of view of a system. Ideally, the scan mirrors should be optically relayed to one another using lenses or mirrors [16–18]. However, if only the central field of view is needed, the scan mirrors may be placed very close together, provided that the intermirror spacing is much smaller than the focal length of the scan lens [15].

Galvanometer-based scan mirrors provide relatively fast scanning and can respond to arbitrary command waveforms within their operating envelope, with smaller mirrors offering faster response times. Resonant scanners can operate about four to ten times faster than galvanometer-based scanners. However, that speed comes at a price. The scan amplitude can be adjusted, but that's it. The scan speed cannot be changed. Unfortunately, the sinusoidal scan pattern is fastest (in degrees per second, or microns scanned per second) in the middle of the field of view and slowest at the edges—exactly the opposite of what would be ideal for imaging [19]. Linear scanners are usually set to scan the field of view at a constant rate and then speed up at the edges to get ready for the next line. Linear scanners also use a variable scan offset to "pan" around the field of view. Resonant scanners cannot be used to "pan" in this way, the scanned region will always be centered in the same place, in the center of the resonant axis. To address this limitation, another linear scanner can be added (e.g., resonant x → linear x → linear y), and this offers great flexibility, at the cost of increased complexity.

It is also possible to add on rapid z-scanning abilities via tunable lenses [20], piezo-actuated objective positioners [21], or adaptive optics [22, 23]. Tunable lenses are the simplest (and least expensive) to implement, but they add aberrations that impair imaging when focusing far from the objective's natural imaging depth. Piezo-actuated objective positioners are more expensive, and involve considerable motion next to the sample, but can offer lower aberration imaging. Adaptive optics offer low aberration imaging, fast response times, and no objective motion. However, they are the most expensive and involved of the three options mentioned above. Rapid scanning can also be supported by spinning mirrors [24], acousto-optical deflectors [25–27], and ultrasound lenses [28]. Moreover, spatial light modulators and holographic techniques can be used to reposition the excitation as well [29–31]. Pushing the limits of scan engine designs often requires careful optical design to minimize optical aberrations [18, 32]. Indeed, scan engines can quickly become rather complex, and we have only touched on the variety here [33]. Moreover, the scan pattern itself can be optimized for a particular measurement. For example, instead of raster scanning an entire image plane, the beam could be directed along an arbitrary path to sample from the key regions of interest within the field of view. The desired scan pattern can influence the scan engine design in turn. Scan engine design is certainly a rich area for creative optical engineering.

2.3   The Objective

Two-photon excitation works best with high numerical aperture (NA) objectives for two reasons: the excitation photons are concentrated to a higher intensity, which increases the likelihood of getting excitation events (and resulting fluorescence photons), and high NA systems will collect a large fraction of emitted fluorescence photons.

What you just read isn't completely true in practice. When imaging the brain, the scattering of the tissue degrades the performance of high NA systems more severely than more moderate NA systems [34, 35]. This is partly because the marginal rays for high NA systems have so great a path length through the tissue. Along these greater path lengths, there are more opportunities for scattering and absorption events. Moreover, when imaging hundreds of microns into a sample (where the strength of two-photon excitation is key), tissue-induced aberrations reduce the effectiveness of high NA focusing [36]. Therefore, while higher NA is always better for signal collection, in practice, moderate NA (0.40–0.80) systems can be sufficient for two-photon imaging [18, 32, 37]. Moreover, moderate NA systems can offer ergonomic benefits including longer working distances, more flexibility on immersion media (e.g., air objectives, which often have lower NA than water-dipping objectives), and access angle to position instruments including electrodes within the field of view of the objective.

High NA objectives are sensitive to many parameters that are not appropriately constrained in many neurobiology experiments. For example, high NA objectives (>1.0 NA) are designed to be used at a specific temperature, and their performance can drop precipitously over just a few degrees Celsius from their design temperature. Also, high NA objectives are designed to be used with a particular type of coverslip, but some neurobiological preparations use no coverslip while others use stacks of three coverslips on top of each other (few objectives are designed for such use). Consult with the manufacturers to determine what imaging conditions their nominal performance is specified for. The good news is that high NA objectives can still be used in nonideal circumstances. Even if they are not offering their nominal optical performance, their residual performance can exceed that of moderate NA systems. Relatedly, objectives of any NA must be overfilled at their back aperture to use all of their excitation NA and offer the best possible resolution [38]. However, underfilling objectives, particularly high NA objectives, can still yield acceptable results.

The NA governs the resolution of the system, among other factors [39], and the required resolution varies by application. Sometimes relatively small excitation volumes (related to the point spread function, PSF, [40]) are required, while in other applications this requirement can be relaxed, sometimes deliberately so. In calcium imaging, several techniques have been used to engineer expanded PSF for specific applications. Extending the excitation volume axially [41–43] can provide faster scanning of a volume since fewer imaging planes are scanned, at the cost of reduced axial resolution. For sparsely labeled samples, that trade-off is often acceptable because there are fewer structures above and below the region of interest that yield contaminating signals. Axially extending the excitation volume can also help make imaging less sensitive to axial shifts which can occur during movement of the preparation or session-to-session misalignments. Small transverse drifts or misalignments can often be corrected offline because the data is there, just in a different pixel location. However, with high-resolution optical sectioning, axial shifts cannot be corrected for (unless multiple z planes are imaged at each time point). Extending the excitation volume laterally, with temporal focusing [44], can support faster scanning, as long as the trade-off in spatial resolution is tolerable.

For most applications, water dipping objectives are preferred. These systems can provide good optical performance and facilitate the simultaneous use of glass micropipettes for visualized patch-clamp recording. Air objectives can offer ergonomic benefits in some situations. When working with objectives positioned at large angles (off of vertical), or with preparations that are rapidly switched in and out (e.g., automated head-fixing systems [45, 46]), not needing immersion media can simplify the instrumentation significantly. A drawback to air objectives is the relatively large mismatch in index of refraction, n, between the glass of the objective and air, and between air and the physiological sample. This leads to spherical aberration and alters z measurements [47]. These aberrations can be minimized with moderate NA objectives (aberration effects are greater with higher NA systems) and compensated for with adaptive optics.

Objectives have several parameters that should be optimized for two-photon imaging. They should, of course, have appropriate antireflective (AR) coatings. The scan engine needs appropriate AR coatings as well, but only for the infrared excitation wavelengths. The objective must pass visible wavelengths well, with the infrared wavelengths a close second priority. Transmission rates over 80% are typical for objectives across the visible and infrared range. Objectives should also be relatively achromatic across the infrared wavelengths used to minimize pulse distortions [48], and these can be determined using standard raytracing software during design or modeling [49]. However, obtaining such specifications from manufacturers can be difficult if not impossible. The simplest approaches are to test a candidate objective on a working two-photon imaging system, or to use commercial objectives that have already been proven to work well for two-photon imaging. When testing optics, be sure to test on a specimen that is similar or identical to the specimens on which measurements will be made. Some samples, like pollen grains, can be very forgiving and provide excellent looking images when actual system performance is far from ideal.

Again, we are only scratching the surface of the parameter space that can be explored. Most two-photon microscopes use off-the-shelf components, especially the objective. In those cases, the rest of the system is built around those off-the-shelf components. An alternative approach is to start completely from scratch and design end-to-end custom optics, optimized for a specific application [17, 18]. This approach affords even greater creativity.

2.4   The Detector

Collection optics gather fluorescence photons from the objective and direct them to a detector. Theoretically, it can be difficult to design an optimal collection system, in part because scattered light cannot be focused, by definition. Large diameter collection optics can be helpful in collecting scattered light [50, 51]. Simple two-lens schemes function well in many setups [52], and in practice, the position tolerance of the collection is often not tight (i.e., translating individual lenses ± a few millimeters along the optical axis has minimal effects on signal collection in some systems).

Again here, there is room for great creativity in the design of these optics, including supplemental collection [53], oil immersion optics [17], and coupling to liquid light guides [54].

The detector should be several square millimeters in area, have a high bandwidth (~1 GHz), high quantum efficiency, and low noise. Multi-alkali PMTs, GaAsP PMTs, and hybrid photodetectors [55] offer this combination of features. These devices, which are all single pixel detectors, may seem like a slow way to image, but it's actually the ideal solution for multiphoton imaging deep in scattering tissue. Fluorescence photons should only be created within the excitation volume. Some of these photons will make it out of the tissue without being scattered, and these are called ballistic photons, and they not only reflect signal intensity, but also can be used to determine where in the tissue the fluorescence emission occurred (based on the location and angle at which the photon hits the detector). However, the single pixel detector discards the spatial information and only records the presence of the photon. The spatial information is redundant, because the scan engine already determines where the signal could come from. Other fluorescence photons will be scattered one or more times before exiting the tissue, but they can still make it to the detector. These photons can contribute to overall signal intensity, but the angle and location at which they hit the detector cannot be used to determine where in the tissue the fluorescence emission occurred. Again, this is fine because the scan engine is controlling the location of the excitation volume, so spatial information from the fluorescence photons is not needed. Instead, both ballistic and scattered photons contribute to signal intensity.

Area detectors, or cameras, can be used with two-photon imaging. However, because cameras have integration times (>1 ms) that are much longer than typical pixel dwell times in two-photon imaging (<1 µs), they will detect photons from multiple excitation locations during the same integration window. This can be acceptable when almost all of the photons are ballistic (not scattered), and it opens the door to multibeam scanning and other techniques to more rapidly scan excitation energy across the preparation, potentially improving the overall frame rate [56]. However, when there is significant scattering (e.g., ≥400 µm deep in mouse neocortex), the images become blurred. This is because cameras rely on ballistic photons to form a high-resolution image and scattered photons arrive at the "wrong" pixel locations and blur the image. Thus, the use of area detectors defeats one of the key advantages of two-photon imaging: resolving structures deep in scattering tissue. Systems that use a camera for detection can only be used when imaging at very shallow depths in tissue, similar to 1-photon approaches.

Digitization of signals from the detector generally falls into two categories: photon counting and analog integration. The former can be closer to optimal under some imaging regimes, including very dim samples. However, it requires high bandwidth signal processing, and some implementations sacrifice dynamic range. Analog integration is a more conventional approach and is sufficient for many experiments. Again, there are variations of these approaches including lock-in sampling (at a fixed delay from the pulses of the laser). The relative merits of these different approaches are the subject of much discussion [57] and offer yet more opportunities for creativity and interesting engineering trade-offs.

3   Methods

3.1 Design Constraints

The design of any imaging system is constrained by engineering trade-offs. A design is often trying to simultaneously maximize several competing parameters: frame rate, resolution, and field of view. Dynamics in neurobiology often play out at the subsecond time scale, with micrometer resolution, across millimeters to centimeters of neural circuitry. Systems cannot be designed to meet arbitrary specifications, and thus a good deal of creativity in design comes from deciding which compromises to make.

A major limiting factor for the design of a two-photon imaging system is the expected yield of detected fluorescence photons per laser pulse. This is a function of the properties of the fluorescent indicator, and the laser power the sample can tolerate. Brighter samples and higher laser power will increase the photons per pulse, of course. Too much laser power will cause damage to the preparation. These design factors, excitation power and signal photon levels, are sometimes called the "photon budget."

Generally, the detected yield per laser pulse of photons from fluorescence is less than one. In a study by Driscoll and colleagues, their imaging system detected about ten fluorescence photons per pixel, when using 383 laser pulses per pixel, and thus an average of ~0.03 photons per excitation laser pulse [57]. This is a typical value for two-photon imaging in neuroscience. This is true even for the intense pulses used in three-photon imaging, as values from a functional imaging study show [13]. There were ~100 pixels per cell body in that study, and ~100 photons per second per cell. Converting that to photons/pulse, we get: 100 photons/(8.49 frames/s × 100 pixels per cell × 3 pulses per pixel) = 0.04 photons per pulse. Thus, across a range of imaging system parameters, the signal photons per laser pulse remain typically <<1.

This is important to keep in mind during system design. For example, when using a 12 kHz resonant scanner, each line is scanned in 42 μs (two lines per cycle). In that amount of time, an 80 MHz laser pulses 3360 times. If the line is split into 512 pixels, then there are <10 pulses per pixel. Given the numbers above, there could be <1 photon per pixel even in bright regions.

Therefore, when scanning rapidly, temporal averaging is often necessary to clearly make out the image. There are a lot of pixel samples that are zero (or just noise).

Since two-photon imaging is a point-scanning technique, this low number of signal photons per laser pulse places a limit on the pixel acquisition rate (and by extension, the frame acquisition rate and/or frame size), given a minimal signal-to-noise specification. For example, given a particular resolution, as the field of view and/or depth (i.e., overall imaging volume) increases, the number of pixels per frame increases, and the time required to acquire a frame increases. Thus, it is often preferable to image portions of the full image and/or use multiple imaging beams to sample from multiple regions of an imaging volume [17, 18, 32, 58–62].

Originally, two-photon imaging systems used mostly conventional microscope parts from widefield and confocal microscopes. These parts often suffice in practice, even though they were not designed specifically for two-photon microscopy. In recent years, the market options for components for two-photon imaging systems have expanded considerably. Today, for most systems, commercial off-the-shelf (COTS) components will be sufficient to perform the measurements needed for a particular experiment. For advanced applications, custom optics can be explored when COTS components cannot suffice [17, 18, 32, 63]. Still, even custom optics are subject to the design constraints discussed above.

3.2 Software

Once the system is constructed, it must be connected up to a computer to synchronize beam scanning, signal digitization, image generation, live display, and other processes. Software for two-photon imaging typically grows in complexity as features are added, but the core essentials are simple. The computer needs to output scan mirror commands (e.g., sawtooth voltage commands, where the x frequency is N times the y frequency, where N is the number of scan lines per imaging frame), digitize the detector signals, and construct the data stream into images. There is a lot of optimization and additional features that can be added, but these essentials can be coded quite compactly. Sophisticated image processing is not necessary for the data acquisition stage, particularly in the case of linear galvo scanning. In the case of resonant scanning, the speed of the fast axis varies nonlinearly across the field of view, and so the data need to be processed to compensate for that aspect. The nonlinearity is predictable and relatively stationary, and thus the compensation not typically complex. Simple software can be developed in a day or two, sufficient for basic operation and testing. Subsequently, it typically grows into a labyrinthine suite as more functionality and features are added to support experiments.

4   Troubleshooting and Further Resources

4.1   Tips and Tricks

Here are some brief practical tips building a two-photon microscope. This is to supplement the resources mentioned above.

"Imaging" the Scan Commands

When first setting up a new system, particularly when writing or customizing the software, it can be useful to "image" the scan commands. That is, run the analog command voltages for the x and y mirrors into two acquisition channels. The channel corresponding to the x command should have a gradient from left to right, and the channel for the y command should have a gradient from top to bottom. This quick spot check lets the operator troubleshoot the instrument control and image generation routines without having to actually image anything.

Aligning a Two-Photon Imaging System

The individual lenses within a commercial objective lens are aligned by the manufacturer to high precision. The same goes for the individual lenses within a sophisticated scan lens. However, the relative spacing between the main components of a laser scanning microscope (e.g., scan lens, tube lens, objective) does not typically need to be highly precise—a difference of a few millimeters often makes little difference in performance. There is one exception: the distance between the scan mirrors and the scan lens is critical. This distance has a relatively strong influence on how stable the beam will be at the objective, and the ultimate imaging quality. This distance needs to be carefully adjusted, and fortunately it is relatively easy to do so using visual feedback. First, lay out all of the scan optics with their designed spacings. Activate the scanning in software, at the maximum scan angle. Next, view the beam at the back focal plane of the objective and adjust the scan mirror-to-scan lens distance to make the beam relatively stationary during scanning. In addition, shearing interferometers are useful tools for measuring changes in beam quality as the beam passes through elements. Another trick is to let the beam project to a distant target and examine the beam shape at different points during scanning. For example, removing a mirror can let the beam travel several meters after the scan lens and hit a paper target on a laboratory wall. As the scan coordinates change, the beam shape should remain circular. Any distortions of the spot shape during scanning could indicate aberrations and/or clipping.

Imaging Pollen Grains and Subresolution Beads

The first target for a new imaging system should be a uniform fluorescent sample, for example, a fluorescent piece of plastic or a tube of fluorescin. This provides an opportunity to fine tune the collection optics (maximizing the signal) and ensuring that excitation light is generally making it to the preparation, even if it the PSF is not yet optimized through fine alignment. After that step, it is time to focus on a sample with structure. Imaging a slide with pollen grains is handy, but remember that they are massively forgiving. This preparation is bright and sparse. It is easy to optimize a system using pollen grains and experience abject failure when imaging a living biological specimen. Use it only as a rough starting point. Then move to subresolution (smaller than the expected PSF) beads (preferably embedded in agar or some tissue phantom), and adjust the imaging system to obtain bright signals and a small PSF. Then move to your actual preparation and make further adjustments. Note that biological measurements are not always well served by having the smallest PSF possible. In some cases, slightly underfilling the objective and extending the z resolution can facilitate measurements, in other cases that can lead to problematic contamination from neuropil or other structures.

Tilt-Tip Optimization

One of the easier ways to dramatically degrade your imaging quality is to tilt the coverglass relative to the objective. This tilt causes optical aberrations that expand the PSF and result in lower 2P excitation efficiency and resolution. To deal with this, make sure that it is possible to make fine adjustments to the relative tip/tilt of the objective or the coverglass of the preparation. One solution is to mount the preparation on a tip-tilt stage. Note that while many microscopes offer tilting objectives, they typically only tilt around one axis, and thus are not sufficient to compensate for arbitrary tip-tilt.

4.2 Resources

For detailed and practical discussions of building two-photon microscopes, see work by Philbert Tsai and David Kleinfeld [15] and work by Jeff Squire's group [64]. In addition, Labrigger (labrigger.com) is an online resource with a series of technical notes and other information relevant to two-photon imaging.

References

1. Chalfie M, Tu Y, Euskirchen G, Ward WW, Prasher DC (1994) Green fluorescent protein as a marker for gene expression. Science 263(5148):802–805

2. Heisenberg W (1927) Über den anschaulichen Inhalt der quantentheoretischen Kinematik und Mechanik. Zeitschrift für Physik 43(3–4):172–198

3. Göppert-Mayer M (1931) Über Elementarakte mit zwei Quantensprüngen. Ann Phys (Leipzig) 9:273–294

4. Yang MH, Abashin M, Saisan PA, Tian P, Ferri CG, Devor A, Fainman Y (2016) Non-degenerate 2-photon excitation in scattering medium for fluorescence microscopy. Opt Express 24(26):30173–30187. https://doi.org/10.1364/OE.24.030173

5. Denk W, Strickler JH, Webb WW (1990) Two-photon laser scanning fluorescence microscopy. Science 248(4951):73–76

6. Masters BR, So PT (2004) Antecedents of two-photon excitation laser scanning micros-

copy. Microsc Res Tech 63(1):3–11. https://doi.org/10.1002/jemt.10418

7. So PT, Dong CY, Masters BR, Berland KM (2000) Two-photon excitation fluorescence microscopy. Annu Rev Biomed Eng 2:399–429. https://doi.org/10.1146/annurev.bioeng.2.1.399

8. Denk W, Svoboda K (1997) Photon upmanship: why multiphoton imaging is more than a gimmick. Neuron 18(3):351–357

9. Moulton PF (1986) Spectroscopic and laser characteristics of Ti:Al2O3. OSAB: Opt Phys 3:125–133

10. Podgorski K, Ranganathan G (2016) Brain heating induced by near-infrared lasers during multiphoton microscopy. J Neurophysiol 116(3):1012–1023. https://doi.org/10.1152/jn.00275.2016

11. Kalies S, Kuetemeyer K, Heisterkamp A (2011) Mechanisms of high-order photobleaching and its relationship to intracellular ablation. Biomed Opt Express 2(4):805–816. https://doi.org/10.1364/BOE.2.000816

12. Macias-Romero C, Zubkovs V, Wang S, Roke S (2016) Wide-field medium-repetition-rate multiphoton microscopy reduces photodamage of living cells. Biomed Opt Express 7(4):1458–1467. https://doi.org/10.1364/BOE.7.001458

13. Ouzounov DG, Wang T, Wang M, Feng DD, Horton NG, Cruz-Hernandez JC, Cheng YT, Reimer J, Tolias AS, Nishimura N, Xu C (2017) In vivo three-photon imaging of activity of GCaMP6-labeled neurons deep in intact mouse brain. Nat Methods 14(4):388–390. https://doi.org/10.1038/nmeth.4183

14. Fu W, Wright LG, Sidorenko P, Backus S, Wise FW (2018) Several new directions for ultrafast fiber lasers [Invited]. Opt Express 26(8):9432–9463. https://doi.org/10.1364/OE.26.009432

15. Tsai PS, Kleinfeld D (2009) In vivo two-photon laser scanning microscopy with concurrent plasma-mediated ablation. In: Frostig R (ed) Methods for in vivo optical imaging, vol 3. CRC Press, Boca Raton, pp 59–115

16. Sharafutdinova G, Holdsworth J, van Helden D (2010) Improved field scanner incorporating parabolic optics. Part 2: experimental verification and potential for volume scanning. Appl Opt 49(29):5517–5527. https://doi.org/10.1364/AO.49.005517

17. Sofroniew NJ, Flickinger D, King J, Svoboda K (2016) A large field of view two-photon mesoscope with subcellular resolution for in vivo imaging. Elife 5. https://doi.org/10.7554/eLife.14472

18. Stirman JN, Smith IT, Kudenov MW, Smith SL (2016) Wide field-of-view, multi-region, two-photon imaging of neuronal activity in the mammalian brain. Nat Biotechnol 34(8):857–862. https://doi.org/10.1038/nbt.3594

19. Tian X, Xu L, Li X, Shang G, Yao J (2010) Geometric distortion correction for sinusoidally scanned atomic force microscopic images. Paper presented at the IEEE International Conference on Imaging Systems and Techniques, Thessaloniki, 1–2 July 2010

20. Grewe BF, Voigt FF, van 't Hoff M, Helmchen F (2011) Fast two-layer two-photon imaging of neuronal cell populations using an electrically tunable lens. Biomed Opt Express 2(7):2035–2046. https://doi.org/10.1364/BOE.2.002035

21. Gobel W, Kampa BM, Helmchen F (2007) Imaging cellular network dynamics in three dimensions using fast 3D laser scanning. Nat Methods 4(1):73–79. https://doi.org/10.1038/nmeth989

22. Zurauskas M, Barnstedt O, Frade-Rodriguez M, Waddell S, Booth MJ (2017) Rapid adaptive remote focusing microscope for sensing of volumetric neural activity. Biomed Opt Express 8(10):4369–4379. https://doi.org/10.1364/BOE.8.004369

23. Shain WJ, Vickers NA, Goldberg BB, Bifano T, Mertz J (2017) Extended depth-of-field microscopy with a high-speed deformable mirror. Opt Lett 42(5):995–998. https://doi.org/10.1364/OL.42.000995

24. Kim KH, Buehler C, So PT (1999) High-speed, two-photon scanning microscope. Appl Opt 38(28):6004–6009

25. Reddy GD, Kelleher K, Fink R, Saggau P (2008) Three-dimensional random access multiphoton microscopy for functional imaging of neuronal activity. Nat Neurosci 11(6):713–720. https://doi.org/10.1038/nn.2116

26. Kirkby PA, Srinivas Nadella KM, Silver RA (2010) A compact Acousto-Optic Lens for 2D and 3D femtosecond based 2-photon microscopy. Opt Express 18(13):13721–13745. https://doi.org/10.1364/OE.18.013720

27. Katona G, Szalay G, Maak P, Kaszas A, Veress M, Hillier D, Chiovini B, Vizi ES, Roska B, Rozsa B (2012) Fast two-photon in vivo imaging with three-dimensional random-access scanning in large tissue volumes. Nat Methods 9(2):201–208. https://doi.org/10.1038/nmeth.1851

28. Kong L, Tang J, Little JP, Yu Y, Lammermann T, Lin CP, Germain RN, Cui M (2015) Continuous volumetric imaging via an optical phase-locked ultrasound lens. Nat Methods

12(8):759–762. https://doi.org/10.1038/nmeth.3476

29. Anselmi F, Ventalon C, Begue A, Ogden D, Emiliani V (2011) Three-dimensional imaging and photostimulation by remote-focusing and holographic light patterning. Proc Natl Acad Sci U S A 108(49):19504–19509. https://doi.org/10.1073/pnas.1109111108

30. Paluch-Siegler S, Mayblum T, Dana H, Brosh I, Gefen I, Shoham S (2015) All-optical bidirectional neural interfacing using hybrid multiphoton holographic optogenetic stimulation. Neurophotonics 2(3):031208. https://doi.org/10.1117/1.NPh.2.3.031208

31. Quirin S, Jackson J, Peterka DS, Yuste R (2014) Simultaneous imaging of neural activity in three dimensions. Front Neural Circuits 8:29. https://doi.org/10.3389/fncir.2014.00029

32. Tsai PS, Mateo C, Field JJ, Schaffer CB, Anderson ME, Kleinfeld D (2015) Ultra-large field-of-view two-photon microscopy. Opt Express 23(11):13833–13847. https://doi.org/10.1364/OE.23.013833

33. Ji N, Freeman J, Smith SL (2016) Technologies for imaging neural activity in large volumes. Nat Neurosci 19(9):1154–1164. https://doi.org/10.1038/nn.4358

34. Dunn AK, Wallace VP, Coleno M, Berns MW, Tromberg BJ (2000) Influence of optical properties on two-photon fluorescence imaging in turbid samples. Appl Opt 39(7):1194–1201

35. Tung CK, Sun Y, Lo W, Lin SJ, Jee SH, Dong CY (2004) Effects of objective numerical apertures on achievable imaging depths in multiphoton microscopy. Microsc Res Tech 65(6):308–314. https://doi.org/10.1002/jemt.20116

36. Schwertner M, Booth M, Wilson T (2004) Characterizing specimen induced aberrations for high NA adaptive optical microscopy. Opt Express 12(26):6540–6552

37. Ohki K, Chung S, Kara P, Hubener M, Bonhoeffer T, Reid RC (2006) Highly ordered arrangement of single neurons in orientation pinwheels. Nature 442(7105):925–928. https://doi.org/10.1038/nature05019

38. Helmchen F, Denk W (2005) Deep tissue two-photon microscopy. Nat Methods 2(12):932–940. https://doi.org/10.1038/nmeth818

39. Sheppard CJR, Castello M, Tortarolo G, Vicidomini G, Diaspro A (2017) Image formation in image scanning microscopy, including the case of two-photon excitation. J Opt Soc Am A Opt Image Sci Vis 34(8):1339–1350. https://doi.org/10.1364/JOSAA.34.001339

40. Zipfel WR, Williams RM, Webb WW (2003) Nonlinear magic: multiphoton microscopy in the biosciences. Nat Biotechnol 21(11):1369–1377. https://doi.org/10.1038/nbt899

41. Lu R, Sun W, Liang Y, Kerlin A, Bierfeld J, Seelig JD, Wilson DE, Scholl B, Mohar B, Tanimoto M, Koyama M, Fitzpatrick D, Orger MB, Ji N (2017) Video-rate volumetric functional imaging of the brain at synaptic resolution. Nat Neurosci 20(4):620–628. https://doi.org/10.1038/nn.4516

42. Theriault G, Cottet M, Castonguay A, McCarthy N, De Koninck Y (2014) Extended two-photon microscopy in live samples with Bessel beams: steadier focus, faster volume scans, and simpler stereoscopic imaging. Front Cell Neurosci 8:139. https://doi.org/10.3389/fncel.2014.00139

43. Dufour P, Piche M, De Koninck Y, McCarthy N (2006) Two-photon excitation fluorescence microscopy with a high depth of field using an axicon. Appl Opt 45(36):9246–9252

44. Prevedel R, Verhoef AJ, Pernia-Andrade AJ, Weisenburger S, Huang BS, Nobauer T, Fernandez A, Delcour JE, Golshani P, Baltuska A, Vaziri A (2016) Fast volumetric calcium imaging across multiple cortical layers using sculpted light. Nat Methods 13(12):1021–1028. https://doi.org/10.1038/nmeth.4040

45. Scott BB, Brody CD, Tank DW (2013) Cellular resolution functional imaging in behaving rats using voluntary head restraint. Neuron 80(2):371–384. https://doi.org/10.1016/j.neuron.2013.08.002

46. Murphy TH, Boyd JD, Bolanos F, Vanni MP, Silasi G, Haupt D, LeDue JM (2016) High-throughput automated home-cage mesoscopic functional imaging of mouse cortex. Nat Commun 7:11611. https://doi.org/10.1038/ncomms11611

47. Visser TD, Oud JL (1994) Volume measurements in three-dimensional microscopy. Scanning 16:198–200

48. Bor Z (1989) Distortion of femtosecond laser pulses in lenses. Opt Lett 14(2):119–121

49. Busing L, Bonhoff T, Gottmann J, Loosen P (2013) Deformation of ultra-short laser pulses by optical systems for laser scanners. Opt Express 21(21):24475–24482. https://doi.org/10.1364/OE.21.024475

50. Oheim M, Beaurepaire E, Chaigneau E, Mertz J, Charpak S (2001) Two-photon microscopy in brain tissue: parameters influencing the imaging depth. J Neurosci Meth 111(1):29–37

51. Singh A, McMullen JD, Doris EA, Zipfel WR (2015) Comparison of objective lenses for

multiphoton microscopy in turbid samples. Biomed Opt Express 6(8):3113–3127. https://doi.org/10.1364/BOE.6.003113

52. Zinter JP, Levene MJ (2011) Maximizing fluorescence collection efficiency in multiphoton microscopy. Opt Express 19(16):15348–15362. https://doi.org/10.1364/OE.19.015348

53. Engelbrecht CJ, Gobel W, Helmchen F (2009) Enhanced fluorescence signal in nonlinear microscopy through supplementary fiber-optic light collection. Opt Express 17(8): 6421–6435

54. Ducros M, van 't Hoff M, Evrard A, Seebacher C, Schmidt EM, Charpak S, Oheim M (2011) Efficient large core fiber-based detection for multi-channel two-photon fluorescence microscopy and spectral unmixing. J Neurosci Meth 198(2):172–180. https://doi.org/10.1016/j.jneumeth.2011.03.015

55. Michalet X, Cheng A, Antelman J, Suyama M, Arisaka K, Weiss S (2008) Hybrid photodetector for single-molecule spectroscopy and microscopy. Proc SPIE Int Soc Opt Eng 6862(68620F). https://doi.org/10.1117/12.763449

56. Fittinghoff D, Wiseman P, Squier J (2000) Widefield multiphoton and temporally decorrelated multifocal multiphoton microscopy. Opt Express 7(8):273–279

57. Driscoll JD, Shih AY, Iyengar S, Field JJ, White GA, Squier JA, Cauwenberghs G, Kleinfeld D (2011) Photon counting, censor corrections, and lifetime imaging for improved detection in two-photon microscopy. J Neurophysiol 105(6):3106–3113. https://doi.org/10.1152/jn.00649.2010

58. Amir W, Carriles R, Hoover EE, Planchon TA, Durfee CG, Squier JA (2007) Simultaneous imaging of multiple focal planes using a two-photon scanning microscope. Opt Lett 32(12): 1731–1733

59. Cheng A, Goncalves JT, Golshani P, Arisaka K, Portera-Cailliau C (2011) Simultaneous two-photon calcium imaging at different depths with spatiotemporal multiplexing. Nat Methods 8(2):139–142. https://doi.org/10.1038/nmeth.1552

60. Chen JL, Voigt FF, Javadzadeh M, Krueppel R, Helmchen F (2016) Long-range population dynamics of anatomically defined neocortical networks. Elife 5. https://doi.org/10.7554/eLife.14679

61. Lecoq J, Savall J, Vucinic D, Grewe BF, Kim H, Li JZ, Kitch LJ, Schnitzer MJ (2014) Visualizing mammalian brain area interactions by dual-axis two-photon calcium imaging. Nat Neurosci 17(12):1825–1829. https://doi.org/10.1038/nn.3867

62. Terada SI, Kobayashi K, Ohkura M, Nakai J, Matsuzaki M (2018) Super-wide-field two-photon imaging with a micro-optical device moving in post-objective space. Nat Commun 9(1):3550. https://doi.org/10.1038/s41467-018-06058-8

63. Negrean A, Mansvelder HD (2014) Optimal lens design and use in laser-scanning microscopy. Biomed Opt Express 5(5):1588–1609. https://doi.org/10.1364/BOE.5.001588

64. Young MD, Field JJ, Sheetz KE, Bartels RA, Squier J (2015) A pragmatic guide to multiphoton microscope design. Adv Opt Photonics 7(2):276–378. https://doi.org/10.1364/AOP.7.000276

Chapter 2

Using Multiphoton Imaging for Targeted Electrophysiological Recording and Live Cell Imaging of Fluorescently Labeled Neurons from Isolated Retinas

Qiang Chen and Wei Wei

Abstract

A central goal of neuroscience is to understand how neural computations are implemented by neural circuits. An excellent model system is the mammalian retina. Besides its important role in visual processing, the retina offers technical advantages for circuit interrogation at the cellular and synaptic levels due to its experimental accessibility and well-defined cell types. Recent development of genetic and molecular tools in mice has made the mouse retina a preferred choice for studying retinal circuitry, since an increasing repertoire of cell types can be specifically labeled by fluorescent proteins. However, measuring the light response of fluorescently tagged retinal neurons is challenging because excitation of fluorophores at visible wavelengths often leads to rapid photopigment bleaching that prevents subsequent recording of light responses from retinal neurons. One way to circumvent this problem is to use multiphoton excitation in the infrared range to visualize fluorescent protein-expressing cells. In this chapter, we describe a detailed protocol for multiphoton-targeted electrophysiological recording from fluorescently labeled retinal neurons while preserving their sensitivity to visual stimulation. This technique also enables live imaging of the three-dimensional morphology of the recorded neurons. With the continued development of cell-specific markers in the mouse retina, this method is expected to be widely used for harnessing the power of genetic and molecular tools in retinal circuit analysis.

Key words Multiphoton microscopy, Retina, Light response, Fluorescence proteins, Patch-clamp recording

1 Introduction

The retina, the neural tissue of the eye, belongs to the central nervous system. Visual processing in the retina is implemented by five major classes of retinal neurons that are organized into three cellular layers interconnected by two synaptic layers [1] (Fig. 1). In the vertical pathway, visual inputs are relayed and transformed by photoreceptor–bipolar cell–ganglion cell connections. Importantly, this forward signaling is profoundly modified by lateral connections made by horizontal and amacrine cells. The complexity of retinal

Espen Hartveit (ed.), *Multiphoton Microscopy*, Neuromethods, vol. 148,
https://doi.org/10.1007/978-1-4939-9702-2_2, © Springer Science+Business Media, LLC, part of Springer Nature 2019

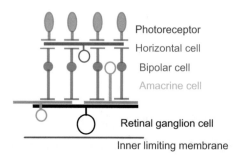

Fig. 1 Schematic side view of the major cell types in the mammalian retina

circuitry is reflected in its diverse cell types (over 100 distinct types according to the current estimate [1, 2]. Through precise wiring between specific neuronal types, visual inputs are processed in parallel by >30 retinal circuits [3]. Each circuit extracts a specific visual feature, which is represented as the spiking output of a retinal ganglion cell type [4]. Together, the axons of >30 types of retinal ganglion cells exit the retina at the optic disc and convey processed visual information to multiple brain targets such as thalamus, superior colliculus, accessory optic system, and suprachiasmatic nucleus [5]. One well-known example of feature extraction in the retina is implemented by the direction-selective circuit (reviewed in [6, 7]), which we use as an example to illustrate the method described in this chapter. The output neurons of the circuit, the direction-selective ganglion cells, fire action potentials maximally to motion in their preferred direction, but minimally to motion in the opposite, "null" direction. The direction-selective ganglion cells consist of multiple types that project their axons to distinct visual nuclei. The On direction-selective ganglion cells innervate the accessory optic system and mediate the optokinetic reflex [8–10], while the On–Off type mainly innervates the superior colliculus and the dorsal lateral geniculate nucleus and is involved in motion processing in these nuclei and the primary visual cortex [11–14]. The diverse set of retinal circuits that perform parallel visual processing makes the retina an intriguing place to study synaptic, cellular, and network level mechanisms of feature detection. Understanding retinal computations will have broad implications for the general principles of information processing by the brain.

Technically, the retina is a highly accessible system for experimental manipulations and recording. Since the retina receives minimal feedback inputs from the rest of the brain [15], the neural circuitry in the isolated retina is largely intact. When protected from visible light, acutely isolated retina survives and remains light-responsive for a sufficiently long period to enable electrophysiological experiments on visually evoked neuronal responses. For this purpose, retinal neurons are traditionally visualized using infrared optics and distinguished by morphological characteristics of their somas such as size, shape, and retinal location [16]. While a few cell types can be efficiently targeted by this approach, it poses a significant challenge for targeting the rest of the retinal cell types for focused and extensive circuit analysis. Recent progress in genetically engineered mouse lines and viral vectors offers an unparalleled opportunity to record from specific and sparse subpopulations of retinal neurons [1]. Selective labeling of one or several types of retinal neurons can be achieved by expressing a fluorescent protein directly under a cell type-specific promoter, or by using a binary system with "driver" and "reporter" genes such as the Cre-loxP system [17]. Genetic targeting leads to greatly improved efficiency of finding cells of interest. Importantly, stable expression of genetic markers offers the unique advantage of studying the early development of visual circuits even before the retina becomes light-responsive [7, 18].

While cell type-specific labeling greatly simplifies the initial targeting step, recording light-evoked responses from fluorescently labeled retinal neurons is subject to the "observer effect" in which the process of observation affects the observed results. In this case, the visible light source used to excite fluorophores also potently stimulates the photoreceptors in the retina, and therefore contaminates the neuron's response to the visual stimulus of interest. Even more problematic is that the visible light used to excite the fluorophores often bleaches the photoreceptors rapidly and prevents subsequent experiments using visual stimulation. For cells with bright fluorescence, this caveat can be alleviated by minimizing the intensity and duration of the excitation light exposure to the retina. However, this approach is not applicable to cells with low level of fluorescence.

Two-photon microscopy, thus, becomes a preferred choice for targeting retinal neurons labeled with green fluorescent protein (GFP) or other two-photon excitable fluorophores, because the infrared wavelengths used in two-photon excitation cause much weaker single-photon absorption for mammalian retinal photopigments. It is notable that the infrared laser used during two-photon imaging still causes activation of the photoreceptors, primarily due to two-photon excitation of the photopigments [19]. However, for most practical purposes, two-photon excitation of photoreceptors at typical laser intensities (~ 5 mW) does not cause a significant bleaching effect and therefore does not pose serious problems for subsequent recording experiments due to the following two factors. First, the imaging process for identifying a fluorescently labeled neuron is usually short in duration (e.g., 1–5 min). Second, the cells most suitable for electrophysiological recording are located in the retinal ganglion cell layer near the surface of the whole mount retina preparation. Due to the good optical sectioning of two-photon microscopy, two-photon excitation is spatially restricted to the illumination focal point at the ganglion cell layer, which is ~200 µm above the outer segments of the photoreceptors where the photopigments are located. Therefore, the activation and photobleaching of the photoreceptors is further minimized. In addition, the other general advantages of two-photon microscopy also apply to the isolated retina preparation, including good depth penetration and three-dimensional resolution, and minimized photodamage to the living tissue [20]. Together, two-photon targeting of fluorescently labeled retinal neurons allows for the acquisition of both electrophysiological recordings of a cell's light response and detailed live morphology of the cell's dendritic arbors.

In this chapter, we describe the equipment and procedure to perform two-photon targeted recording and imaging of fluorescent protein-expressing retinal neurons in whole-mount mouse retinas. The retina is first isolated under infrared illumination, and then transferred to a two-photon microscope for fluorescence imaging. The identified cells of interest are then subject to patch-clamp recording aided by infrared optics and an IR-sensitive camera while visual stimuli are presented to the retina through the microscope condenser. Inclusion of a dye in the internal solution during patch-clamp recording allows for subsequent live imaging of the recorded cell in three dimensions.

2   Materials

2.1   Reagents

1. Transgenic mice expressing two-photon excitable fluorophores in retinal neurons (all procedures need to be performed in accordance with ethical and safety guidance of relevant institutions and authorities). When retinal orientation is important, for example, during investigation of direction-selective or orientation-selective circuits, or when the fluorescence labeling pattern is nonuniform across the retina, a transgenic mouse line from a pigmented background such as C57/BL6 is recommended, because they have more distinguishable landmarks on the choroid than albino strains.

2. Ames' medium (pH 7.36; e.g., Sigma, Cat. No. A1420-10X1L).

3. Intracellular solution for patch-clamp recording, prepared according to requirements of experiments being performed (e.g., cesium-based internal solution containing: 110 mM $CsMeSO_4$, 2.8 mM NaCl, 4 mM EGTA, 5 mM TEA-Cl, 4 mM adenosine 5′-triphosphate (magnesium salt), 0.3 mM guanosine 5′-triphosphate (trisodium salt), 20 mM HEPES, 10 mM phosphocreatine (disodium salt), and 5 mM N-ethyllidocaine chloride (QX-314 chloride, to block voltage-gated Na^+ channels), pH 7.25). Fluorescent dyes like Alexa

Fluor 594 or 488 hydrazide (ThermoFisher Scientific) can be added to the internal solution to fill the recorded cells for subsequent two-photon imaging.

4. Isoflurane (e.g., Phoenix Pharmaceuticals, Cat. No. NDC 57319-507-05) for anesthesia.

2.2 Equipment

2.2.1 Retina Dissection

1. Black mixed cellulose ester membrane filter paper (0.45 μm; e.g., Millipore, Cat. No. HABG01300) for mounting the isolated retina.

2. White filter paper (e.g., Whatman Cat. No. 1001090) for holding the eyeballs in step 3 of section 3.1.

3. Surgical razor blade (e.g., Feather Safety Razor Co., Cat. No. GRF-2976#11).

4. Pyrex Petri dish (100 mm × 15 mm; e.g., Fisher Scientific, Cat. No. 08-747C).

5. Dissection tools (fine dissection scissors and forceps) (e.g., Roboz Surgical Instrument Inc.).

6. Dissection microscope (e.g., Olympus SZ61).

7. Infrared light source for the dissection microscope (e.g., CMVision IR30 illuminator).

8. Red LED headlamp for ambient room illumination and during dissection (e.g., Energizer, Model No. HD33AIEN).

9. Two infrared (IR)-sensitive CCD cameras (e.g., Watec, Model No. WAT-902H) for visualization under infrared optics during retina dissection and patch-clamp recording, respectively.

10. Two video monitors (e.g., Sanyo, Model No. DP1B41B) for visualization under infrared optics during retina dissection and patch-clamp recording, respectively.

2.2.2 Visual Stimulus

1. White organic light-emitting display (OLED; Emagin Corporation, model no. 100100-01 with glass faceplate) for presenting visual stimuli to the retina.

2. The "Design Reference Kit" (Emagin Corporation) for connecting the OLED to a VGA port of a PC for visual stimulation.

3. Custom-made OLED holder for use with Thorlabs 30 mm cage systems.

4. Cage plate (Thorlabs, Cat. No. CP02).

5. XY translating lens mount (Thorlabs, Cat. No. HPT1).

6. Right-angle kinematic mount for elliptical mirrors (Thorlabs, KCB2EC).

7. Protected silver-coated elliptical mirrors (Thorlabs PFE20-P01).

8. Cage assembly rods 8″ (Thorlabs, Cat. No. ER8).

9. Manual rotation base (Thorlabs, QRP02).

10. Dual-port high-performance video card for the OLED (Nvidia GeForce, 9500 GT 512 MB).

11. Computer monitor for viewing the visual stimuli on the OLED (e.g., Dell 1704FPT). Both this computer monitor and the OLED are connected to a PC with a dual-port video card. This allows the user to conveniently see the pattern of the visual stimulus shown on the OLED during experiments.

2.2.3 Electrophysiology

1. Amplifier with headstage(s) (e.g., Molecular Devices, model no. Multiclamp 700B).

2. Analog-to-digital converter (e.g., Molecular Devices, Digidata 1440A).

3. Micromanipulator (e.g., Sutter Instrument, MPC-200).

4. Borosilicate glass capillaries (1.5 mm outer and 1.10 mm inner diameters, 7.5 cm length, e.g., Sutter Instrument, Cat. No. BF150-110-7.5).

5. Glass microelectrode puller (e.g., Narishige PC-100).

6. Custom-made grounding wires for the headstages.

2.2.4 Two-Photon Microscopy

1. Two-photon microscope with dual-channel external multi-alkali detectors and a translatable stage. We have verified this protocol with multiphoton systems from Bruker (previously "Prairie") and Scientifica.

2. Femtosecond, tunable IR laser source (e.g., 730–1080 nm, Coherent Chameleon Ultra II or Spectra-Physics Mai Tai).

3. Water-immersion objectives (Olympus LUMPlan Fl/IR 60×/0.90NA Water Objective).

4. 5× objectives for bright-field observation (Olympus MPlan N 5×/0.10NA Microscope Objective).

5. Bandpass filter (optional, e.g., Chroma ET470/40x), for placing in front of the OLED for simultaneous visual stimulation and functional calcium imaging.

2.2.5 Other Equipment

1. Recording chamber (e.g., Warner Instruments, RC-26GLP).

2. Perfusion pump (e.g., Watson Marlow 120S, Cat. No. 14-284-202).

3. In-line solution heater (e.g., Warner, Model No. 64-0102) for warming solutions flowing into recording chamber. Check the temperature of the bath with the thermistor probe regularly to ensure that the bath temperature is stable.

4. Computers for two-photon imaging, electrophysiological recording, and visual stimulation.

5. Stand-alone breakout board for a parallel port cable (Winford Engineering, BRK25F-R-FT) for sending computer-generated TTL trigger signals to the patch-clamp device.

6. MATLAB software (Mathworks) with Psychophysics Toolbox installed for generating visual stimuli.

2.3 Experimental Setup

Schematics of overall equipment layout are shown in Fig. 2.

1. Visual stimulation

The OLED is secured by a custom-made plastic holder that fits into the Thorlabs 30 mm cage system that can be swung in under the condenser for visual stimulation or out to allow infrared light illumination of the tissue for patch-clamp recording (Fig. 2d). Images from the OLED are reflected by a silver mirror below the condenser mounted on the 30 mm cage system and projected and focused onto the photoreceptors through the condenser lens. In our setup, the area of retina stimulated by the OLED is 330 µm in diameter. The size of the stimulated area can be adjusted by adjusting the light path distance between OLED and condenser. Custom visual stimuli are generated using MATLAB and the Psychophysics Toolbox (http://psychtoolbox.org) [21].

There are alternative designs in which visual stimulation is delivered through the objective [19, 22]. In the current protocol, visual stimulation is delivered through the condenser since this configuration requires no custom modification for most commercial upright microscopes. The only requirement is that the distance between the bottom of the condenser and the transmitted light source is sufficiently large to accommodate the Thorlabs 30 mm cage system. We have successfully incorporated this 30 mm cage system in the commonly used Olympus BX51WI and Scientifica SliceScope. If this distance in an existing microscope is not large enough (e.g., ~10 cm), the Thorlabs 16 mm cage system can be used, or the microscope stage can be raised.

2. Synchronize visual stimuli with electrophysiological recording

The onset of the visual stimulus is accompanied by a transistor–transistor logic (TTL) pulse. Details on generating TTL pulse triggers can be found at the MATLAB Psychophysics Toolbox website (https://github.com/Psychtoolbox-3/Psychtoolbox-3/wiki/FAQ:-TTL-Triggers-in-Windows). Briefly, a TTL pulse is generated by the Psychophysics Toolbox in MATLAB in the visual stimulation computer, which is then sent to the patch-clamp devices through a parallel cable and a

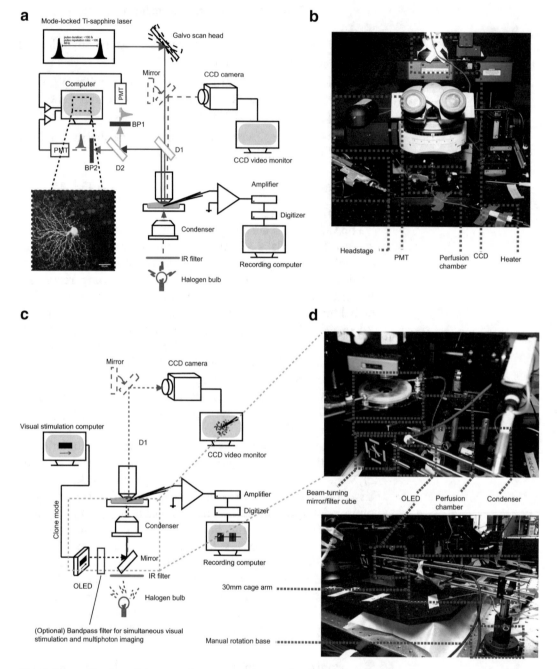

Fig. 2 Schematic view of the setup for targeted electrophysiological recording of light responses from retinal neurons expressing fluorescent proteins. (**a**) Schematic diagram of the two-photon microscope configured for targeted recording. Once a fluorescently labeled neuron is identified, the cell is targeted for recording aided by transmitted infrared (IR) illumination (cyan) and an IR-sensitive CCD camera. D1, dichroic filter (e.g., 695 nm split); D2, dichroic filter (e.g., 585 nm split); BP1, band pass filter (e.g., 500–550 nm); BP2, band pass filter (e.g., 600–660 nm). The detailed specifications of the optical filters depend on the choice of the fluorophores and the vendor of the multiphoton systems. (**b**) Example layout based on a Bruker Ultima multiphoton system. (**c**) Schematic diagram of the two-photon microscope configured for visual stimulation. Once a successful recording has been established, visual stimuli from the OLED are delivered to the retina through the condenser. (**d**) Example layout of a visual stimulation module built around the Thorlabs 30 mm cage system. (Modified from Wei et al. (2010) with permission from Springer Nature [22])

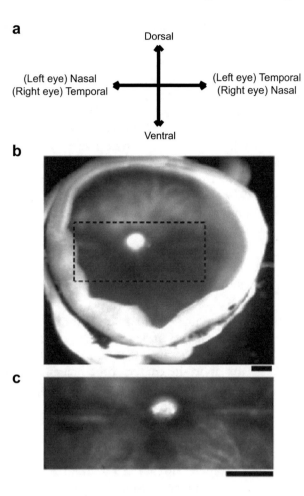

Fig. 3 The landmarks in the choroid for marking the orientation of the retina. Under a dissection microscope, a retina attached to the pigment epithelium is positioned with the ganglion cell layer facing up and dorsal side facing upward (**a, b**). The nasal-temporal axis is aligned with a horizontal stripe running beneath the optic nerve, with a dark-appearing patch in the ventral side (**c**). The nasal/temporal side is opposite for the left and right eyes (**a**). Scale bars, 250 μm (**b, c**). (Modified from Wei et al. (2010) with permission from Springer Nature [22])

breakout board to trigger electrophysiological recording (Fig. 3).

3. Alignment of the OLED to the image-forming center of the objective

First, the images on the OLED need to be centered to the field of view through the objective. To do this, a whole-mount retina sample is placed into the recording chamber perfused with oxygenated Ames' solution. Then, focus the objective on the outer segment of the photoreceptors. Next, move the microscope stage in the x-y plane so that the retina preparation is outside the field of view and an area containing unobstructed coverslip bottom of the imaging chamber is under the objective. Swing in the visual stimulus arm below the condenser and project a centered circle and crosshair pattern on the OLED. Without moving the objective in the z-axis, adjust the condenser focus knob below the microscope stage until the crosshair pattern is focused under the objective. Next, without adjusting the condenser focusing screw, move the adjusting screws of the silver mirror mount in the visual stimulation arm so that the center of the crosshair pattern in the OLED coincides with the crosshair center of the eyepiece (imaging-forming center of the objective). It is recommended that the alignment and focus of the visual stimulation is checked daily before experiments.

4. Create a reference point to align the field of view under laser scanning with that under transmitted light illumination.

5. Fill a glass electrode with a two-photon excitable dye such as Alexa Fluor 488 or Alexa Fluor 594. Focus the tip of the dye-filled pipette under the 60x objective. First, image the dye-filled pipette tip with the two-photon microscope, and position the tip in the center of the imaging window. Mark the center position on the monitor of the imaging computer with a piece of tape. Gentle positive pressure should be applied to the pipette so that the tip is clearly visible during imaging due to the constant flow of the dye out of the pipette. Next, stop laser scanning, and without moving the pipette position, visualize the pipette tip on the TV monitor using the IR-sensitive CCD camera under transmitted light illumination. Mark the position of the tip on the TV screen with another piece of tape. Now, a reference point has been created to represent the image-forming center of the objective in both the CCD video monitor and two-photon imaging monitor. It is important to maintain the two-photon imaging window in a fixed position on the imaging computer monitor. It is also necessary to image the fluorescently labeled neuron before and after recording and dye filling to verify the correct targeting.

3 Methods

3.1 Preparation of Acutely Isolated Retina Samples

1. Cut black membrane filter paper into pieces that fit into the recording chamber. Cut a hole (~1 mm × 2 mm) at the center of the filter paper with a sharp razor blade. When the orientation of the retina is important, mark the filter paper with the razor blade so that the dorsal, ventral, nasal, and temporal directions can be recognized.

2. Adapt mice to darkness for at least 1 h before euthanization. Anesthetize mice with isoflurane, and then decapitate the mice. Enucleate the eyeballs rapidly under dim red-light illumination (all animal handling and euthanization procedures need to be performed according to the ethical guidelines of the relevant institution and authorities).

3. Under infrared illumination, place the eyeball on a piece of white filter paper. Note the left and right eyeballs. Make an incision through the cornea with a sharp surgical blade or a 15 gauge needle.

4. Transfer the eyeball to a glass Petri dish filled with oxygenated (95% O_2, 5% CO_2) and buffered Ames' medium. Remove the cornea, the lens and the vitreous from the eye under a dissection microscope under infrared illumination. The vitreous body needs to be removed completely for patch-clamp recording.

5. Identify the dorsal/ventral side of retina according to the landmarks in the choroid (Fig. 3). Mark the dorsal-ventral axis by making small cuts at the periphery of the retina.

6. Carefully peel the pigmented epithelial layer, choroid, and sclera from the retina using a pair of fine forceps. Cut the isolated retina into dorsal and ventral halves along the nasal-temporal axis.

7. Under dim red light, mount the isolated retina pieces onto filter paper prepared in step 1 with ganglion cell layer facing up. Make sure that the retina fully covers the hole in the middle of the filter paper. Note down the orientation of the retina on the filter paper.

To maintain the health of the tissue, transfer the retina to a new Petri dish with freshly oxygenated Ames' medium every 10 min during step 4.1.1–4.1.7.

8. Keep the mounted retina in the continuously bubbled (95% O_2, 5% CO_2) incubation chamber with Ames' medium in darkness at room temperature before transferring to the microscope's bath chamber for recording and imaging. The mounted retina samples sink to the bottom of the chamber, and do not need to be held in place by additional tools.

3.2 Two-Photon Targeted Recording of Fluorescently Labeled Neurons

1. Place a retina sample with ganglion cell layer facing up into the microscope chamber continuously perfused with oxygenated Ames' medium at physiological temperature (34–36 °C).

2. Check the condenser position so that the images from the OLED are focused on the outer segments of the photoreceptors in the filter paper hole (see Sect. 2.3).

3. Focus the 60x objective at the ganglion cell layer within the filter-paper hole with infrared optics and the CCD camera.

4. Switch the microscope to the laser scanning/imaging mode (Fig. 2a). Depending on the fluorescent probe expressed in the cells of interest, tune the wavelength of the infrared laser accordingly. Adjust the x-y position of the retina so that the labelled cell is positioned at the reference point (see Sect. 2.3).

5. Switch back to transmitted infrared illumination (Fig. 2c) and use the CCD camera to visualize the targeted cell body at the reference point on the TV screen. Use a glass patch electrode to carefully remove the inner limiting membrane around the cell body, exposing the targeted cell while minimizing damage to the surrounding tissue.

6. Perform patch-clamp recording from the target cell using a new glass pipette filled with appropriate internal solution and an intracellular dye.

3.3 Recording Light Responses from the Targeted Neuron

1. After the recording is established, turn off all transmitted light sources and swing in the OLED stimulus arm under the condenser. Allow the retina to adapt (e.g., ~10 s) to the background light intensity of the OLED before presenting visual stimuli.

2. Present visual stimuli to the retina and record the cell's light response.

3. At the end of the recording, acquire a z-stack of the dye-filled neuron using the two-photon microscope.

3.4 Representative Results

We have used this protocol to obtain light responses and morphology of fluorescently labeled retinal neurons such as direction-selective ganglion cells and starburst amacrine cells. The acute retina preparation remains healthy and light-responsive for at least 8 h when incubated in oxygenated Ames' medium at room temperature. We have successfully targeted GFP-expressing neurons from a transgenic mouse line Drd4-GFP with low levels of GFP expression (Fig. 4) [23, 24]. Such low fluorescence levels are barely detectable with a standard CCD camera under full-field UV-illumination from xenon arc lamps. This method can be used in combination with pharmacology to address fundamental questions in retinal circuitry. It is important to note that the onset of the laser scanning at typical intensity (~5 mW) still activates the photoreceptors [20]. Despite using infrared wavelengths and the local nature of multiphoton excitation that alleviates the photobleaching effects, it is good practice to use minimal laser power and laser exposure time when locating fluorescent-protein-expressing neurons.

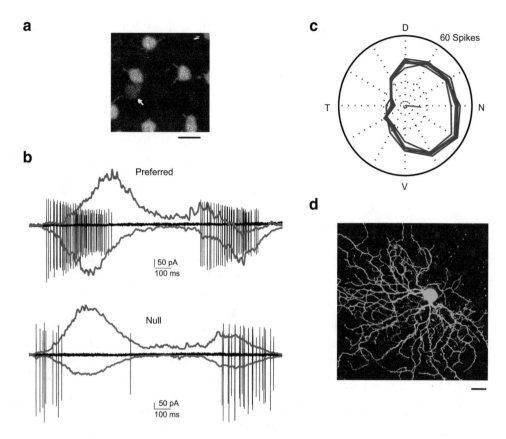

Fig. 4 Example of two-photon imaging and targeted recording from a GFP-expressing on–off direction-selective ganglion cell. (**a**) An image of a retina from a transgenic mouse line that labels an on–off direction-selective ganglion cell with low level of GFP (green, arrow) and starburst amacrine cells with td-Tomato (red). Scale bar, 15 μm. (**b**) Spiking (black), inward excitatory (red), and outward inhibitory (blue) postsynaptic currents measured in an on–off direction-selective ganglion cell evoked by a bar moving in the preferred and null directions. Spiking response was measured by loose-attached recording before breaking into the cell. Excitatory and inhibitory currents were measured by holding the cell at −60 mV and 0 mV, respectively, in the whole-cell patch-clamp configuration. On and off responses were evoked by the leading and trailing edges of the bright bar, respectively. (**c**) Polar plot of the total spike count (3 repetitions, same cell as (**b**) as a function of bar direction. Black lines indicate individual repetitions and the red line indicates the mean response. The vector sum of spike counts is shown as the red line in the middle. D, dorsal; N, nasal; V, ventral; T, temporal. (**d**) Top view (maximum intensity projection) of the dendritic arbors of the Alexa Fluor 488-filled direction-selective ganglion cell (recorded in **b**, **c**). Scale bar, 20 μm

4   Troubleshooting and Further Directions

To obtain high-quality electrophysiological recordings of visually evoked responses from retinal neurons, the multiphoton rig should be equipped with a Faraday cage to reduce electromagnetic interference. The power supply unit of the OLED is a major source of noise and should be positioned outside the Faraday cage. To obtain high-quality images of the labeled neurons, the visual stimulation module should be swung away from below the condenser during multiphoton imaging.

Several factors contribute to the health and responsiveness to light of the isolated retina samples. The rate of circulation of oxygenated Ames' medium is recommended to be greater than 10 ml/min. The bath temperature needs to be above 32 °C to obtain good visual responses. Before recording, the retina samples should be examined by the infrared optics and the CCD camera to ensure that the vitreous body is completely removed from the retina. Incomplete removal of the vitreous body can prevent successful electrophysiological recordings from healthy retinal neurons.

In the present protocol, multiphoton imaging of the fluorescent proteins is performed first, followed by simultaneous electrophysiological recording and visual stimulation. Therefore, the white light from the OLED needs to be blocked from the retina before the start of multiphoton imaging. However, this setup can be easily modified for simultaneous multiphoton imaging and visual stimulation without major modification of the microscope hardware (Fig. 2). A narrow bandpass filter can be placed in front of the OLED screen so that the wavelengths permitted for detecting fluorescence at the photomultiplier tubes (PMTs) are well separated from those of the visual stimuli (see Sect. 2.3). We have used this approach to record light-evoked calcium transients from retinal neurons expressing the genetically encoded calcium sensor GCaMP6 [23].

References

1. Sanes JR, Masland RH (2015) The types of retinal ganglion cells: current status and implications for neuronal classification. Annu Rev Neurosci 38:221–246

2. Demb JB, Singer JH (2015) Functional circuitry of the retina. Annu Rev Neurosci 1:263–289

3. Baden T, Berens P, Franke K, Roman Roson M, Bethge M, Euler T (2016) The functional diversity of retinal ganglion cells in the mouse. Nature 529:345–350

4. Gollisch T, Meister M (2010) Eye smarter than scientists believed: neural computations in circuits of the retina. Neuron 65:150–164

5. Dhande OS, Stafford BK, Lim J-HA, Huberman AD (2015) Contributions of retinal ganglion cells to subcortical visual processing and behaviors. Annu Rev Neurosci 1:291–328

6. Vaney DI, Sivyer B, Taylor WR (2012) Direction selectivity in the retina: symmetry and asymmetry in structure and function. Nat Rev Neurosci 13:194

7. Wei W, Hamby AM, Zhou KL, Feller MB (2011) Development of asymmetric inhibition underlying direction selectivity in the retina. Nature 469:402

8. Barlow HB, Hill RM, Levick WR (1964) Retinal ganglion cells responding selectively to direction and speed of image motion in the rabbit. J Physiol 173:377–407

9. Barlow HB, Levick WR (1965) The mechanism of directionally selective units in rabbit's retina. J Physiol 178:477–504

10. Simpson JI (1984) The accessory optic system. Annu Rev Neurosci 7:13–41

11. Cruz-Martín A, El-Danaf RN, Osakada F, Sriram B, Dhande OS, Nguyen PL et al (2014) A dedicated circuit linking direction selective retinal ganglion cells to primary visual cortex. Nature 507:358

12. Hillier D, Fiscella M, Drinnenberg A, Trenholm S, Rompani SB, Raics Z et al (2017) Causal evidence for retina-dependent and -independent visual motion computations in mouse cortex. Nat Neurosci 20:960–968

13. Huberman AD, Wei W, Elstrott J, Stafford BK, Feller MB, Barres BA (2009) Genetic identification of an On-Off direction-selective retinal ganglion cell subtype reveals a layer-specific subcortical map of posterior motion. Neuron 62:327–334

14. Shi X, Barchini J, Ledesma HA, Koren D, Jin Y, Liu X et al (2017) Retinal origin of direction selectivity in the superior colliculus. Nat Neurosci 20:550–558

15. Zucker CL, Dowling JE (1987) Centrifugal fibres synapse on dopaminergic interplexiform cells in the teleost retina. Nature 330:166–168

16. van Wyk M, Wassle H, Taylor WR (2009) Receptive field properties of ON- and OFF-ganglion cells in the mouse retina. Vis Neurosci 26:297–308

17. Zeng H, Madisen L (2012) Mouse transgenic approaches in optogenetics. Prog Brain Res 196:193–213

18. Yonehara K, Balint K, Noda M, Nagel G, Bamberg E, Roska B (2011) Spatially asymmetric reorganization of inhibition establishes a motion-sensitive circuit. Nature 469: 407–410

19. Euler T, Hausselt SE, Margolis DJ, Breuninger T, Castell X, Detwiler PB et al (2009) Eyecup scope–optical recordings of light stimulus-evoked fluorescence signals in the retina. Pflugers Arch 457:1393–1414

20. Svoboda K, Yasuda R (2006) Principles of two-photon excitation microscopy and its applications to neuroscience. Neuron 50:823–839

21. Brainard DH (1997) The Psychophysics Toolbox. Spat Vis 10:433–436

22. Wei W, Elstrott J, Feller MB (2010) Two-photon targeted recording of GFP-expressing neurons for light responses and live-cell imaging in the mouse retina. Nat Prot 5:1347–1352

23. Chen Q, Pei Z, Koren D, Wei W (2016) Stimulus-dependent recruitment of lateral inhibition underlies retinal direction selectivity. Elife 5:e21053

24. Pei Z, Chen Q, Koren D, Giammarinaro B, Ledesma HA, Wei W (2015) Conditional knock-out of vesicular GABA transporter gene from starburst amacrine cells reveals the contributions of multiple synaptic mechanisms underlying direction selectivity in the retina. J Neurosci 35:13219–13232

Chapter 3

Two-Photon Neurotransmitter Uncaging for the Study of Dendritic Integration

Alexandra Tran-Van-Minh, Nelson Rebola, Andreas Hoehne, and David A. DiGregorio

Abstract

Neurons transform the information arising from up to thousands of synaptic inputs into specific spiking patterns. Nonlinear dendritic integration is a crucial step in this process and is thought to increase the computational ability of neurons. However, studying how complex spatiotemporal patterns of synaptic inputs drive neuronal spiking is technically challenging. Two-photon neurotransmitter uncaging allows researchers to activate sequences of single synapses with high spatiotemporal precision and thus systematically examine how single and multiple synaptic activation patterns may recruit dendritic nonlinearities. Here, we describe the theoretical and practical considerations of using two-photon uncaging to mimic synaptic activation and monitor the electrical and biochemical signaling in dendrites when evoked by various synaptic patterns.

Key words Two-photon microscopy, Neurotransmitter uncaging, Dendritic integration, Synaptic potentials, Photolysis, Neuronal computations

Abbreviations

[Glut]	Glutamate concentration
2PLSM	Two-photon laser scanning microscopy
ACSF	Artificial cerebrospinal fluid
AMPAR	α-Amino-3-hydroxy-5-methyl-4-isoxazolepropionic acid *receptor*
AOD	Acousto-optic deflector
AOM	Acousto-optic modulator
CDNI-glutamate	4-Carboxymethoxy-5,7-Dinitroindolinyl-Glutamate
dI/O	Dendritic input/output relationship
DiO/DPA	Two-component optical *voltage* sensor made of the neuronal tracer dye $DiOC_{16}(3)$ and dipicrylamine
DNI-glutamate TFA	4-Methoxy-5,7-dinitroindolinyl-L-glutamate trifluoroacetate

Espen Hartveit (ed.), *Multiphoton Microscopy*, Neuromethods, vol. 148,
https://doi.org/10.1007/978-1-4939-9702-2_3, © Springer Science+Business Media, LLC, part of Springer Nature 2019

GM	The molecular two-photon cross-section is usually quoted in the units of Göppert-Mayer (**GM**), where 1 GM is 10^{-50} cm^4 s photon^{-1}
I/O	Input/output relationship
MNI-Glutamate	4-Methoxy-7-nitroindolinyl-caged-L-glutamate
NMDAR	N-Methyl-D-aspartate *receptor*
PSF	Point spread function
PV	Parvalbumin
SLM	Spatial light modulator
u[Glut]	Uncaging-evoked glutamate transient
uEPSC	Uncaging-evoked EPSC
uEPSP	Uncaging-evoked EPSP
VGCC	Voltage-gated calcium channel
θ_B	Bragg angle, the angle between the incident laser beam into an AOM and the diffracted beam

1   Introduction

1.1 Background

Describing how a neuron transforms different synaptic input patterns into action potential output firing is critical to understanding a neuron's computational capabilities. Decades of in vitro and in vivo work have shown that dendrites transform synaptic inputs in a nonlinear fashion. Consequently, for a given number of synaptic inputs, different spatial patterns of inputs produce different postsynaptic potentials, and hence a different probability of generating an action potential. Modeling work performed by Wilfrid Rall in the 1960s demonstrated the importance of dendrites to transform post-synaptic potentials (PSPs) between the synaptic input site and the soma [1]. Rall's work showed that dendrites behave like electrical cables, along which the PSP is progressively attenuated in amplitude and slowed in kinetics. In addition, dendrites contain a multitude of ion channels, the nature and distribution of which differ depending on neuron type, which can further alter PSP amplitude and shape. As a result, the integration of multiple synaptic inputs is highly dependent on the location of those inputs on the dendritic tree.

Dendritic patch-clamp recordings allow for the study of dendritic integration using synaptic-like current injections at single locations within the dendritic tree [2–4]. However, this technique cannot easily mimic the near simultaneous synaptic activation at multiple different locations within the dendritic tree that is thought to occur in vivo. This limitation can be circumvented by the use of a focused laser beam to locally liberate neurotransmitter by photo-uncaging and systematically probing the dendrite's voltage response to different spatial and temporal patterns that mimic synaptic input activity. Caged neurotransmitters are composed of a neurotransmitter molecule made biologically inert by a covalent bond to a photolabile protective group or "cage." The absorption of a photon (or two photons of half energy) by the caging group leads to an excitation to a high energetic state (triplet state) from which there is a high probability that a covalent bond will be destroyed, thereby liberating the active neurotransmitter molecule (Box 1).

Critical properties that caged compounds should display have been described previously [5–7] and include high two-photon (2P) cross section, good solubility at physiological ionic strength and pH, high stability of the caged form at physiological ionic strength and pH in order to ensure that little or no free neurotransmitter is introduced in the experimental preparation, and fast dark reaction rates (the steps between the light-dependent excitation of the caged compound and its conversion to photolysis products). In addition, the caged form of neurotransmitter and its photolysis by-products should be biologically inert. These properties are all essential for mimicking responses to neurotransmitter release from the fusion of single synaptic vesicles. Among the commercially available forms of caged glutamate, nitroindoline-based cages in particular meet these criteria, but suffer from the principal shortcoming that they promiscuously block GABA$_A$ receptors (for MNI-glutamate: 50% block of inhibitory postsynaptic currents (IPSCs) at 500 µM [8] and total block at the working concentration of 2.5 mM [9]). Rubi-based caged compounds also display off-target effects, but thanks to their better absorption cross section, they can be used at lower concentrations, resulting in less GABA$_A$ receptor antagonism than MNI-based compounds; however, their effect is still large enough to be

Box 1 Properties of caged neurotransmitters. Reaction of photolytic release of a typical caged compound, 4-methoxy-7-nitroindolinyl-glutamate (MNI-glutamate)

Ideal properties of caged compounds for 2P uncaging include:

- High 2P cross section.
- Fast dark reaction rate constants (k_{dark}) and excitation rate constants (k_e).
- Biologically inert cage form and by-products.
- High solubility at physiological ionic strength and pH.
- High stability at physiological ionic strength and pH.

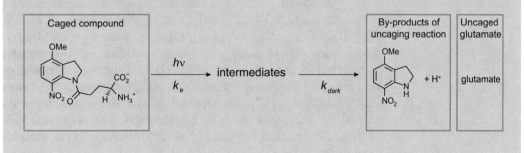

worth noting (50% block of IPSCs at the working concentration 300 µM [9]). Thus, for cases where high concentrations of caged compounds must be used, an antagonist of GABA$_A$ receptors is often added to the bathing solution.

Here, we discuss the technical and biological considerations when using photo-uncaging of neurotransmitter to study synaptic integration in dendrites, with a particular focus on 2P photolysis of caged glutamate. We will highlight recent technical advances that facilitate reliable studies of glutamatergic inputs onto aspiny dendrites, a less-often considered preparation. As examples, we discuss the technical requirements to perform two types of studies: (1) spine-to-dendrite electrical signal transfer and (2) quantification of dendritic integration properties using multispot photolysis.

1.2 Why Use Two-Photon Excitation to Uncage Neurotransmitter?

Like 2P excitation of fluorescent molecules, photolysis of caged compounds occurs when two low-energy photons are simultaneously absorbed, driving electrons of the chromophore into an excited state from which chemical bonds tethering the chemical cage to neurotransmitter are broken. 2P uncaging also benefits from excitation, and hence uncaging, within a small, well-defined focal volume that can be localized deep within scattering tissue, such as a brain slice [10]. Restricted 2P uncaging volumes also reduce out-of-focus photodamage and the accumulation of uncaging by-products. 2P uncaging permits bath-application of the caged compound, which guarantees homogenous concentration of caged neurotransmitter, and thereby permits probing different parts of the dendritic tree over long distances. For 1P uncaging, out-of-focus photolysis can be minimized by local perfusion of the caged compound [11]. However, local perfusion suffers from a heterogeneous spatial distribution in the concentration of caged compound, resulting in a variable amount of neurotransmitter uncaging for a given laser pulse energy at different locations relative to the perfusion pipette.

One long-standing limitation of caged neurotransmitters is the poor 2P cross section of nitroindoline-based caged compounds (~0.06 GM). Despite this limitation, it is nevertheless possible to successfully perform 2P photolysis of these compounds for physiological experiments [12–15]. Recently developed caged glutamate compounds exhibit a much higher 2P cross section (Rubi-glutamate, 0.14 GM [16], and CDNI-glutamate, ~0.24 GM [17], DNI-glutamate TFA [18]), and can be uncaged by 2P excitation using lower average powers than for MNI-glutamate, thereby reducing phototoxicity. Additionally, the improved uncaging efficiency allows the use of short uncaging pulses (10 µs with CDNI-glutamate [17], as opposed to the 0.2–1 ms durations needed for MNI-glutamate to elicit synaptic-like responses [13, 19]). The differences between 1P versus 2P uncaging are summarized in Table 1.

Table 1
Comparison of experimental constraints for uncaging of locally applied MNI-glutamate (4 mM, using perfusion pipette) using 1P (405 nm) or 2P (720 nm) uncaging light

	1P uncaging	2P uncaging
Depth penetration	Limited to most superficial part of the slice (~20 μm)	Limited by penetration of caged glutamate in the slice
Spatial resolution	Out-of-focus excitation above uncaging spot	Focal excitation, uncaging restricted to the PSF volume
Minimal uncaging time	20 μs	100 μs
Power	<1 mW	<20 mW to minimize phototoxicity

2 Materials

2.1 Basic Tools for Dendritic Uncaging

Visualization of individual dendrites and spines with diffraction-limited optical resolution is essential for precise positioning of the uncaging spots. This is achieved by labeling neurons with a fluorescent dye, possibly loaded through the patch pipette, or by sparse expression of genetically encoded fluorescent proteins, combined by a high contrast imaging technique such as confocal or multiphoton microscopy. To achieve synaptic-like uncaging responses, it is recommended to monitor postsynaptic receptor gating using whole-cell patch recordings in either voltage or current clamp. Thus, a fluorescence imaging and photoactivation microscope should also be equipped with micromanipulators and electrophysiology.

Confocal imaging provides a high spatial resolution visualization of dendrites and spines, but the scattering properties of brain slices deteriorate the signal-to-noise ratio of the image beyond 20–50 μm within the tissue. While 2P imaging provides slightly lower spatial resolution (<2×), the longer excitation wavelengths and the use of nondescanned photon detection (i.e., no pinhole) ensure efficient light collection, resulting in high-contrast fluorescence images of dendrites even deep within the scattering brain tissue.

Because imaging and uncaging are generally performed with different excitation wavelengths, independent 2P imaging and uncaging is most effectively implemented on a single microscope by using two optical pathways and two separate femtosecond pulsed lasers. This arrangement allows simultaneous imaging and uncaging. For microscopes with only a single laser, minutes are needed to change between imaging and uncaging wavelengths, thus precluding imaging-based monitoring of neuronal responses to neurotransmitter uncaging. Moreover, the long delays imparted by wavelength changes accentuate the deterioration of uncaging spot localization due to preparation drift. Automated drift correction strategies are also more efficient without wavelength change times.

To minimize unwanted photolysis during imaging, 2P imaging and uncaging wavelengths should, ideally, be well separated (e.g., 720 nm for nitroindoline-based cage compounds, 810–840 nm for excitation of fluorescein-based dyes, and 920–940 nm for excitation of green fluorescent proteins). Short imaging dwell times at low powers also help to minimize unwanted uncaging, and can facilitate the use of caged compounds and fluorescent dyes with unavoidable spectral overlap between uncaging and imaging (e.g., when using Rubi-based caged compounds with green fluorescence dyes). It is important to ensure that the focal plane for imaging and uncaging lasers is similar (parfocal), as differences in laser beam collimation (divergence) from each of the two different sources will alter focal planes for the same objective. The axial location of each laser beam focus can be assessed by imaging a thin fluorescent sample (fluorescent beads or highlighter pen streaks), first with the imaging beam and then the uncaging beam. Divergence of the uncaging laser beam can be adjusted using two lenses with identical focal lengths and placed apart by the distance (approximately 2x their focal distance) required to match the divergence between the two lasers. This simple lens system can be inserted in the uncaging optical path, and then adjusted to match the imaging plane of the uncaging and imaging path.

2.2 Optical Modulators to Control Laser Illumination Intensity and Duration

Sub-millisecond duration laser illumination is needed to generate uncaging-evoked synaptic-like neuronal responses (see below, Sect. 3.1), thus requiring rapid and precise control of light. Mechanical shutters are not suitable to elicit such rapid light pulses, as they have several millisecond open times, and cannot be easily triggered at greater than 1 kHz repetition frequency needed for mimicking multiple synapse activation.

Electro-optical modulators (e.g., Pockels cell) are devices that can control laser illumination with a precision of microseconds. However, light cannot be completely blocked by electro-optical devices. In practice, a 500-fold reduction in the laser intensity (1:500 extinction ratio) is required to minimize unwanted uncaging when eliciting synaptic-like responses. While Pockels cells can achieve such an extinction, small misalignment can deteriorate the extinction ratio and also leads to poor transmission. Thus, Pockels cells are often used in combination with hard shutters to completely block laser light between experimental trials. Pockels cells can also suffer from slow changes in optical transmission, presumably due to transient heating by the 2P laser.

Acousto-optical modulators (AOMs) are a good alternative to Pockels cells as they can achieve transmission changes of tens of nanoseconds (see comparison at https://labrigger.com/

blog/2018/01/14/laser-power-modulation-eoms-and-aoms).
They exhibit greater than 1:3000 extinction ratios (if one uses the
first-order diffraction beam), which is important to minimize
unwanted uncaging between laser pulses, and achieve nearly 90%
transmission. AOMs have a small footprint and can be more
robustly mounted on the optical table than Pockels cells, resulting
in a better overall mechanical stability. When gating 2P laser beams,
AOMs do introduce some group velocity dispersion, but this can
be compensated by commercial dispersion compensation units
(e.g., DeepSee unit, Spectra-Physics). The principal disadvantage
of AOMs is that the angle between the incident and diffracted
beam, given by the Bragg angle ($\theta_B = \dfrac{\lambda F}{2v}$, where F is the acoustic
frequency and v the acoustic velocity in the crystal), depends on
the wavelength. Thus, adjustments in alignment are required if the
experimenter wishes to use multiple wavelengths.

2.3 Devices for Uncaging Laser Spot Positioning

Characterization of how dendrites integrate activation of multiple
synapses requires the rapid movement of the uncaging laser beam
between multiple locations. The most common way to achieve this
is to use a dual optical path scan-head with two sets of fast-deflecting
galvanometric mirrors: one for imaging and one for the uncaging
light path. This configuration, which is commercially provided by
several major microscope manufacturers, enables simultaneous
excitation and imaging within the same focal plane. However, the
point-to-point displacement time (within a single xy focal plane) is
limited to ~100–200 μs by the inertia of the mirrors. Given the
additional 200 μs spent at each point for photolysis of the caged
compound, the total point-to-point cycle is ~300–400 μs.

Acousto-optic deflectors (AODs) do not require mechanical
movement to deflect the laser beam and can thus exhibit faster
point-to-point displacement time (~30 μs for total point-to-point
cycle). One main advantage is that they allow random-access pointing of the laser beam in non-neighboring locations (in contrast,
see Sect. 4.2.3 for a discussion of the limitations of galvanometric
mirrors), including in different axial planes, thus providing access
to larger fractions of the dendritic tree [20, 21]. Although to date
they have mostly been used for imaging, recent improvements in
their light transmission properties over a large range of wavelengths
[20] make them amenable to uncaging experiments.

Another possibility is to use a spatial light modulator (SLM),
a liquid crystal device that generates holographic patterns of
illumination to form multiple focal points or arbitrary shapes [22,
23]. The main advantage of this method is that it can produce truly
simultaneous illumination at multiple locations. The limit to the
number of uncaging locations is determined by the laser power per
spot necessary to achieve synaptic-like responses. Illumination spot
placement can also be achieved in three dimensions [24]. The main
limitation is the refresh rate of current SLMs (50–200 Hz) due to

relaxation of the liquid crystal [25] (see also www.meadowlark.com). The main experimental constraints for the three different systems are summarized in Table 2.

2.4 Essential Hardware Requirements for 2P Neurotransmitter Uncaging

To perform simultaneous 2P imaging and uncaging, we use an Ultima in vivo dual scan-head (Bruker Corporation, www.bruker.com) equipped with an imaging path and a photostimulation path, mounted on an Olympus BX61WI microscope equipped with a 60× (1.1 NA) water-immersion objective (Olympus, www.olympus-lifescience.com). The microscope is controlled using the PrairieView software (Bruker Corporation), which drives two sets of galvanometric mirrors, each of which scans separate laser beam paths. For excitation, we use two pulsed Ti:Sapphire lasers (Mai Tai, Spectra-Physics, www.spectra-physics.com) with pulse widths of less than 120 fs. We use a Pockels cell (350-50-BK 02, Conoptics, www.conoptics.com) to modulate the laser intensity on the imaging path, and an AOM (MT110-B50A1.5-IR-Hk, AA Optoelectronic, www.aaoptoelectronic.com) to modulate laser intensity on the uncaging path. For electrophysiological recordings, we use motorized stage and micromanipulators (Luigs and Neumann, www.luigs-neumann.com), a Multiclamp 700B amplifier (Molecular Devices, www.moleculardevices.com), and a multifunction input/output board (NI USB-6259, National Instruments, www.ni.com) for digitization.

Other commercially available systems with compatible dual-galvanometric scan heads are the LASU system (Scientifica,

Table 2
Comparison of constraints on the multisite spatial pattern using different optical systems

	Galvo mirrors	AODs	SLM
Point-to-point move time (μs)	100–200	~25	Simultaneous
Maximal inter-point distance (μm)	~5% of field of view	Limited by field of view (up to hundreds of μm) and resolution	Limited by field of view (up to hundreds of μm) and resolution
Maximum number of points visited for I/O experiment	<10 (3 locations/ms)	<100 (~30 locations/ms)	Limited by power/point
Maximum spatial spread of uncaging pattern (μm)	<50 μm	Limited by field of view	Limited by field of view
Uncaging on multiple focal planes	No	Yes	Yes

www.scientifica.uk.com) and the FemtoS–Galvo system (Femtonics, femtonics.eu). For custom and open-source systems, ScanImage software (Vidrio Technologies, scanimage.vidriotechnologies.com) is capable of controlling a second laser path and galvanometric mirror pair for photostimulation [26]. Complete liquid crystal SLMs, controllers, and software packages can be purchased from Intelligent Imaging Innovations (www.intelligent-imaging.com) or Bruker Corporation.

For those experimentalists who are just getting started, we provide here a checklist of essential hardware/software requirements to perform synaptic-like 2P neurotransmitter uncaging (Table 3) using commercial turnkey or custom-built microscopes. Generally, we have found that due to the requirement of microsecond synchronization of uncaging spot positioning, laser illumination, fluorescence detection, and electrophysiological recordings, commercial systems are the most rapid solutions to implement. As these systems can be rather expensive, it may be preferable to perform modifications to existing commercial two-

Table 3
Essential hardware requirements and manufacturers for an uncaging setup

Function	Typical device	Vendors
High contrast microscope for scattering tissue	2P scanning microscope (galvanometer mirror-based)	Bruker, Thorlabs, Femtonics, Intelligent Imaging Innovations, Scientifica,
Simultaneous imaging and uncaging	Dual-path scan head	Bruker, Scientifica, Femtonics
2P laser	Ti:Sapphire femtosecond laser	Coherent, Spectra-Physics
Rapid and accurate control of laser illumination	Pockels cells	Conoptics (www.conoptics.com)
	AOM	Gooch and Housego (www.goochandhousego.com); AA Opto-Electronic (www.aaoptoelectronic.com)
Software to define, synchronize and calibrate laser illumination, galvanometer mirror positioning and electrophysiology or fluorescence	Vendor software, ScanImage	Vidrio Technologies (ScanImage)
Optimize laser divergence for parfocality	1x telescope beam expander	Parts from Thorlabs
Low-volume bath recirculation	Peristaltic pump (Minpuls 3)	Gilson (www.gilson.com)
Patch-clamp	Axon instruments multi-clamp 700B	Molecular devices (www.moleculardevices.com)

photon scanning microscopes or custom-built systems. For hardware control in the latter case, we recommend a generic scanning software package, such as ScanImage from Vidrio (www.vidrio-technologies.com).

2.5 Brain Slice Preparation Protocol

Acute brain slices are prepared from postnatal day 30–60 mice (F1 cross from BalbC and C57/Bl6J strains). Mice are quickly decapitated and whole brains are then quickly transferred to an ice-cold solution containing (in mM): 2.5 KCl, 0.5 $CaCl_2$, 4 $MgCl_2$, 1.25 NaH_2PO_4, 24 $NaHCO_3$, 25 glucose, and 230 sucrose, bubbled with 95% O_2 and 5% CO_2. Slices are cut using a vibratome (Leica VT1200S) and incubated for 30 min in a solution containing (in mM) 85 NaCl, 2.5 KCl, 0.5 $CaCl_2$, 4 $MgCl_2$, 1.25 NaH_2PO_4, 24 $NaHCO_3$, 25 glucose, and 75 sucrose, bubbled with 95% O_2 and 5% CO_2. Slices are subsequently transferred to a solution containing (in mM): 125 NaCl, 2.5 KCl, 0.5 $CaCl_2$, 4 $MgCl_2$, 1.25 NaH_2PO_4, 24 $NaHCO_3$, 25 glucose, and 0.5 L-ascorbic acid, bubbled with 95% O_2 and 5% CO_2 and maintained at room temperature for up to 8 h.

2.6 Preparation of Solutions Containing Caged Compound

Caged glutamate can be purchased from Tocris (MNI-glutamate #1490, and Rubi-glutamate, #3574). MNI-glutamate is prepared in 40 mM stock solutions in HEPES-buffered ACSF buffered to pH 7.4, containing (in mM): 125 NaCl, 2.5 KCl, 0.5 $CaCl_2$, 4 $MgCl_2$, 1.25 NaH_2PO_4, 40 HEPES, 25 glucose, and kept in light-protected aliquots at −20 °C. Rubi-glutamate is directly dissolved in ASCF on a daily basis at a final concentration of 1 mM. To prevent oxidative stress due to high-power light excitation, we add 0.5 mM L-ascorbic acid (Sigma-Aldrich, www.sigmaaldrich.com) in the recording solution. In experiments using MNI-glutamate, we add 10 μM Gabazine (Tocris) to the recording solution in order to avoid confounding results due to partial blockade of $GABA_A$ receptors by the cage compound. For specific recommendations for preparing and handling solutions of caged neurotransmitter, *see* **Note 4.1**.

We would also like to encourage a method for verifying the efficiency of each lot of the caged compound, as purity can influence uncaging efficiency. We recommend recording photolysis-evoked EPSCs during bath application of the cage compound on a particular cell type (preferable a spiny dendrite to facilitate consistent spot positioning). One should use a single laser power and duration and select several proximal spot locations at a consistent depth within the tissue.

3   Methods

One of the principal advantages of the confined excitation of 2P uncaging is the possibility to evoke synaptic-like currents (uncaging-evoked EPSCs, uEPSCs) or voltage responses (uncaging-evoked EPSPs, uEPSPs), as measured at the soma [12, 27]. Rapid EPSCs are a result of the activation of low-affinity AMPA-type glutamate receptors in the postsynaptic density [28] by brief glutamate concentration ([Glut]) transients (<100 μs) produced by synaptic vesicle exocytosis. In order to elicit synaptic-like photolysis-evoked [Glut] transients, one must uncage neurotransmitter molecules over short durations and limited volumes. Thus, the size of the illumination spot and duration of the laser pulse must be kept to a minimum, while ensuring the uncaging reaction and receptor binding are within their linear regimes. Below, we fully elucidate these requirements.

3.1   Theoretical Considerations for Producing Fast Photolysis-Evoked [Glut] Transients

Simple diffusion theory posits that neurotransmitter clearance is slowed by increases in the spatial extent of the source as well as duration of the molecular flux, which for uncaging corresponds to the illumination pulse duration. To illustrate this theory, we used a simple one-dimensional diffusional model to examine how the spatial and temporal profile of glutamate concentration (referred to as [Glut]) in an idealized synaptic cleft would be altered when varying the spatial extent and duration of glutamate flux (Fig. 1). We used an experimentally determined diffusion coefficient for glutamate of 0.33 $\mu m^2/ms$ [30]. An instantaneous injection of molecules at a point results in a rapid dissipation in space and in time of the [Glut] (impulse response; Fig. 1a, left). For point sources, the relationship between the square of the gradient width (standard deviation, σ) and time is linear ($\sigma^2 = 4Dt$, where D is the diffusion coefficient of the molecule; [32]). If the size of the source is extended, the gradient width and decay time (time at which the concentration decreases by 50%) is also increased (Fig. 1a). If the flux duration is increased from 20 to 500 μs, the decay of the [Glut] transient is dramatically slowed for both source widths. Interestingly, within the range of source widths and flux durations corresponding to typical experimental conditions, increased flux durations have a larger impact on decay times than increasing the size of the source (Fig. 1b, c). Altering the uncaging volume, approximated in this case by reducing the lateral dimension from 0.5 to 0.3 μm (40% decrease in width), produced only a minor (<50%) slowing of the decay (Fig. 1c, gray region), whereas adjusting the flux durations between 0.05 and 2 ms (40-fold increase), a range often used in experiments, can lead to up to a ten-fold alteration of the [Glut] decay. This effect of flux duration on molecular

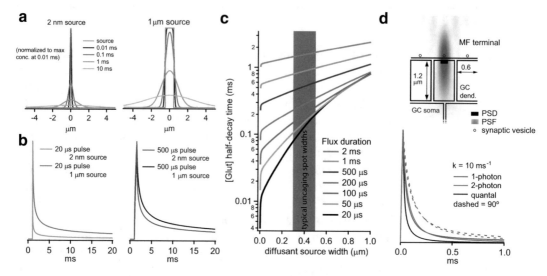

Fig. 1 Simple diffusion theory shows that temporal decay of a diffusant depends on the spatial extent and flux duration of the source. (**a**) One-dimensional simulation ([29], eq. 2.5) of the relative diffusant (glutamate) concentration at different times (after termination of the glutamate flux) for a 2 nm and 1 μm source size. The diffusion coefficient was 0.33 $μm^2/ms$ [30], and the flux duration was 20 μs. (**b**) The same simulation as (**a**) but showing the decay time course for either 20 or 500 μs flux duration. (**c**) Half-decay time of simulated [Glut] transients using the model from (**a**) for source sizes varying from 2 nm to 1 μm, with a flux duration varying between 20 μs and 2 ms. (**d**) A three-dimensional simulation of glutamate diffusion in a model synaptic cleft representing a cerebellar mossy-fiber (MF) to granule cell (GC) synapse in response to a 20 μs simulated laser pulse illumination over a 1P PSF (red trace) or a 2P PSF (blue trace). The black trace is a diffusion simulation of quantal glutamate release using an instantaneous flux over a 5 nm point source. All [Glut] transients are normalized. Dashed line is a simulation with the illumination spot rotated 90 degrees, parallel with the synaptic cleft. The cartoon at the top shows a cross section (x–z plane) of the synaptic space overlaid with the 1P PSF (blue trace). (Reproduced from DiGregorio et al. [31] with permission from the Society for Neuroscience)

clearance is due to the non-exponential nature of the solution to the diffusion equation, which also contributes a significant slow decay component that is not captured by a half-decay metric. Thus, in comparison with optimizing the focal spot size, minimizing the pulse duration is the most important consideration for achieving synaptic-like [Glut] transients. Practical considerations to this effect are described in **Note 4.2.1**.

The fusion of neurotransmitter-containing synaptic vesicles is thought to be nearly instantaneous (< 50 μs) at fast synapses [30] and occurs over a dimension much less than the 40 nm vesicle diameter, thus approximating a point source. Diffusion theory predicts that the decay of [Glut] transients evoked from spatially extended sources, on the order of the microscope Point Spread Function (PSF, ~300 nm), are more than two-fold slower than the [Glut] decay following instantaneous release of neurotransmitter into the synaptic cleft (Fig. 1c). The time course of [Glut] transients in a more realistic experimental setting, within a morphological model of the extracellular space and evoked by a volume corresponding to the spatial extent of a PSF, has been examined theoretically [31]. In Fig. 1 we reproduce these simulated uncaging-evoked [Glut] transients (u[Glut]). The [Glut] was simulated in response to a point source release of glutamate molecules followed by diffusion in a complex extracellular space (Fig. 1d, top). To emulate the [Glut] driving AMPA receptor opening, we spatially averaged the [Glut] over the cleft volume adjacent to an average postsynaptic density (0.2 × 0.2 μm). The simulation results show that neither 1P or 2P uncaging can reproduce [Glut] transients as fast as those expected for vesicle fusion ([31]; Fig. 1d, bottom). However, given that the half-decay of the [Glut] transient for the 2P simulation is only about twofold slower than that predicted for quantal release (Fig. 1d, bottom), it is conceivable that slow AMPARs, such as those generally found in pyramidal cells ([28]), might still be activated by uncaging in a manner similar to fusion of a synaptic vesicle, provided that the flux duration (i.e., the laser illumination pulse) is sufficiently short. The effect of glutamate receptor type on uEPSP and uEPSCs will be further discussed in Sect. 3.2.3.

3.2 Practical Considerations for Achieving Brief u[Glut] Transients

3.2.1 Uncaging Reaction Must Be Faster than Glutamate Receptor Gating

First and foremost, the caged compound must exhibit a fast uncaging reaction. The photophysical rates should be on the order of a microsecond or less, which is the case for most modern caged compounds, in particular the nitroindoline-based ones [5]. If there is a significant delay between photon absorption and complete transmitter liberation, then diffusion will effectively increase the size of the uncaging volume. Second, the caged compound should exhibit a high photolysis efficiency, thereby enabling the use of brief, relatively low-power illumination pulses. Indeed, laser illumination is the most convenient light source because (1) the coherent source can be focused to a diffraction-limited spot and (2) the resulting power density of the focused beam is very high and thus well-suited for efficient absorption by the caged compound. The trade-off is the potential photodamage, which is higher within the focal spot when using 2P excitation compared to 1P excitation [33]. Thus, the lower photodamage threshold and the relatively poor 2P cross sections of nitroindoline-based cages limit the efficiency of short laser pulses [32]. Therefore, in order to achieve physiological receptor occupancy (binding), the duration of the laser pulse must be increased, which, in turn, slows the rise and fall of the u[Glut] (see Fig. 1).

3.2.2 Minimize Light-Evoked Nonlinear Processes

In order to produce the high [Glut] (>1 mM) necessary to achieve the high glutamate receptor occupancies achieved during synaptic transmission (50%, [31]), high concentrations of caged compound (>2 mM) and high laser power are needed. Other factors that can greatly influence the time course and spatial specificity of single-spot photolysis are nonlinearities of the (1) photolysis reaction and

(2) receptor binding. The uncaging reaction can be approximated as a simple first-order reaction in which the rate of photolysis (k) is dependent on laser power. As the value of k increases, the amount of photolysed compound increases during the laser pulse. Experimentally estimated values of k suggest it is possible to deplete the caged compound within the focal volume. For large values of k, the concentration of liberated glutamate saturates in the center but can continue to increase more laterally within the PSF, thus effectively increasing the lateral and axial extent of the uncaging volume. This power-dependent broadening of the apparent photo-reaction volume is observed for fluorescence after photobleaching experiments. Watt Webb and colleagues provided a mathematical description of the relationship between the bleaching rate (analogous to photolysis rate) and the bleached spot size, as well as fluorescence recovery time courses [34]. These relationships have also been modeled specifically for uncaging experiments [35].

We also simulated the effect of different photolysis rates on the decay of the u[Glut] transient for our measured 1P and 2P PSFs, but within a synaptic cleft. We observed that indeed, for the estimated 1P uncaging efficiencies, the photolysis volume and [Glut] decay are broadened and slowed, respectively, for high rates of uncaging (Fig. 2a, b), while lower photolysis rates (<100 ms^{-1}) did not exhibit much broadening. Given the poor efficiency of 2P photolysis of most caged glutamate compounds, light-dependent alterations in the time course or spatial dependence of [Glut] transients may not be a major concern. However, care should be taken to ensure linearity of the uncaging reaction by minimizing uncaging duration and laser power (see Sects. 4.2.1 and 4.2.2). In Sect. 3.2.4, we examine and discuss the implications of the slow u[Glut] time course for the study of dendritic integration in spiny and aspiny neurons.

If high concentrations of [Glut] are generated by uncaging, then the resulting high occupancy (i.e., saturation of binding) of the glutamate receptors by agonist could produce an apparent increase in the spatial extent of the uncaging reaction, even in the absence of any change in the photolysis volume. This nonlinear relationship between [Glut] and receptor activation occurs when high concentrations of the caged compound and high laser powers are used. For example, using 1P photolysis of MNI-glutamate (10 mM), we observed a broadening (four-fold) of the spatial dependence of uEPSCs with increasing laser power (ten-fold, Fig. 2c–f, from [31]). Because the broadening is more than we expected for the estimated photolysis rates, we suggested that the receptor nonlinearities likely played a dominant role in the dependence of the effective uncaging volume on laser power. Because high receptor occupancies can also be achieved with 2P uncaging [27], receptor nonlinearities would also be expected to contribute to alterations in the effective uncaging resolution during 2P excita-

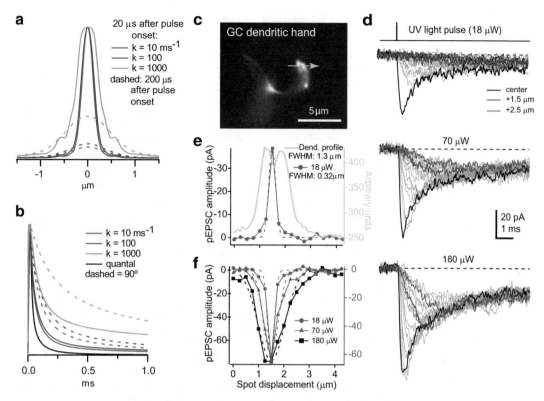

Fig. 2 High laser power and receptor nonlinearities can slow uEPSC decay. (**a, b**) Effect of varying the rate of photolysis on the spatial and temporal profile of [Glut] decay, for a 20 μs pulse and 5 nm point source. Times are relative to the start of the simulated laser pulse. (**c–f**) Effect of increasing the uncaging light intensity on uEPSCs spatial profile at the mossy-fiber to granule cell synapse. (**c**) Image of a granule cell dendritic hand visualized with Alexa Fluor 594. The green arrow indicates the length (4 μm) and location over which uncaging responses were elicited. (**d**) uEPSCs evoked with a brief (20 μs) diffraction-limited UV illumination spot during local perfusion of 10 mM MNI-glutamate at three different light intensities elicited at different locations (250 nm apart) across the dendritic hand. Colored traces corresponding to three locations are highlighted: black, at the center of the dendritic hand, blue, 1.5 μm from the center, red, 2.5 μm from the center. Gray traces are responses recorded at all other spot locations tested (see **e**). (**e**) Isochronal plot of uEPSC amplitudes (data, red trace) at the time of the peak of the largest EPSC, showing a hotspot of AMPARs within the dendrite (18 μW laser power). The dashed line is a Gaussian fit to the isochronal amplitude plot. The green trace is the line profile of dendritic fluorescence along green arrow in (**c**). (**f**) Isochronal plots of uEPSC amplitudes showing a broader spatial dependence (reduced resolution) for increasing laser power. 10 mM MNI-glutamate was locally perfused for all experiments presented. Dashed lines are Gaussian fits. (Reproduced from DiGregorio et al. [31] with permission from the Society for Neuroscience)

tion. These alterations in the size of the effective uncaging volume would also limit the ability to mimic synapse-evoked conductances and increase the chance of activating extrasynaptic receptors.

3.2.3 Influence of AMPAR Kinetics on the Ability to Mimic Synapse-Evoked EPSCs

It is well known that different neurons can express AMPARs which display different intrinsic kinetic responses to brief application of glutamate, depending on their subunit composition [28], which can thus influence the time course of uncaging responses. In particular, GABAergic interneurons typically express glutamatergic synaptic conductances with very fast kinetics [36]. We illustrate this here, using the same uncaging protocol for cortical L2/3 pyramidal neurons (PN) and PV positive interneurons. The interneuron's uEPSCs were much faster rising (<0.5 ms rise time for PV+ cells vs ~1 ms for L2/3 PN, Fig. 3a–d) and decaying (half-decay: 1.6 ms for PV+ cells vs 6.5 ms for L2/3 PN). The distance of the uncaging site from the soma was comparable in the two cases, and the dendritic diameter is thinner for the interneurons, thus it seems possible that channel kinetics could underlie the different uEPSC kinetics, which is consistent with differences in kinetic properties of somatic receptors in pyramidal neurons and interneurons [28]. These differences also imply that diffusion limitations arising from the finite illumination volume, described above, might have a stronger impact on synapses with faster receptor kinetics, making it potentially easier to mimic synapse-evoked responses in pyramidal neurons, for example.

Another hurdle to mimicking the time course of synapse-evoked EPSCs or EPSPs is cable filtering. Only in the case where channel closing kinetics are on the order of cable filtering or slower can the time course of uEPSCs be tuned to approximate synaptic-like responses. Direct dendritic electrical recording is also a robust strategy to measure unfiltered synaptic and uncaging-evoked responses [37, 38]. In CA1 pyramidal cells, dendritic patch recordings revealed that local rise time is under 0.2 ms (time constant, τ_{rise}) and the decay is 4 ms (τ_{decay}) [39]. These values contrast EPSC kinetics recorded at the soma of PV+ basket cells, with $\tau_{rise} < 0.1$ ms and τ_{decay} on the order of 0.2 ms (Fig. 3e–f) [40]. Thus, for interneurons, very brief illumination durations are necessary to approach the time course of synaptic events. In summary, intrinsic receptor kinetics and cable filtering should be considered when adjusting illumination parameters (duration and amplitude) and caged compound concentrations to mimic synapse-evoked responses.

3.2.4 Influence of Uncaging Spot Location on the Speed of u[Glut] Transients

Because of the sharp spatial decay of the [Glut] gradient over submicron scales (Fig. 1a), the photolysis volume must be well aligned with a PSD containing AMPARs in order to efficiently and rapidly activate them. In many pyramidal neurons, the PSDs are systematically located on dendritic spines, small membranous protrusions from dendrites [41]. Thus, the use of fluorescence microscopy (2P in tissue) to identify spines greatly facilitates optimum photolysis volume placement along the dendrite. Progressive displacement of the uncaging laser spot away from dendritic spines reveals a steep distance dependence of both amplitude and rise time of light-evoked postsynaptic currents, with only a nominal response when the uncaging spot is placed at a distance of 1 μm from the spine head (Fig. 3a, b). The steep spatial dependence highlights the need for precise positioning of the uncaging laser beam close to the targeted synapse. Thus, when spines can be used as a guide, laser spot placement should be within less than 1 μm from the spine head (*see* **Note 4.2.3**).

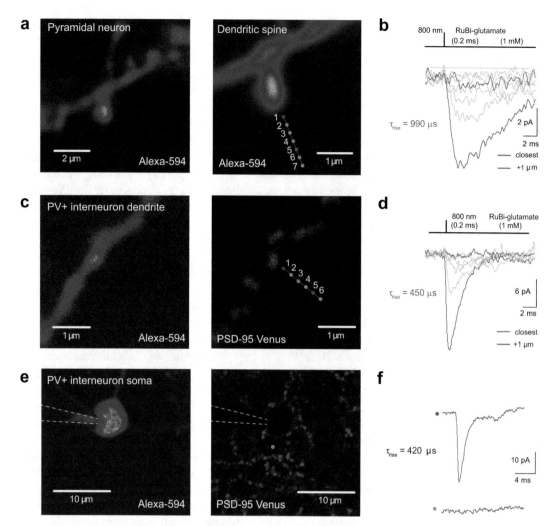

Fig. 3 Spatial dependence of photolysis-evoked synaptic currents using 2P glutamate uncaging in excitatory and inhibitory neurons. (**a**) 2PLSM image of a dendrite of layer 2/3 pyramidal cell loaded with the fluorescent marker Alexa Fluor 594 via the patch pipette. A spine was identified and 2P glutamate uncaging was performed at increasingly longer distances from the spine head (steps 250 nm). (**b**) Steep spatial dependence, of both rise time and peak amplitude, of uEPSCs evoked from locations indicated in (**a**) (colored traces were recorded from locations indicated by colored dots in **a**). (**c**) Dendrite of a neocortical PV+ interneuron loaded with Alexa Fluor 594 via the patch pipette. Conditional expression of the Venus-tagged PSD-95 protein using PSD95-ENABLED mice allowed identification of synapse location along interneuron dendrites. (**d**) uEPSCs get progressively slower and smaller as the uncaging spot is moved away from a fluorescent cluster. Colored traces were recorded from locations indicated by colored dots in (**c**). (**e, f**) Conditional expression of the Venus-tagged PSD-95 protein can also be used to identify somatic glutamatergic synapses in PV+ interneurons, at which location rapid uEPSCs could be observed. White dashed lines in (**e**) indicate the position of the patch pipette. Uncaging near a PSD-95 spot (**e**, right, indicated by a red dot) evoked a somatic uEPSC (**f**, top), while uncaging away from PSD-95 fluorescence (**e**, right, indicated by a gray dot) did not evoke a somatic uEPSC (**f**, bottom). All uEPSCs in this figure were evoked with a 0.2 ms long 800 nm laser pulse while the bath was perfused with 1 mM RuBi-glutamate

The positioning of the uncaging spot along dendrites of aspiny neurons (e.g., most GABAergic interneurons) is more of a challenge. When the experimenter does not have a distinct morphological correlate of synapse location, there is a reliance on trial-and-error placement of photolysis volumes along the dendrite. To precisely align the photolysis volume on PSDs, one must systematically displace it along both the x and y dimensions within the focal plane of the dendrite and then identify locations where uEPSCs are the largest and have the fastest rise time [31, 42]. This iterative approach increases substantially the duration and complexity of uncaging experiments, as well as potential photodamage, particularly for the identification of multiple uncaging locations. It should be noted that the absolute optimal rise time of uPSCs is likely to vary in rise times and peak amplitudes at different distances along dendrites, due to the electrical cable properties, and between cell types (Fig. 2c, d, f). This trial-and-error placement of photolysis volumes, therefore, requires a priori knowledge on the distance-dependence of synaptic currents amplitude and kinetics in the cell type studied.

Recently, a transgenic mouse line was created, allowing conditional expression of Venus-labeled PSD-95 in Cre-expressing cells [43]. Since PSD-95 is present in the majority of glutamatergic synapses [41], it is therefore possible to fluorescently label sites of excitatory synapses within specific cell types, thereby allowing for precise positioning of the photolysis volume even in aspiny neurons [43]. An example is shown in Fig. 3c–f, where glutamatergic synapses in PV interneurons were easily identified, both at the soma as well as the dendrites, by imaging conditionally expressed PSD-95-Venus. As expected, uncaging glutamate in locations close to PSD-95 fluorescent spots produced the fastest photolysis-evoked AMPAR currents (Fig. 3c–f). Although the use of transgenic mouse models represents an important advantage to visualize excitatory synapse location in aspiny neurons, care must be taken to validate that the transgenic mouse model used has minor impact on the composition and properties of synaptic receptors. One shortcoming of the PSD-95 Venus transgenic line is that its spectrum overlaps with green calcium and voltage indicators.

3.3 Typical Experiments Requiring 2P Uncaging

3.3.1 Estimating Voltage Attenuation Across Spine Necks

Glutamate uncaging is a powerful method to examine systematically the voltage propagation from spine head to parent dendrite. The spine head is connected to the parent dendrite by a spine neck whose diameter varies from 100 to 250 nm [44], but whether these narrow cable-like structures contribute to a difference in voltage between the spine head and parent dendrite is still controversial [45, 46]. Spine heads are too small to directly record synaptic potentials using standard patch-clamp techniques and can be difficult to stimulate in isolation by extracellular stimulation of the presynaptic neuron due to complex axonal arborisations of the input cells. These limitations can be circumvented by combining

2P glutamate uncaging and voltage imaging, in order to optically record a local depolarization and its propagation to the parent dendrite in response to a reliable synaptic-like stimulus. See Note 4.3 for technical considerations when performing simultaneous uncaging and fluorescence imaging.

We show the results of such an experiment in Fig. 4, where we used the optical reporter of voltage DiO/DPA [47, 48]. Following a brief laser pulse in the presence of RuBi-Glutamate, the optically recorded uEPSP detected in the parent dendrite was strongly attenuated as compared to the fluorescence recording of the uEPSP in the spine head. This result is consistent with prominent electrical filtering by the spine neck. Following near-synchronous multi-spine activation using uncaging, the optically recorded voltage responses recorded from the spine head and dendrite are more similar in amplitude, due to the summed depolarization within the parent dendrite, and perhaps as well as with the recruitment of voltage-dependent channels. This experiment shows that uncaging provides a more reliable, flexible, and systematic examination of synaptic integration following single and multiple spine activation, which would not be possible with electrical stimulation of the presynaptic neuron.

Cable theory predicts that amplitude attenuation (filtering) of brief depolarizations can depend on the time course of the depolarization. Generally, fast voltage transients are more sensitive to cable filtering than slow-voltage transients. Therefore, a caveat of the experiment proposed above is that because the uEPSCs are generally slower than synapse-evoked EPSCs (Sect. 3.1), the corresponding uEPSPs may also be slower than their physiological counterpart, thus underestimating the impact of spine neck filtering. We tested this possibility with numerical simulations of synaptic and uncaging conductance waveforms using a model of layer 5 pyramidal neuron and examining synaptic conductance injection onto spines at three different locations on the basal dendrite, and tested for a range of neck resistances (R_{neck} between 25 and 500 MΩ). For all spine neck resistances, varying the decay of the synaptic conductance between 0.5 and 2 ms resulted in corresponding increases in the decay of local spine and dendritic EPSPs, but with very little alteration in their peak amplitude (Fig. 5a–c). This implies that the local recruitment of voltage-dependent nonlinearities would be only modestly affected by the slower decay of uncaging-evoked conductances. However, the slow decay of voltage responses in spines and dendrites resulting from longer uncaging pulses could contribute to enhanced temporal summation [50] in response to high-frequency train stimuli.

The voltage propagation from spine head to dendrite was insensitive to variations in the synaptic conductance decay time constant up until 2 ms. If the synaptic conductance decay is further increased (up to 10 and 100 ms), $V_{dendrite}/V_{spine}$ also increases, provided the

Uncaging Single Spine

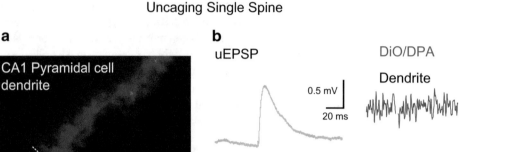

a

CA1 Pyramidal cell
dendrite

5 µm

b

uEPSP

DiO/DPA

Dendrite

0.5 mV

20 ms

Spine

0.1
ΔF/F

uncaging

1 ms

20 ms

Uncaging Multiple Spines

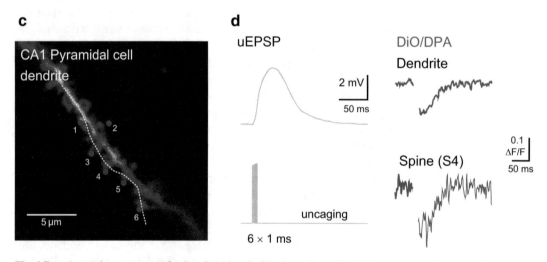

c

CA1 Pyramidal cell
dendrite

1   2

3
4
5

5 µm

6

d

uEPSP

DiO/DPA

Dendrite

2 mV

50 ms

Spine (S4)

0.1
ΔF/F

50 ms

uncaging

6 × 1 ms

Fig. 4 Experimental assessment of spine-dendrite electrical coupling using glutamate uncaging and voltage imaging. (**a**, **b**) Estimating spine-dendrite electrical coupling using 2P glutamate uncaging (2 mM MNI-glutamate, bath applied) and line scan imaging of fluorescent voltage reporter DiO/DPA. (**a**) 2PLSM image of a CA1 pyramidal cell dendrite, loaded with DiO via the patch pipette. The uncaging location is indicated by the red spot next to the spine head. The white dashed line indicates the position of the line scan. (**b**) Somatic uEPSP (left) and photolysis-evoked depolarization in the dendrite and in the spine head, visualized with DiO/DPA. The uncaging pulse produced a transient fluorescence artifact which was blanked from the DiO/DPA traces. Traces are averages of 10 trials. In this example, the peak amplitude of the optically recorded uEPSPs was larger in the spine than in the dendrite, as expected for a voltage drop across the spine neck. (**c**, **d**) The same as (**a**, **b**), but six spines (indicated by red spots in **c**) were stimulated in rapid succession (interpulse interval: 200 µs). Traces are averages of 10 trials. In this example, the peak amplitude of the optically recorded uEPSPs in the spine head and in the dendrite were similar

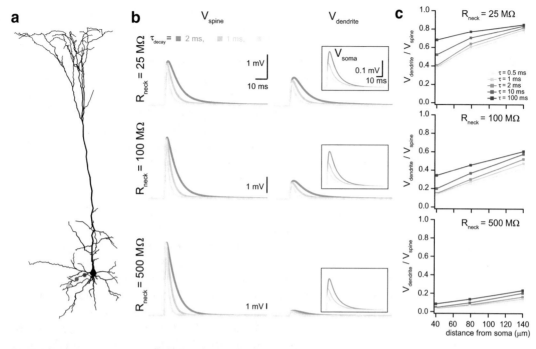

Fig. 5 Time course of the spine uEPSC has little influence on spine-to-dendrite voltage transfer. Simulation of the effect of varying the spine EPSP time course within a cable model of layer 5 pyramidal neuron. The simulation was performed using NEURON (www.neuron.yale.edu) using morphology and passive parameters from [49]. (**a**) Model morphology with colored spots indicating the location of synapses (located 40, 79, and 140 μm from the soma, respectively). Spine neck resistance R_{neck} was 25, 100, or 500 MΩ. Synaptic conductances modeled as alpha functions with decay time constants (τ_{decay}) varying between 0.5 and 100 ms were injected at each spine head, and the resulting EPSPs measured in the spine head (V_{spine}), the parent dendrite ($V_{dendrite}$), or at the soma (V_{soma}). (**b**) Example traces of V_{spine} (left), $V_{dendrite}$ (right), and V_{soma} (insets) for the synapse located 79 μm from the soma (indicated by a red spot in **a**) and for a synaptic conductance with a τ_{decay} of 0.5, 1, or 2 ms. (**c**) Spine-to-dendrite voltage attenuation for a range of synaptic EPSP kinetics, distances, and spine neck resistances

neck resistance is less <100 MΩ. Thus, the small differences in the time course of the [Glut] transients produced by synaptic and uncaging-evoked conductances described above is unlikely to alter voltage propagation between the spine head and the parent dendrite. However, for fast-gating AMPA receptors (see Sect. 3.2.3), the local uEPSP time course in the spine and dendrite would be slower than that of a synapse-evoked EPSP, which could differentially activate local voltage-gated ion channels and modify the gain of the I/O relationship. Finally, these simulations provide further confirmation that the somatic EPSP time course is not always the best indicator of the local EPSP, suggesting that tuning uncaging pulses according to somatic recordings can be misleading.

In aspiny neurons, dendritic summation of synaptic or uEPSPs is more straightforward; however, depending on the thickness of the dendrites, the local depolarization can be large and the decay exceptionally fast, due to rapid charge redistribution down the dendrite [15].

3.3.2 Using 2P Uncaging to Mimic Synaptic Activity Patterns Evoked by Multiple Synapses

The morphological and intrinsic properties of dendrites can influence how multiple synaptic potentials summate [1]. Synaptic potentials can either summate *linearly* (if the response amplitude equals the arithmetic sum of the individual synaptic responses), *sublinearly* (if the response amplitude is less than the sum of the individuals), or *supralinearly* (if the response amplitude is greater than the sum of the individuals). This dendrite-specific mathematical operation has been proposed to greatly enhance the computational capacity of neurons [51], allowing neurons to improve the precision of information transfer [52], discriminate temporal input sequences [53], detect sparse input patterns [42], enhance coincidence detection [54], influence orientation tuning [55], store input features (memories) [56], contribute to active tactile sensation of rodent whiskers [57], and are likely critical for grid cell formation [58]. Convincing evidence also indicates that clustering of synaptic inputs carrying similar information determines output feature selectivity [59, 60]. Thus, the concerted action of select subsets of synapses is critical, since only specific spatiotemporal patterns of synaptic activation are likely to engage nonlinearities.

As discussed above, the restricted excitation volume achieved with 2P uncaging is well suited to mimic the activation of a single synapse. The use of rapid optical deflectors or spatial light modulators allows to perform synchronous activation of multiple synapses, even located on distant dendrites. Moreover, the flexibility of 2P uncaging enables the systematic examination of the number of synapses required to recruit nonlinearities. Specifically, this is studied by examining the *dendritic input/output relationship* (dI/O) (Fig. 6). Such relationships can be characterized by systematically uncaging at multiple synapses, then comparing the arithmetic sum of uncaging responses to individual sites ("expected") to compound responses evoked by "simultaneous" multisite uncaging protocols ("observed"). Nonlinear operations are those that deviate above or below a linear relationship between expected and observed compound EPSPs.

The dI/O is physiologically relevant when the uncaging-evoked local dendritic voltage amplitude and time course match those for synaptic activation; however, this is not always the case (see Sect. 3.1). Because dendritic nonlinearities are often mediated by NMDARs [14, 37], long uncaging pulses or large spots could result in excessive NMDAR activation and a confounding interpretation of the number of active synapses required to recruit nonlinearities. Thus, optimizing uncaging pulse durations and amplitudes is also an important step to achieve similar recruitment of NMDARs as for synaptic activation (*see* **Notes 4.2.1** and **4.2.2**).

To obtain the time course of synapse-evoked EPSCs, and guide tuning of uncaging parameters, quantal EPSCs evoked at the same electrotonic distance as the dendritic segment should be used as a

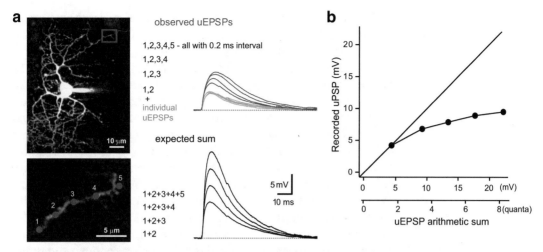

Fig. 6 Using 2P uncaging to examine dendritic synaptic input/output (dI/O) relationships. (**a**) Left, 2P image of a cerebellar stellate cell filled with Alexa Fluor 594 (top) and the dendrite where glutamate uncaging was elicited (area in red rectangle enlarged in the bottom image). Uncaging locations are indicated by red spots (1–5). Right, the top traces show uEPSPs recorded at the soma in response to near-simultaneous multisite uncaging at an increasing number of locations (blue traces) or individual site uncaging (gray traces). At the bottom are the expected EPSP waveforms, constructed by linearly summing the individual uEPSPs (black traces). (**b**) Dendritic I/O relationship resulting from (**a**) showing sublinear summation of uEPSPs. The lower x-axis represents the uEPSP amplitude transformed into the number of quanta, which was estimated from the quantal EPSP (~2.5 mV). (Figure is modified with permission from Abrahamsson et al. [42])

reference. If the circuit anatomy allows electrical stimulation of the presynaptic axons without excitation of other collaterals, as is the case for parallel fiber inputs in the cerebellum, this can be obtained by performing minimal extracellular stimulation of presynaptic inputs [42]. By lowering the extracellular [Ca²⁺] to a range in which release events are observed in <10% of the trials, the resulting average of the successful trials represents a good approximation of the locally evoked quantal EPSC [42]. Another strategy is to locally puff a high-osmolarity solution (such as sucrose) on a presynaptic terminal to evoke exocytosis of presynaptic vesicles [37]. Once the average quantal EPSC is determined, then amplitude and time course can be compared to uEPSCs. Using either a comparison of the amplitude or total synaptic charge, one can then calibrate the equivalent number of synaptic responses per uEPSC (Fig. 6 from [42]).

Synaptic activation can generate both electrical and chemical responses locally within dendrites. Electrical signals are driven by the total ion flux through neurotransmitter- and voltage-gated channels, whereas local intracellular [Ca²⁺] changes are driven by Ca²⁺ influx through local AMPAR, NMDAR, or VGCC. [Ca²⁺] transients can be recorded using 2P imaging of calcium indicators (e.g., Fluo-5F, ThermoFisher Scientific) and are often more spatially restricted than their electrical counterparts due to endogenous calcium buffering and geometrical constraints such as spine necks or dendritic branch-points. The local [Ca^{2+}] change can drive signal transduction mechanisms within restricted regions of the dendrite, for example driving a local activity-dependent regulation of synaptic plasticity [15, 61, 62]. We have shown previously that the combination of glutamate uncaging, whole-cell somatic patch-clamp electrophysiology, and Ca^{2+} imaging allowed us to determine electrical and chemical operations simultaneously during dI/O experiments (Fig. 7a–e; *see* also **Note 4.3**). We found that the synaptic ion flux that produces measurable local [Ca^{2+}] changes does not necessarily translate into significant voltage changes at the soma, due largely to high surface-to-volume ratio. In cerebellar stellate cells, the small volume imparted by thin-diameter (400 nm) dendrites results in a supralinear increase in [Ca^{2+}] in response to the combined activation of neighboring synapses, but the response involves only a small number of ion channels and thus a small Ca^{2+} flux, which contributes little to the local depolarization (Fig. 7f–h [15]). As a consequence, local synaptic plasticity can be triggered by [Ca^{2+}] changes without affecting the overall dendritic voltage operation [15].

In summary, the restricted photolysis volume provided by 2P excitation, combined with flexible and rapid positioning of the uncaging spot, allows for a flexible biophysical examination of dendritic integration of both electrical and chemical signals under near-physiological conditions.

4   Notes

4.1   Handling Caged Compounds

In order for an uncaging experiment to be effective, it is essential that the caged solution applied to the tissue is completely devoid of any bioactive uncaging products, i.e., free neurotransmitter. Good storage conditions help prevent the spontaneous hydrolysis of the caged compound. Stock solutions must generally be prepared in a HEPES-buffered ACSF at physiological pH, in the dark, aliquoted in order to minimize freeze/thaw cycles, and stored at −80 °C or −20 °C in light-protected tubes. During the experiment, the solution must be protected from exposure to light at wavelengths that could cause undesired uncaging. Any lamps used for ambient light and microscope trans-illumination used during approach of the patch-clamp electrode must be covered with UV and/or yellow filters, typically purchased from photography vendors (e.g., www.leefilters.com: 738 JAS Green or 010 Medium Yellow).

The high concentration of caged compound required to achieve a synaptic-like uncaging response, combined with the price of the compound, requires that the volume of extracellular solution containing the caged compound be reduced to a minimum. For bath perfusion of bicarbonate-buffered ACSF solutions, one strategy is to recirculate small volumes (10–15 ml) using a peristaltic

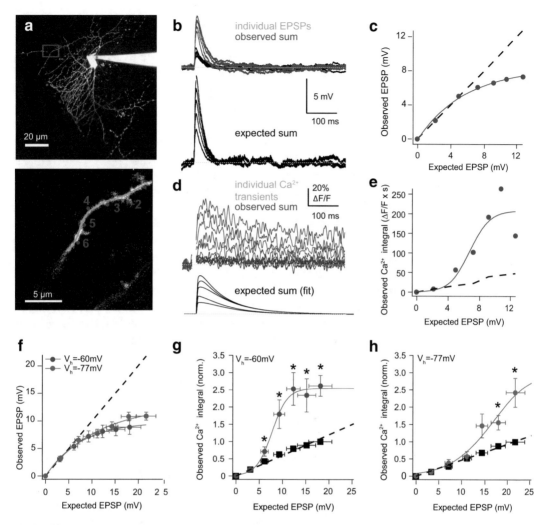

Fig. 7 Differential integration of uncaging-evoked electrical and biochemical responses in cerebellar interneuron dendrites. (**a**) 2P image of a cerebellar stellate cell filled with Alexa Fluor 594 (top) and dendrite (area in red rectangle enlarged in the bottom image) with uncaging locations indicated by red spots, and line scan location indicated by the blue line. (**b**) Top, somatic uEPSPs recorded in response to multisite uncaging (blue traces) or individual site uncaging (gray traces). Bottom, expected EPSPs constructed by summing the individual uEPSPs. (**c**) dI/O relationship resulting from data in (**b**). The dashed line is the linear relationship. (**d**) Same as in (**b**), but with corresponding uncaging-evoked dendritic Ca^{2+} transients (300 µM Oregon Green Bapta 5N) imaged using fast line scans. Multiexponential fits of Ca^{2+} transients are plotted below calcium transients. (**e**) Plot of the integral of fitted Ca^{2+} transients over a 590 ms window after the uncaging pulse, versus the expected EPSP amplitude. Dashed line is the plot of the arithmetic sum of the expected Ca^{2+} transient integral, against the expected EPSP amplitude. Blue and gray traces are the average of 10 trials. (**f**) Population average dI/O relationship ($n = 10$ cells) shows that EPSP summation is sublinear at a holding potential of either $V_h = -60$ mV or $V_h = -77$ mV. The dashed line is a linear extrapolation between the origin and the first data point of the d(I/O) when $V_h = -77$ mV. (**g, h**) The dI/O relationship for local Ca^{2+} summation, on the other hand, is supralinear, and its gain and threshold are both dependent on the holding potential (blue symbols). Black symbols represent the arithmetic sum of the Ca^{2+} transient integral versus the expected EPSP amplitude, and the dashed line is a linear fit to the data. $*p < 0.05$, paired Wilcoxon signed-rank test between observed and expected Ca^{2+} integral. (Figure is modified with permission from Tran-Van-Minh et al. [15])

pump (*MINIPULS 3* multichannel peristaltic pump, Gilson), short tubing, and a small reservoir. Bath perfusion of high concentrations of the caged compound requires that ACSF be prepared at slightly higher concentrations to account for the addition of the stock solution of caged compound, which includes 10 mM Na-HEPES and a maximum of 40 mM caged compound for MNI-glutamate (according to solubility). During the course of the experiment, the experimenter should monitor the change in volume of the extracellular solution due to evaporation, and top-up regularly the solution with an appropriate volume of water, to prevent unphysiological changes in osmolarity. This is easy to monitor if extracellular solution is being kept in a graduated tube (e.g., Falcon, 15 ml) or any other narrow graduated reservoir. The tube or cylinder should be covered with parafilm (except for small holes for CO_2/O_2 bubbling tubing). If recordings are performed at near-physiological temperatures (33–35 °C), we suggest an in-line heater (e.g., SF-28, Warner Instruments, www.warneronline.com). At these temperatures, we estimate an evaporation rate of 1 ml/h.

For local application of the caged compound on the dendrite of interest, we suggest using a glass perfusion pipette kept under constant pressure. Coupling a manometer to the perfusion pipette is useful to ensure that the pressure (and thus flow of solution out of the pipette) remains constant. An advantage of this method is that it requires smaller volumes of caged compounds and allows the local application of higher concentrations of caged compounds, therefore requiring shorter duration and lower power uncaging pulses. Another advantage is that because the extracellular solution does not need to be recycled, the accumulation of uncaging by-products during the course of the experiment is avoided. On the other hand, the cage solution is extruded from the perfusion pipette as a plume of limited size and inevitably produces a gradient of caged glutamate concentration. This restricts the area of the dendritic tree that can be tested at one given time to about 30 μm of dendrite (assumed orthogonal to the perfusion pipette). For optimal application of the highest concentration of caged glutamate on the dendrite of interest, it is important to position the perfusion pipette in the same focal plane. This can be performed using fluorescence imaging of the dendrite of interest and the perfusion pipette, both filled with a different fluorescent dye, or using IR-Dodt contrast (Luigs and Neumann) to visualize the perfusion pipette. Insertion of the perfusion pipette within the slice generally causes mechanical distortion of the tissue. To minimize this problem, we restrict the pipette placement to the most superficial 20 μm of the slice, and allow the tissue to relax for a few minutes after positioning of the pipette. It is generally preferable to avoid placement of the perfusion pipette in the vicinity of the patch pipette to prevent direct pressure and thereby possible loss of the patched cell or reduction of recording quality.

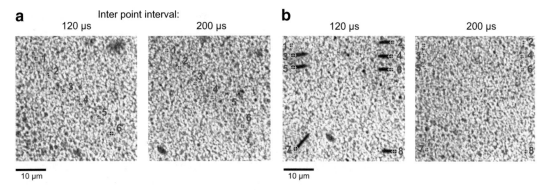

Fig. 8 Estimation of minimal point-to-point galvanometer mirror-based focal spot displacement interval. 2P images of a fluorescent slide (whiteboard marker on a microscope slide) imaged at 840 nm and bleached by rapid multisite laser illumination at 720 nm. To characterize the spatiotemporal accuracy of the uncaging system, we fixed the uncaging pulse duration to 200 μs but varied the interpulse duration and distance between illumination locations and imaged the region where fluorescence was bleached (black areas). Intended bleaching locations are indicated by yellow crosshairs and blue numbers. (**a**) Bleaching patterns obtained for interpoint intervals of 120 and 200 μs, and 7 μm apart. Bleached regions indicate the effective area which is illuminated and show an estimate of a putative region of uncaging. (**b**) Increasing the distance between uncaging points revealed a substantial elongation ("smearing") of the bleached regions for the 120 μs interpoint interval, but not for the 200 μs interpoint interval, showing that the minimal galvanometer move time that does not produce smearing is longer for larger distances between points

4.2 Calibrating Laser Illumination Pulse (Duration, Power, and Position)

4.2.1 Laser Pulse Duration

Nonlinear dendritic integration is highly dependent on the temporal synchrony of the synaptic inputs. Lengthening the intersite uncaging interval results in a reduced synchrony of neighboring synapses and thus in a reduction of the gain of the I/O function [63, 64]. In radial oblique dendrites of CA1 pyramidal neurons, increasing the laser pulse duration from 0.1 to 1 ms results in a transformation of the original supralinear I/O relationship to a linear I/O relationship. This results from asynchronous uEPSPs that fail to produce sufficient depolarization to activate the voltage-dependent conductances responsible for the generation of a dendritic spike [13]. It is therefore critical, when using scanning devices, to minimize the time needed to sequentially uncage at multiple locations in order to maintain near simultaneous synaptic activation. Moreover, the total duration of this protocol must remain consistent between cells/dendrites to allow for direct comparison of the I/O relationships.

Point-to-point deflection time is a critical parameter influencing the delay between uncaging locations and thus the degree of synchrony during a dI/O relationship experiment. The minimum displacement time of uncaging mirrors between two uncaging locations is fixed for a given uncaging system and is on the order of ~100–200 μs (Fig. 8). The duration of the photolysis laser pulse then adds to the total move time. Because 0.1–2 ms laser illumination durations are necessary to achieve physiologically relevant peak amplitudes of uEPSC/Ps, the temporal synchronicity of the multi-point protocol becomes notably reduced. Therefore, to achieve higher receptor activation, it is preferable to keep the laser pulse duration constant and change the power to increase the u[Glut]. This is commonly done to compensate for the loss of power due to tissue scattering deep in the tissue.

4.2.2 *Laser Pulse Power*

After the quantal size has been characterized (as described in Sect. 3.3.2), the power delivered at each uncaging location should be calibrated so that uEPSPs recorded in response to uncaging on an individual synapse approaches the quantal response for that location/region of the dendrite (Fig. 6b). We find that for 1 ms pulses, at locations less than 50 μm deep in the tissue, synaptic-like responses can be evoked by <20 mW (measured on the imaging side of the objective). In this case, it is also possible to normalize the amount of glutamate delivered at each synapse using fractional bleaching of fluorescence of the dye loaded via the patch pipette to visualize the dendrites and spines (e.g., Alexa Fluor dye) [65]. Ideally, the software should provide easy modulation of amplitude of the voltage pulse controlling the laser at each location (e.g., PrairieView, ScanImage) and visualization of the uEPSC/P.

4.2.3 *Laser Spot Positioning Accuracy*

Even in imaging systems optimized for fast focal spot displacement, the inertia of the galvanometric mirrors puts a lower bound to the minimal interpoint delay. When moving from one point to the next, each mirror needs to accelerate and decelerate again. If the galvanometric mirror is still moving during the uncaging light pulse, uncaging reactions can occur at undesired locations along the mirrors' displacement path. In order to correctly evoke synaptic-like responses at multiple uncaging locations, it is important to ensure that the laser pulse occurs during the time when the mirrors are stationary. In practice, this can be visualized by bleaching a fluorescent slide using the uncaging laser pulses (either a commercial one, e.g., FluorCal optical calibration slide from Valley Scientific, valley-scientific.com, or a microscope slide covered with whiteboard marker; Fig. 8). For a protocol that produces an uncaging laser pulse only at the desired location, one expects to observe discrete, round bleached spots, matching perfectly the target uncaging spots (Fig. 8a). If the interpulse interval is too short and the galvanometer mirror is not immobile for the duration of the laser pulse, one will observe elongated bleached spots, often looking "smeared" between intended spot locations (Fig. 8b).

Another strategy to confirm synchronized galvanometric mirror movements and laser illumination is to record the galvanometer position signals from the scan card, as well as timing of the laser light pulse using a microscope slide to reflect a small fraction of the illumination to a fast photodetector (e.g., PDA100A2, Thorlabs, www.thorlabs.com). With this configuration, it is possible to check that the light pulse occurs only when both mirrors are immobile. In addition, while current commercial microscopes are capable of parking the illumination spots reliably and precisely according to a reference image, it is important to confirm the accuracy of positioning (dependent on in-software calibration) and to regularly confirm that the tissue is not drifting during the experiment. During an experiment, drift of the tissue can be checked by frequently acquiring a high-magnification image of the dendrite on which uncaging spots are positioned and comparing with an image taken at the time when uncaging spots were positioned. Both drift in the x-y plane and of the focal plane should be corrected for, so that the current image and initial image are matching, either manually or via automatic software drift correction [26].

4.3 Technical Considerations for Simultaneous Uncaging and Calcium or Voltage Imaging

Combining fluorescence imaging of dendritic Ca^{2+} or voltage with an uncaging experiment requires synchronization of fast imaging with the delivery of the uncaging pulses. This can be achieved using fast imaging along a single line that traverses the longitudinal axis of the dendrite (linescan) and provide a temporal resolution of 1–1.5 kHz [15]. However, imaging is limited to a single focal plane and thereby restricts imaging of dendrites that project in three dimensions to length segments shorter than the total dendritic length. Random-access imaging of the dendrites using AODs can be combined with uncaging and can greatly extend the fraction of the dendritic tree that can be studied [18].

Combining Ca^{2+} or voltage imaging with uncaging often involves the use of dyes that can be excited by uncaging light. For example, fluorescein or Oregon Green BAPTA-based calcium indicators are partially excited at 720 nm (wavelength used for uncaging MNI-glutamate) and strongly excited at 800 nm (used for uncaging Rubi-glutamate). This causes a fluorescence artifact during the uncaging pulse [12, 15, 62]. It is important to check the decay kinetics of this artifact to make sure it does not contaminate the photolysis-evoked fluorescence transient that follows the uncaging pulse. This can be controlled by delivering the uncaging pulse in an area of the slice devoid of the dye. If the artifact is strictly limited to the duration of the laser pulse, then it can be safely blanked.

References

1. Rall W (1967) Distinguishing theoretical synaptic potentials computed for different soma-dendritic distributions of synaptic input. J Neurophysiol 30:1138–1168

2. Magee JC, Johnston D (1995) Characterization of single voltage-gated Na^+ and Ca^{2+} channels in apical dendrites of rat CA1 pyramidal neurons. J Physiol 487:67–90. https://doi.org/10.1113/jphysiol.1995.sp020862

3. Stuart GJ, Sakmann B (1994) Active propagation of somatic action potentials into neocortical pyramidal cell dendrites. Nature 367:69–72. https://doi.org/10.1038/367069a0

4. Williams SR, Mitchell SJ (2008) Direct measurement of somatic voltage clamp errors in central neurons. Nat Neurosci 11:790–798. https://doi.org/10.1038/nn.2137

5. Corrie JET (2005) Photoremovable protecting groups used for the caging of biomolecules. Dyn Stud Biol 1999:1–94

6. Lester HA, Nerbonne JM (1982) Physiological and pharmacological manipulations with light flashes. Annu Rev Biophys Bioeng 11:151–175. https://doi.org/10.1146/annurev.bb.11.060182.001055

7. Sarkisov DV, Wang SS-H (2007) Combining uncaging techniques with patch-clamp recording and optical physiology. In: Walz W (ed) Neuromethods, Patch-Clamp analysis, vol 38, 2nd edn. Humana Press, Totowa, NJ, pp 149–168. https://doi.org/10.1007/978-1-59745-492-6_5

8. Palma-Cerda F, Auger C, Crawford DJ et al (2012) New caged neurotransmitter analogs selective for glutamate receptor sub-types based on methoxynitroindoline and nitro-phenylethoxycarbonyl caging groups. Neuropharmacology 63:624–634. https://doi.org/10.1016/j.neuropharm.2012.05.010

9. Fino E, Araya R, Peterka DS et al (2009) RuBi-Glutamate: two-photon and visible-light photoactivation of neurons and dendritic spines. Front Neural Circuits 3:1–9. https://doi.org/10.3389/neuro.04.002.2009

10. Zipfel WR, Williams RM, Webb WW (2003) Nonlinear magic: multiphoton microscopy in the biosciences. Nat Biotechnol 21:1369–1377. https://doi.org/10.1038/nbt899

11. Trigo FF, Corrie JET, Ogden D (2009) Laser photolysis of caged compounds at 405 nm: photochemical advantages, localisation, phototoxicity and methods for calibration. J Neurosci Meth 180:9–21. https://doi.org/10.1016/j.jneumeth.2009.01.032

12. Carter AG, Sabatini BL (2004) State-dependent calcium signaling in dendritic spines of striatal medium spiny neurons. Neuron 44:483–493. https://doi.org/10.1016/j.neuron.2004.10.013

13. Losonczy A, Magee JC (2006) Integrative properties of radial oblique dendrites in hippocampal CA1 pyramidal neurons. Neuron 50:291–307. https://doi.org/10.1016/j.neuron.2006.03.016

14. Branco T, Häusser M (2011) Synaptic integration gradients in single cortical pyramidal cell dendrites. Neuron 69:885–892. https://doi.org/10.1016/j.neuron.2011.02.006

15. Tran-Van-Minh A, Abrahamsson T, Cathala L, DiGregorio DA (2016) Differential dendritic integration of synaptic potentials and calcium in cerebellar interneurons. Neuron 91:837–850. https://doi.org/10.1016/j.neuron.2016.07.029

16. Salierno M, Marceca E, Peterka DS et al (2010) A fast ruthenium polypyridine cage complex photoreleases glutamate with visible or IR light in one and two photon regimes. J Inorg Biochem 104:418–422. https://doi.org/10.1016/j.jinorgbio.2009.12.004

17. Ellis-Davies GCR, Matsuzaki M, Paukert M et al (2007) 4-Carboxymethoxy-5,7-Dinitroindolinyl-Glu: an improved caged glutamate for expeditious ultraviolet and two-photon photolysis in brain slices. J Neurosci 27:6601–6604. https://doi.org/10.1523/JNEUROSCI.1519-07.2007

18. Chiovini B, Turi GF, Katona G et al (2014) Dendritic spikes induce ripples in parvalbumin interneurons during hippocampal sharp waves. Neuron 82:908–924. https://doi.org/10.1016/j.neuron.2014.04.004

19. Gasparini S, Magee JC (2006) State-dependent dendritic computation in hippocampal CA1 pyramidal neurons. J Neurosci 26:2088–2100. https://doi.org/10.1523/JNEUROSCI.4428-05.2006

20. Fernandez-Alfonso T, Nadella KMNS, Iacaruso MF et al (2014) Monitoring synaptic and neuronal activity in 3D with synthetic and genetic indicators using a compact acousto-optic lens two-photon microscope. J Neurosci Meth 222:69–81. https://doi.org/10.1016/j.jneumeth.2013.10.021

21. Katona G, Szalay G, Maák P et al (2012) Fast two-photon in vivo imaging with three-dimensional random-access scanning in large tissue volumes. Nat Methods 9:201–208. https://doi.org/10.1038/nmeth.1851

22. Lutz C, Otis TS, DeSars V et al (2008) Holographic photolysis of caged neurotransmitters. Nat Methods 5:821–827. https://doi.org/10.1038/nmeth.1241

23. Hernandez O, Papagiakoumou E, Tanese D et al (2016) Three-dimensional spatiotemporal focusing of holographic patterns. Nat Commun 7:12716. https://doi.org/10.1038/ncomms12716

24. Anselmi F, Ventalon C, Begue A et al (2011) Three-dimensional imaging and photostimulation by remote-focusing and holographic light patterning. Proc Natl Acad Sci USA 108:19504–19509. https://doi.org/10.1073/pnas.1109111108

25. Bovetti S, Fellin T (2015) Optical dissection of brain circuits with patterned illumination through the phase modulation of light. J Neurosci Meth 241:66–77. https://doi.org/10.1016/j.jneumeth.2014.12.002

26. Smirnov MS, Evans PR, Garrett TR et al (2017) Automated remote focusing, drift cor-

rection, and photostimulation to evaluate structural plasticity in dendritic spines. PLoS One 12:1–14. https://doi.org/10.1371/journal.pone.0170586

27. Matsuzaki M, Ellis-Davies GC, Nemoto T et al (2001) Dendritic spine geometry is critical for AMPA receptor expression in hippocampal CA1 pyramidal neurons. Nat Neurosci 4:1086–1092. https://doi.org/10.1038/nn736

28. Jonas P (2000) The time course of signaling at central glutamatergic synapses. News Physiol Sci 15:83–89. https://doi.org/10.1113/jphysiol.1992.sp019417

29. Crank J (1975) The mathematics of diffusion, 2nd edn. Oxford University Press, Oxford

30. Nielsen TA, DiGregorio DA, Silver RA (2004) Modulation of glutamate mobility reveals the mechanism underlying slow-rising AMPAR EPSCs and the diffusion coefficient in the synaptic cleft. Neuron 42:757–771. https://doi.org/10.1016/j.neuron.2004.04.003

31. DiGregorio DA, Rothman JS, Nielsen TA, Silver RA (2007) Desensitization properties of AMPA receptors at the cerebellar mossy fiber granule cell synapse. J Neurosci 27:8344–8357. https://doi.org/10.1523/JNEUROSCI.2399-07.2007

32. Kiskin NI, Chillingworth R, McCray JA et al (2002) The efficiency of two-photon photolysis of a "caged" fluorophore, o-1-(2-nitrophenyl)ethylpyranine, in relation to photodamage of synaptic terminals. Eur Biophys J 30:588–604. https://doi.org/10.1007/s00249-001-0187-x

33. Koester HJ, Baur D, Uhl R, Hell SW (1999) Ca²⁺ fluorescence imaging with pico- and femtosecond two-photon excitation: signal and photodamage. Biophys J 77:2226–2236. https://doi.org/10.1016/S0006-3495(99)77063-3

34. Axelrod D, Koppel DE, Schlessinger J et al (1976) Mobility measurement by analysis of fluorescence recovery kinetics. Biophys J 16:1055–1069

35. Kiskin NI, Ogden D (2002) Two-photon excitation and photolysis by pulsed laser illumination modelled by spatially non-uniform reactions with simultaneous diffusion. Eur Biophys J 30:571–587. https://doi.org/10.1007/s00249-001-0186-y

36. McBain CJ, Fisahn A (2001) Interneurons unbound. Nat Rev Neurosci 2:11–23. https://doi.org/10.1038/35049047

37. Nevian T, Larkum ME, Polsky A, Schiller J (2007) Properties of basal dendrites of layer 5 pyramidal neurons: a direct patch-clamp recording study. Nat Neurosci 10:206–214. https://doi.org/10.1038/nn1826

38. Magee JC, Cook EP (2000) Somatic EPSP amplitude is independent of synapse location in hippocampal pyramidal neurons. Nat Neurosci 3:895–903

39. Smith MA, Ellis-Davies GCR, Magee JC (2003) Mechanism of the distance-dependent scaling of Schaffer collateral synapses in rat CA1 pyramidal neurons. J Physiol 548:245–258. https://doi.org/10.1113/jphysiol.2002.036376

40. Geiger JRP, Lübke J, Roth A et al (1997) Submillisecond AMPA receptor-mediated signalling at a principal neuron-interneuron synapse. Neuron 18:1009–1023

41. Cane M, Maco B, Knott G, Holtmaat A (2014) The relationship between PSD-95 clustering and spine stability in vivo. J Neurosci 34:2075–2086. https://doi.org/10.1523/JNEUROSCI.3353-13.2014

42. Abrahamsson T, Cathala L, Matsui K et al (2012) Thin dendrites of cerebellar interneurons confer sublinear synaptic integration and a gradient of short-term plasticity. Neuron 73:1159–1172. https://doi.org/10.1016/j.neuron.2012.01.027

43. Fortin DA, Tillo SE, Yang G et al (2014) Live imaging of endogenous PSD-95 using ENABLED: a conditional strategy to fluorescently label endogenous proteins. J Neurosci 34:16698–16712. https://doi.org/10.1523/JNEUROSCI.3888-14.2014

44. Tønnesen J, Katona G, Rózsa B, Nägerl UV (2014) Spine neck plasticity regulates compartmentalization of synapses. Nat Neurosci 17:678–685. https://doi.org/10.1038/nn.3682

45. Popovic MA, Carnevale N, Rozsa B, Zecevic D (2015) Electrical behaviour of dendritic spines as revealed by voltage imaging. Nat Commun 6:8436. https://doi.org/10.1038/ncomms9436

46. Acker CD, Hoyos E, Loew LM (2016) EPSPs measured in proximal dendritic spines of cortical pyramidal neurons. eNeuro 3:1–13. https://doi.org/10.1523/ENEURO.0050-15.2016

47. Bradley J, Luo R, Otis TS, DiGregorio DA (2009) Submillisecond optical reporting of membrane potential in situ using a neuronal tracer dye. J Neurosci 29:9197–9209. https://doi.org/10.1523/JNEUROSCI.1240-09.2009

48. Fink AE, Bender KJ, Trussell LO et al (2012) Two-photon compatibility and single-voxel, single-trial detection of subthreshold neuronal activity by a two-component optical voltage sensor. PLoS One 7. https://doi.org/10.1371/journal.pone.0041434

49. Palmer LM, Stuart GJ (2009) Membrane potential changes in dendritic spines during

action potentials and synaptic input. J Neurosci 29:6897–6903. https://doi.org/10.1523/JNEUROSCI.5847-08.2009

50. Krueppel R, Remy S, Beck H (2011) Dendritic integration in hippocampal dentate granule cells. Neuron 71:512–528. https://doi.org/10.1016/j.neuron.2011.05.043

51. Poirazi P, Brannon T, Mel BW (2003) Pyramidal neuron as two-layer neural network. Neuron 37:989–999. https://doi.org/10.1016/S0896-6273(03)00149-1

52. Hu H, Martina M, Jonas P (2010) Dendritic mechanisms underlying rapid synaptic activation of fast-spiking hippocampal interneurons. Science 327:52–58. https://doi.org/10.1126/science.1177876

53. Branco T, Clark BA, Häusser M (2010) Dendritic discrimination of temporal input sequences in cortical neurons. Science 329:1671–1675. https://doi.org/10.1126/science.1189664

54. Agmon-Snir H, Carr CE, Rinzel J (1998) The role of dendrites in auditory coincidence detection. Nature 393:268–272. https://doi.org/10.1038/30505

55. Mel BW, Ruderman DL, Archie KA (1998) Translation-invariant orientation tuning in visual "complex" cells could derive from intradendritic computations. J Neurosci 18:4325–4334. https://doi.org/10.1098/rspb.1986.0060

56. Losonczy A, Makara JK, Magee JC (2008) Compartmentalized dendritic plasticity and input feature storage in neurons. Nature 452:436–441. https://doi.org/10.1038/nature06725

57. Xu N, Harnett MT, Williams SR et al (2012) Nonlinear dendritic integration of sensory and motor input during an active sensing task. Nature 492:247–251. https://doi.org/10.1038/nature11601

58. Schmidt-Hieber C, Toleikyte G, Aitchison L et al (2017) Active dendritic integration as a mechanism for robust and precise grid cell firing. Nat Neurosci 20:1114–1121. https://doi.org/10.1038/nn.4582

59. Wilson DE, Whitney DE, Scholl B, Fitzpatrick D (2016) Orientation selectivity and the functional clustering of synaptic inputs in primary visual cortex. Nat Neurosci 19:1003–1009. https://doi.org/10.1038/nn.4323

60. Scholl B, Wilson DE, Fitzpatrick D (2017) Local order within global disorder: synaptic architecture of visual space. Neuron 96:1127–1138. https://doi.org/10.1016/j.neuron.2017.10.017

61. Soler-Llavina GJ, Sabatini BL (2006) Synapse-specific plasticity and compartmentalized signaling in cerebellar stellate cells. Nat Neurosci 9:798–806. https://doi.org/10.1038/nn1698

62. Weber JP, Andrásfalvy BK, Polito M et al (2016) Location-dependent synaptic plasticity rules by dendritic spine cooperativity. Nat Commun 7:11380. https://doi.org/10.1038/ncomms11380

63. Silver RA (2010) Neuronal arithmetic. Nat Rev Neurosci 11:474–489. https://doi.org/10.1038/nrn2864

64. Tran-Van-Minh A, Cazé RD, Abrahamsson T et al (2015) Contribution of sublinear and supralinear dendritic integration to neuronal computations. Front Cell Neurosci 9:1–15. https://doi.org/10.3389/fncel.2015.00067

65. Higley MJ, Soler-Llavina GJ, Sabatini BL (2009) Cholinergic modulation of multivesicular release regulates striatal synaptic potency and integration. Nat Neurosci 12:1121–1128. https://doi.org/10.1038/nn.2368

Chapter 4

Two-Photon Glutamate Uncaging to Study Structural and Functional Plasticity of Dendritic Spines

Ivar S. Stein, Travis C. Hill, Won Chan Oh, Laxmi K. Parajuli, and Karen Zito

Abstract

The activity-dependent structural remodeling of dendritic spines in response to sensory experience is vital for the dynamic regulation of neuronal circuit connectivity that supports complex behavior. Here, we discuss how the two-photon glutamate uncaging technique can be applied to study the mechanisms that drive activity-dependent structural and functional plasticity in individual dendritic spines. Our goal is to provide the reader with the key background for this technique and to present guidelines, practical details, and potential caveats associated with its implementation.

Key words Glutamate uncaging, Two-photon imaging, Dendritic spine, Synaptic plasticity, Glutamate receptor

1 Introduction

Experience- and activity-dependent modification of the number and strength of synaptic connections drives the refinement of neuronal circuits during learning. Most of the excitatory synapses in the mammalian cerebral cortex occur on dendritic spines, microscopic membranous protrusions from neuronal dendrites [1, 2]. Precise regulation of the growth, stabilization, and elimination of dendritic spines is necessary for learning [3–5]. Spine volume is also dynamically regulated and highly correlated with the number of AMPA-type glutamate receptors (AMPARs), which mediate fast excitatory synaptic transmission; thus, spine size is tightly linked to synaptic function [6]. Indeed, increased synaptic strength through the induction of long-term potentiation (LTP) is associated with spine enlargement [7, 8], while decreased synaptic strength via the induction of long-term depression (LTD) is associated with spine shrinkage or loss [8, 9]. Dysregulation of spine development and plasticity mechanisms can lead to alterations in dendritic spine

Espen Hartveit (ed.), *Multiphoton Microscopy*, Neuromethods, vol. 148,
https://doi.org/10.1007/978-1-4939-9702-2_4, © Springer Science+Business Media, LLC, part of Springer Nature 2019

morphologies and densities and is associated with cognitive impairments [10]. Thus, study of the mechanisms that regulate spine morphogenesis and stability is critically important for understanding the cellular basis of learning and also how these mechanisms are disrupted in disease.

In this chapter, we describe how photolysis of caged glutamate using two-photon excitation can be implemented to study the mechanisms that drive the structural and functional plasticity of dendritic spines. The development of caged neurotransmitters with adequate two-photon cross sections has revolutionized our ability to study and manipulate single synapses. Stimulation of individual synapses with light-induced neurotransmitter release offers many advantages over traditional electrophysiological approaches. First, light can be delivered with exceptional spatial precision and can be rapidly redirected to stimulate different locations, enabling precise spatiotemporal control compared to stimulation with microelectrodes or perfusion with pharmacological agents. Second, light can be delivered to synapses deep in tissue relatively noninvasively, as compared to electrode placement. Third, unlike electrical stimulation of afferent fibers, uncaging allows the stimulation of single synapses with a single neurotransmitter, permitting study of the effects of that specific neurotransmitter. Fourth, uncaging stimulation bypasses the presynaptic terminal and thus permits investigation of the effect of pharmacological manipulations exclusively on postsynaptic transmission and plasticity mechanisms. Finally, two-photon glutamate uncaging can be combined with calcium imaging and electrophysiological recording to study the functional properties of single synapses. Altogether, two-photon glutamate uncaging provides tremendous advantages for the study of single synapse structural and functional plasticity mechanisms.

1.1 Caged Neurotransmitters

Caged compounds are defined as biologically active molecules that have been rendered inert through chemical modification with a photolabile protecting group. Photostimulation with the appropriate wavelength and intensity of light breaks the covalent bond connecting the protective group, thereby "uncaging" and releasing the active biomolecule. The first successful applications of caged compounds involved the addition of ortho-nitrobenzyl (NB) protecting groups to ATP [11] and cAMP [12], which paved the way for the development of other NB-caged compounds, including neurotransmitters like glutamate [13–15]. However, NB photochemistry is not suitable for two-photon uncaging due to poor two-photon absorption cross section [16]. In this chapter, we focus specifically on the caged glutamates available for studying structural and functional plasticity of dendritic spine synapses. Many different caging strategies and syntheses have been used to produce caged glutamate which can be used with two-photon

Table 1
Properties of caged glutamate compounds suitable for two-photon uncaging

Cage	1P λ_{max} (nm)	Quantum Yield (QY)	Extinction coefficient (ε, M^{-1} cm^{-1})	$\varepsilon \times$ QY	2P λ_{max} (nm)	Rate (s^{-1})	2P action cross section (GM)	Commercially available
MNI-Glu	340	0.08	4500	383	730	10^5	0.06	Yes
CDNI-Glu	330	0.50	6400	3200	720	ND	0.06	No
MDNI-Glu	350	0.47	8600	4042	730	ND	0.06	Yes
RuBi-Glu	450	0.13	5600	728	800	$>10^5$	0.14	Yes
DEAC450-Glu	450	0.39	43,000	16,800	900	ND	0.50	No
PMNB-Glu	317	0.10	9900	990	800	ND	0.45	No
PENB-Glu	317	0.10	9900	990	740	ND	3.20	No

Abbreviations: *1P* one-photon, λ_{max} absorption maximum, *2P* two-photon, *ND* not determined
The properties of MNI-Glu [6], CDNI-Glu [17], MDNI-Glu [18, 19], RuBi-Glu [20], DEAC450-Glu [21], PMNB-Glu [22], and PENB-Glu [23] have been described. Suppliers of the commercially available glutamate uncaging reagents include, e.g., Tocris, Hello Bio, and Femtonics

excitation, including nitroindoline (NI) derivatives, coumarin derivatives, and novel inorganic compounds, such as ruthenium-bipyridine (RuBi) (Table 1) [6, 17, 20, 21, 24].

In order to be experimentally useful, caged compounds must exhibit several fundamental properties. Most importantly, photorelease or "uncaging" should be efficient in response to the uncaging light and must rapidly lead to the production of the biologically active molecule. The efficiency of photorelease depends on (1) the *extinction coefficient of a molecule*, or how well the caged compound absorbs light, (2) the *quantum yield*, or how often light absorption will lead to photorelease, and (3) the *rate at which light absorption leads to photorelease*. The overall quality of the caged compound is often assessed using the product of the extinction coefficient and the quantum yield at a particular wavelength, which reflects both the efficiency of light absorption and the probability that light absorption will result in successful photorelease.

To best mimic synaptic glutamate release, the rate of photorelease of active glutamate must be considerably faster than the rise time of synaptic glutamate receptor currents. AMPARs are responsible for the fast component of excitatory synaptic currents, which display rise times of 100–500 μs. Photorelease of 4-methoxy-7-nitroindolinyl (MNI)-glutamate occurs within 10 μs, and therefore uncaging of MNI-glutamate can be used to mimic glutamate release and to study excitatory postsynaptic currents [6]. The two-photon action cross section of MNI-glutamate is adequate at 0.06

GM (Göppert-Mayer unit, 1 GM = 10^{-50} cm^4 s/photon) at 730 nm [6]; however, it is ~1000-fold less efficient than its absorption of near-UV light, requiring the use of much higher concentrations of caged reagent (mM) and higher intensity light for successful two-photon uncaging [6, 25].

The successful biological application of caged neurotransmitters also requires *chemical stability* (no hydrolysis at physiological pH and temperature) and *biological inertness* of both the caged compound and the released photolabile protecting group. No hydrolysis of MNI-glutamate can be detected after 8 h at room temperature (pH 7.4) [26] and MNI-caged glutamate has been shown to be inert with no activation of glutamate receptors at concentrations up to 10 mM [6]. Surprisingly (and disappointingly), MNI-glutamate is not biologically inert in that it is a strong antagonist of GABA$_A$ receptors (GABA$_A$Rs) at concentrations commonly utilized for two-photon uncaging [20, 21, 25, 27, 28]. Thus, in the absence of tetrodotoxin (TTX) to block action potentials, bathing a cultured slice in MNI-glutamate can lead to epileptiform-like activity, which can cause problems for cellular and circuit level studies. This antagonistic GABA$_A$R side effect appears to be less dramatic, but still an issue with the more recently developed caged glutamate compounds CDNI-, RuBi-, and DEAC450- (a 7-diethylaminocoumarin derivative) glutamate [17, 20, 21], which are reported to have better two-photon action cross sections and higher extinction coefficients compared to MNI-glutamate (Table 1) and therefore can be used at lower concentrations, reducing the antagonism of GABA$_A$Rs.

Of the caged glutamates with adequate two-photon cross sections (Table 1), MNI-glutamate has been commercially available for the longest time and is well-studied and chemically stable, making it the most widely applied caged glutamate for two-photon uncaging (**see Note 1**). MNI- and CDNI-glutamate have two-photon absorption maxima at 720 nm, whereas RuBi-glutamate and DEAC450-glutamate are red-shifted. The two-photon absorption maximum for RuBi-glutamate is at 800 nm, which is close to the peak of the maximum power output of the Ti:sapphire lasers and therefore could be advantageous when splitting the uncaging beam with a spatial light modulator (SLM) [29, 30]. However, this wavelength is also closer to the wavelengths often used for simultaneous calcium imaging, and therefore care must be taken that the caged compound is not uncaged during imaging. DEAC450-glutamate is even further red-shifted with maximum two-photon photolysis at 900 nm [21]. Thus, DEAC450-caged compounds can be used in combination with other nitroindolinyl-cages like MNI or CDNI, which have their excitation peaks at 720 nm, for two-color, two-photon uncaging. Indeed, another coumarin derivative named N-DCAC-GABA already has been used in combination with CDNI-glutamate for simultaneous two-photon uncaging of glutamate and GABA to study how excitatory and inhibitory inputs shape dendritic integration [31]. Additional caged glutamate compounds suitable for two-photon uncaging that have been less widely tested include 3-(2-propyl)-4′-methoxy-4-nitrobiphenyl (PMNB)-glutamate and 3-(2-propyl-1-ol)-4′-tris-ethoxy(methoxy)-4-nitrobiphenyl (PENB)-glutamate (Table 1) [22, 23].

1.1.1 Applications of Two-Photon Glutamate Uncaging in the Study of Dendritic Spines

The development of caged compounds suitable for two-photon glutamate uncaging opened the door for the detailed study of synaptic plasticity mechanisms at the single spine level. Uncaging of MNI-glutamate was first used to map how the functional properties of individual spines relate to their structural properties and to demonstrate that spine volume is tightly correlated with the amplitude of excitatory postsynaptic current [6]. Matsuzaki and colleagues further pioneered studies on the structural plasticity of individual dendritic spines, showing that repetitive glutamate uncaging (usually at 0.5–2 Hz) under nominally Mg^{2+}-free conditions, or when coupled with postsynaptic depolarization to 0 mV, induces a rapid and selective enlargement of the stimulated spine, a single-spine structural long-term potentiation (sLTP) [7].

Several groups, including our own, have implemented two-photon glutamate uncaging to study in detail the structural and functional plasticity of spiny synapses. An initial study characterized the functional development of nascent dendritic spines. Notably, new spines expressed glutamate-evoked currents that were indistinguishable from those of mature spines of comparable size, demonstrating that the formation and growth of new spines is tightly coupled to formation and strengthening of glutamatergic synapses [32]. A subsequent study went on to show that these newly formed spines are stabilized long-term by high-frequency glutamate uncaging leading to the induction of LTP of synaptic transmission (Fig. 1a–d) [33]. In contrast, prolonged low-frequency (0.1 Hz) uncaging of glutamate at single spines resulted in spine shrinkage and synaptic depression (Fig. 1e–g) [34]. Intriguingly, NMDA-type glutamate receptor (NMDAR) activation can drive spine shrinkage in the absence of NMDAR-dependent ion flow and Ca^{2+} influx (Fig. 1h–l) [35]. Furthermore, a strong, high-frequency (5 Hz) glutamate uncaging stimulus can also drive the de novo growth of spines from the dendrite (Fig. 1m) [36–38]. These studies give a taste of how simultaneous two-photon imaging and glutamate uncaging can be used in combination with calcium imaging and electrophysiological recordings to study the synaptic plasticity mechanisms at single synapses that underlie learning.

This chapter provides detailed methods on how two-photon imaging can be combined with two-photon glutamate uncaging to

I. High-Frequency Uncaging (HFU) of glutamate increases uEPSC amplitude and size of new spines

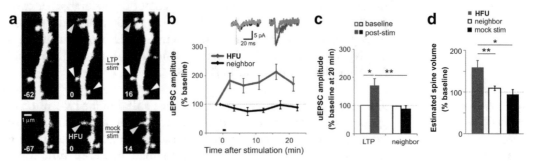

II. Low-Frequency Uncaging (LFU) of glutamate leads to synaptic depression and spine shrinkage

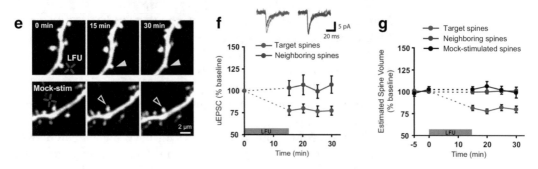

III. Block of ion flow through the NMDAR converts uncaging-induced spine enlargement into shrinkage

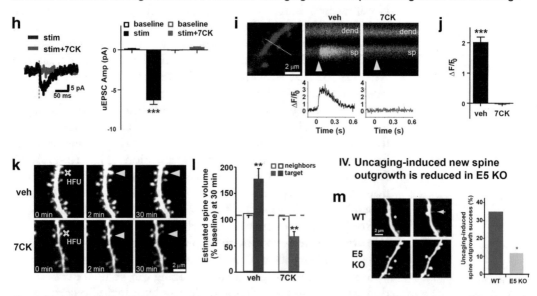

Fig. 1 Examples of plasticity induced by two-photon glutamate uncaging. (**a**) New spines (yellow arrowhead), but not neighbors, enlarge following high-frequency glutamate uncaging (HFU). (**b, c**) HFU leads to an increase in the uEPSC of the stimulated spine (red), but not of neighboring spines (black). (**d**) Mock stimulation (black) in the absence of MNI-glutamate does not induce spine growth. (**e**) Target spines shrink following low-frequency uncaging (LFU), but not following mock stimulation (red crosses). LFU but not mock stimulation leads to a decrease in uEPSC amplitude (**f**) and size (**g**) of target spines (red circles), but not unstimulated neighbors (blue circles). (**h**) NMDAR uEPSCs (black) are blocked

study the mechanisms that drive structural plasticity of dendritic spines on pyramidal neurons of the hippocampus.

2 Materials

2.1 Buffers/ACSF/ Drugs

Simultaneous two-photon imaging and glutamate uncaging experiments are performed in perfusing artificial cerebrospinal fluid (ACSF; in mM: 127 NaCl, 25 $NaHCO_3$, 1.2 NaH_2PO_4, 2.5 KCl, 25 D-glucose, aerated with 95% O_2/5% CO_2, ~310 mOsm, pH 7.2), typically containing 0 Mg^{2+}, 2 mM Ca^{2+}, 1 μM TTX (to block spontaneous spiking activity) and 2.5 mM MNI-glutamate (Tocris) (see Notes 1 and 2). Note that there can be batch-to-batch variations in the efficacy of MNI-glutamate (see Methods Sect. 3.6.1).

2.2 Two-Photon Microscope Setup

Figure 2a shows a typical layout of a two-photon microscope for simultaneous imaging and photolysis of caged compounds. For two-photon imaging, power of the infrared (IR) laser (mode-locked Ti:sapphire laser; e.g., Coherent [Santa Clara, CA] or Spectra-Physics [Mountain View, CA]) beam is controlled with a Pockels Cell (e.g., Model 350-80, Conoptics [Danbury, CT]), and xy position is controlled with galvanometers (e.g., Model 6210H, Cambridge Technology [Lexington, MA]). The beam is directed through a scan lens into a conventional upright microscope (e.g., Olympus [Tokyo, Japan]) and through a tube lens and a water-immersion objective designed to readily transmit IR light. To maximize detection of emitted photons, which originate almost exclusively from the focal volume, photomultiplier tubes (PMTs; e.g., R3896, Hamamatsu Photonics [Hamamatsu City, Japan]) both above and below the sample are used and the signals are summed electronically. A second pulsed IR laser beam for uncaging photostimulation is combined with the first beam using a polarizing beam-splitting cube (Thorlabs) and scanned either simultaneously with the imaging beam (Fig. 2a) or independently using a second set of scan mirrors. Alignment of the imaging and

---

Fig. 1 (Continued) by 7-chlorokynurenic acid (7CK, red). (**i**) Image of a dendritic segment from a cell transfected with a red cell fill (DsRedExpress) and the Ca^{2+} indicator GCaMP6 (green). Overlays of red and green fluorescence line-scan images of the target spines (sp) and dendrite (dend) from the region indicated by the white dashed line in the absence or presence of 7CK before and after glutamate uncaging (yellow arrowhead). (**i, j**) NMDAR Ca^{2+} transients (black) are completely blocked by 7CK (red). (**k, l**) HFU (yellow crosses) in the presence of 7CK (red) results in shrinkage of the target spine instead of enlargement as seen in vehicle conditions (blue). (**m**) A new spine (yellow arrow) forms following HFU (yellow circle) near a thickened, low-spine density dendritic segment from WT, but not Ephexin 5 (E5) KO mutant mice. The success rate of HFU-induced new spine outgrowth is higher in WT (~35%) than in E5 KO (10%). Panels (**a–d**) are adapted from Hill and Zito (2013) [33] with permission from the Society for Neuroscience. Panels (**e–g**) are adapted from Oh et al. (2013) with permission from the National Academy of Sciences [34] and panels (**h–l**) are adapted from Stein et al. (2015) [35] with permission from the Society for Neuroscience. Panel (**m**) is adapted from Hamilton et al. (2017) [36] with permission from Elsevier

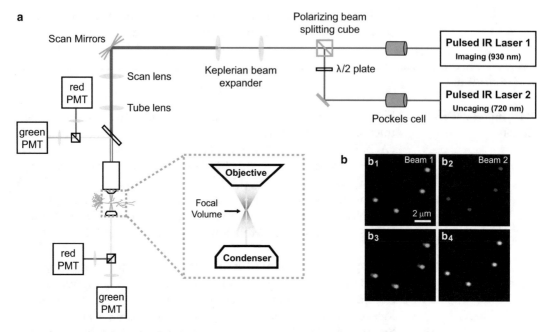

Fig. 2 Microscope setup for simultaneous two-photon imaging and glutamate uncaging. (**a**) For two-photon imaging, the power of the pulsed IR laser is controlled via a Pockels cell and the x-y position is controlled using scan mirrors. Two-photon excitation is restricted to the focal volume (gray inset) and because of this localization of excitation, all emitted photons contribute to the signal and can be collected. Top and bottom PMTs are used to maximize photon collection and the signals are summed electronically. A second pulsed IR laser is used for glutamate uncaging. In this schematic, both the imaging and the uncaging laser beams are controlled by the same set of scan mirrors, but for faster and independent control of the uncaging stimulus the laser beams also can be controlled by two independent sets of scan mirrors and only combined downstream. (**b**) Pseudocolored images of 0.2 μm red fluorescent beads excited with the imaging (b_1, green) or photostimulation (b_2, red) lasers. If the overlay of the bead images shows little overlap (b_3), the angular steering mirrors can be adjusted to bring the two beams in alignment and achieve complete overlap (yellow) of the pseudocolored bead images (b_4)

photostimulation lasers should be regularly monitored and optimized (see Methods Sect. 3.4).

2.3 Software

Data acquisition software needs to be versatile in order to accommodate changes in rig configuration and experimental design. A highly configurable software suitable for two-photon imaging and photostimulation experiments is the open-source software ScanImage and Ephus (http://openwiki.janelia.org/wiki/display/ephus/; [39]; now developed and supported by Vidrio Technologies, LLC), written in MATLAB (The Mathworks [Natick, MA]).

2.4 Electro physiology

Confirming the efficacy of caged neurotransmitters can be accomplished using simultaneous imaging and electrophysiology. Setting up electrophysiology on a two-photon microscope is relatively straightforward. At minimum, the setup consists of an amplifier, a head stage, recording and ground electrodes, micromanipulators, shielding, and a perfusion system. There are a few special considerations when setting up for simultaneous electrophysiology and two-photon microscopy. First, because two-photon imaging experiments are typically performed in vivo or in deep tissue of brain slices, the microscope should be upright (not inverted) and should have a water-immersion objective with a long working distance (providing space for the microelectrodes to access the specimen). We use an Olympus LUMPLFL60×/W, which has a numerical aperture (NA) of 1.0 and a working distance of 2 mm. Second, stepper motors and preamplifiers can be considerable sources of electrical noise, so care should be taken to properly ground these items. Third, shutters can be a considerable source of vibrational noise; anchoring shutters to the optical table using nylon screws can significantly diminish this noise. Finally, it is important to plan ahead so that recording pipettes are positioned in a way that does not interfere with image acquisition and, when using dye-filled pipettes, imaging regions should be distant from the pipette and dye spill should be minimized by approaching the cell quickly and reducing the positive pressure applied to the pipette because excessive dye spill confounds image analysis [40].

3 Methods

3.1 Preparation and Transfection of Organotypic Hippocampal Slice Cultures

Preparation of organotypic hippocampal slice cultures from postnatal day 6–8 (P6-8) rodent brain according to the interface method [41] has been described [42–44]. Biolistic transfection of rat hippocampal slices cultures has been described in detail elsewhere [45].

3.2 Preparation of Acute Hippocampal Slices

Acute slice preparation protocols will vary depending upon species, age, and brain area. Acute hippocampal slice preparation from P16-19 rodents has been described [46, 47]. Acute slices can be prepared from transgenic mice with sparsely labeled neurons, such as Thy-1-GFP-M [48], or from animals that have been transfected using in utero electroporation [49, 50] or transduced by injection with low titers of virus [51]. Alternatively, individual cells from wild-type (WT) slices can be loaded with dyes or calcium indicators by diffusion from patch pipettes [52, 53].

3.3 Two-Photon Time-Lapse Imaging

When imaging the morphology of fluorescently labeled neurons in brain tissue, two-photon microscopy provides distinct advantages due to reduced background fluorescence and decreased phototoxicity compared to wide-field and confocal microscopy [54].

Step-by-Step Instructions:

1. Screen for fluorescently labeled CA1 pyramidal neurons using a dissecting stereomicroscope with fluorescence or an epifluorescence microscope with a 10× air objective. Note that following biolistic transfection, incubate 1–2 days for expression of small cytosolic proteins, such as Enhanced Green Fluorescent Protein (EGFP), and 3–4 days for expression of large or membrane-bound proteins. Viral expression typically takes 1–2 weeks.

2. Place organotypic or acute hippocampal slices in an imaging chamber perfused with recirculating ACSF with custom concentrations of Mg^{2+} and Ca^{2+}. For uncaging spine plasticity experiments, the ACSF often contains nominally 0 Mg^{2+}, 2 mM Ca^{2+}, 0.001 mM TTX, and 2.5 mM MNI-glutamate (see Notes 1 and 2) and is maintained at 30 °C (e.g., heater Model TC-324B, Warner Instruments). Organotypic slices can be held in place with a horseshoe of inert gold wire (Alfa Aesar # 10966-BQ), which is placed to weigh down the attached piece of membrane from the cell culture insert (Millipore # PICM0RG50). Acute slices can be held in place with a harp (Warner Instruments # 64-1421).

3. Neurons situated in a healthy (see Note 3) cell layer typically at depths of 20–50 μm are imaged using a custom two-photon microscope with a pulsed Ti:sapphire laser tuned to 930 nm, which excites GFP and most red fluorophores, simultaneously. To avoid bleaching and phototoxicity, use imaging powers in the range of 0.5–2 mW at the sample (hand-held power meter, e.g., Coherent # 1098293).

4. For each neuron, image stacks (512 × 512 pixels; ~0.02–0.04 μm per pixel) with 1 μm z-steps (typically 10–15 slices) are collected from secondary or tertiary apical or basal dendrites, 40–100 μm from the soma. Each dendritic segment is repeatedly imaged at defined intervals (e.g., every 5 min). Typically, at least two baseline images are taken before glutamate uncaging, and then the dendritic segments are followed for at least 30 min following uncaging stimulation.

3.4 Beam Alignment   Precise alignment of the imaging and uncaging lasers is essential for accurate and successful single spine stimulation. Laser alignment can be checked using subresolution 0.1 or 0.2 μm fluorescent beads (e.g., F8810, F8803, Thermo Fisher Scientific), which can be alternately excited using either the imaging or uncaging laser (Fig. 2b). If the beams are not in perfect alignment, the overlay of the bead images will not overlap completely, and the angular steering mirrors should be adjusted until the two beams are in alignment in xy and there is complete overlap of the bead images excited with the two different lasers. Because the resolution limit in the z-axis is ~2 μm, the slight offset of the uncaging (720 nm) and imaging (930 nm) beams in z-axis due to the different focal points of the distinct wavelengths is typically not of concern.

3.5 Photostimulation

Photostimulation can be accomplished either during scanning or by parking of the uncaging laser beam at a single spot. Photostimulation during scanning allows for simultaneous imaging and real time monitoring of the uncaging beam localization but has the disadvantage that the stimulation time and frequency are constrained by the field of view. Consequently, this configuration usually requires shorter stimulation times and higher laser powers. Beam parking, on the other hand, has the advantage of flexible stimulation times and frequencies but, when used on setups with only one set of scan mirrors, is associated with a slight delay before imaging and does not allow real time monitoring of the uncaging beam or sample drift. The accuracy of photostimulation during beam parking with one set of scan mirrors can be tested by bleaching of fluorescent beads (Fig. 3a–d) or bleaching of specific patterns in a fluorescent slide (e.g., 2273, Ted Pella; Fig. 3e, f).

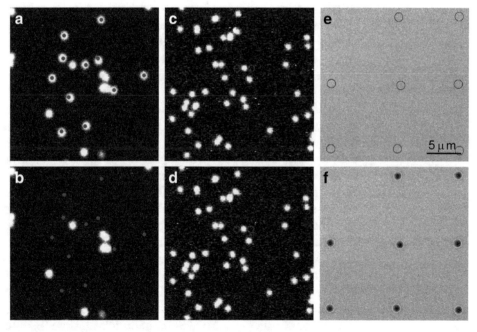

Fig. 3 Testing the accuracy of beam parking for photostimulation. Images of subresolution 0.2 μm red fluorescent beads before (**a**) and after (**b**) photobleaching of a selected subset of the beads (red dots). Due to the localization of excitation of two-photon imaging, targeting of the photostimulation directly next to the fluorescent beads (red circles, **c**), instead of directly on them (**a**, red dots), does not lead to bleaching (**d**). (**e**) Image of a fluorescent slide on which the beam parking targets were defined (red circles) and selectively bleached, indicated as loss of fluorescence signal in the green circles (**f**)

3.6 Glutamate Uncaging to Study the Structural and Functional Plasticity of Dendritic Spines

Two-photon uncaging of MNI-glutamate is achieved through irradiation with a pulse (0.25–4 ms) of 720 nm light from a mode-locked Ti:sapphire laser. The power used to uncage MNI-glutamate will depend on the pulse duration, the depth of the target in the tissue, the concentration of MNI-glutamate in the bath, and the numerical aperture of the objective. The duration and power of the uncaging stimulus can be adjusted so that the kinetics of the uncaging-evoked Excitatory PostSynaptic Currents (uEPSCs) closely mimic those of miniature excitatory postsynaptic currents (mEPSCs) [6]. The spatial resolution of two-photon uncaging permits stimulation of individual dendritic spines [6].

3.6.1 Calibration of the Glutamate Uncaging Stimulus

To calibrate the glutamate uncaging stimulus, visually identified CA1 pyramidal neurons in slice culture (DIV14–18, depths of 20–50 µm) are patched (pipette tip resistances, 5–7 MΩ) in whole-cell configuration (V_{hold} = −65 mV; series resistances, 15–45 MΩ) using Cs-based internal solution (in mM: 135 Cs-methanesulfonate, 10 HEPES, 10 Na_2 phosphocreatine, 4 $MgCl_2$, 4 Na_2-ATP, 0.4 Na-GTP, 3 Na L-ascorbate, and 0.2 Alexa Fluor 488, and ~300 mOsm, pH ~7.25) in standard ACSF containing 2 mM Ca^{2+}, 1 mM Mg^{2+}, 1 µM TTX, and 2.5 mM MNI-glutamate at 25 °C. uEPSCs are evoked using 1 ms laser pulses (720 nm) at the power to be tested (typically around 10 mW at the sample) at five spines per cell within 50 µm of the soma on secondary and tertiary apical and basal dendrites (Fig. 4a). Data acquisition is managed with open-source Ephus software (http://openwiki.janelia.org/wiki/display/ephus/; [55]).

uEPSC amplitudes from individual spines are quantified as the average (5–7 test pulses at 0.1 Hz) from a 2 ms window centered on the maximum current amplitude within 50 ms following pulse delivery. The average uEPSC amplitudes of these 5–7 test pulses should be robust and stable, although individual responses to each single photostimulation are variable (Fig. 4b). Uncaging pulse power should be adjusted with each individual batch of MNI-glutamate to elicit equivalent responses. We aim for an average amplitude of ~10 pA recorded at the soma (at a concentration of 2.5 mM MNI-glutamate and pulse duration of 1 ms) (Fig. 4c), which mimics the amplitude of a quantal response measured at moderate to large synapses [56]. To guarantee stable and comparable uEPSC amplitudes, choose a test pulse frequency that does not lead to changes in the uEPSC amplitude with time (Fig. 4d, e). In addition, the stimulated spine should be greater than 1 µm away from neighboring spines to minimize activation of nearby spines due to glutamate spillover.

It is helpful if the software used for extended time-lapse imaging and glutamate uncaging incorporates a mechanism to correct for drift. In our customized ScanImage software, lateral drift is determined by acquisition of an image immediately before the uncaging stimulation and calculating the cross correlation with an

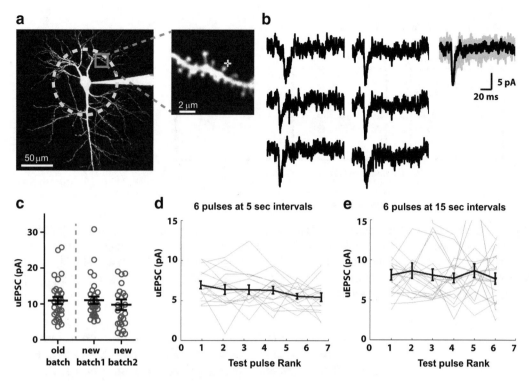

Fig. 4 Testing of individual MNI-glutamate batches. (**a**) Image of a CA1 pyramidal cell filled with Alexa 488 through the recording electrode. Spines are stimulated within a 50 μm radius from the cell body (illustrated as dashed yellow line). Inset shows a dendritic segment with the stimulated spine (yellow cross). (**b**) Individual uEPSC traces evoked by six subsequent uncaging pulses on the same spine (black) and the representative average uEPSC (black) with the six individual responses (gray). (**c**) The average AMPAR uEPSC amplitudes (mean ± SEM) between the old and two new tested batches of MNI-glutamate show no difference between the old and first new batch of glutamate. The second new tested batch of MNI-glutamate has a slightly reduced average AMPAR uEPSC amplitude, which could be adjusted by increasing the power of the uncaging laser pulse. (**d**) Delivery of uncaging test pulses at 5 s intervals on the same spine leads to a reduction of the average uEPSC amplitude (black) over time. (**e**) Increasing the test pulse interval to 15 s results in a stable average uEPSC amplitude (black) over time

initial reference image from the same time series. The drift is then automatically corrected by adjustment of the galvanometer scanning angles (Fig. 5).

The calibration of the MNI-glutamate batch is based on averaging the responses of multiple spines on multiple cells at varying depths in the tissue. It is important to note that, even if the laser power at the objective is held constant, the laser power delivered to the target spine will vary depending on the depth of the target spine in the brain slice and the properties of the tissue above the dendritic region of interest (ROI), which influence how much of the excitation light is scattered and absorbed. Inhomogeneous refraction indices due to tissue structure can affect the degree and volume over which the delivered power is spread. Because it is

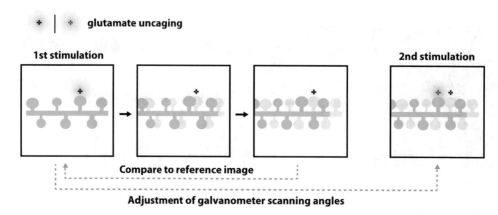

Fig. 5 Drift correction by cross-correlation. Our image acquisition and uncaging software is designed to automatically collect a frame scan image immediately before every uncaging stimulation, and the cross-correlation of this image (green) with a reference image (gray) is used to calculate lateral drift, which is automatically corrected by adjustment of the galvanometer scanning angles so that the uncaging spot (black cross) is redirected to the target spine (red cross)

usually not possible to measure these differences in optical properties of individual brain slices, we typically limit ourselves to a depth range of 20–40 μm and operate under the assumption that variations will average out across multiple cells in different slices from multiple different dissections. Alternatively, others have used a photobleaching protocol to calibrate the laser power for each spine of interest by monitoring bleaching fraction and adjusting power to bleach a constant fraction of the fluorescent dye that fills the spine [57]. While elegant in design, one caveat with this approach is that direct irradiation of the spine with high laser powers can result in cell damage.

3.6.2 Repetitive Glutamate Uncaging for the Induction of Structural Plasticity

The repetitive two-photon uncaging of MNI-glutamate has been shown to induce both input-specific LTP of synaptic transmission and enlargement of spine size (sLTP) [7, 33, 46, 58]. Repetitive glutamate uncaging, at frequencies of 0.5–2 Hz (30–60 stimuli) under low Mg^{2+} conditions, or when paired with postsynaptic depolarization to 0 mV, induces a robust, long-term enlargement of the stimulated spine head and increase in uncaging-evoked spine EPSCs that are stable for at least the next 30 min (Fig. 1a–d, k, l).

In the example illustrated in Fig. 6, the structural plasticity-inducing high-frequency uncaging stimulus consisted of 60 pulses (720 nm, 10 mW at the sample) of 2 ms duration delivered at 2 Hz in 2 mM Ca^{2+}/0 Mg^{2+} ACSF containing 1 μM TTX and 2.5 mM MNI-glutamate ([35, 59] or see similar protocols [7, 33, 46, 60, 61]). During the photostimulation, the uncaging beam (720 nm) is parked at a point ~0.5 μm from the edge of the spine head on the side furthest from the dendrite. Typically, only one dendritic region of interest is stimulated and imaged per cell (*see* **Notes 4 and 5**).

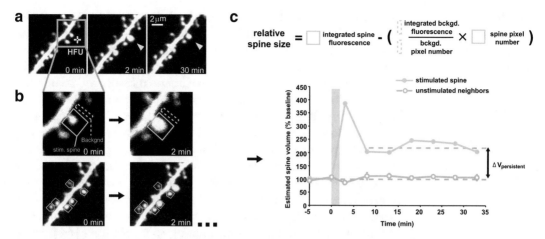

Fig. 6 Analysis of spine size following glutamate uncaging-induced structural plasticity. (**a**) Time-lapse images showing enlargement of the target spine (yellow arrowheads) in response to high-frequency uncaging (HFU, yellow cross) of glutamate. (**b**) 2x magnification of the area around the target spine before (0 min, left) and after (2 min, right) glutamate uncaging. The target region of interest (ROI) drawn around the stimulated spine (solid yellow box) is drawn on the slice in the z-stack in which the analyzed spine is the brightest. A thinner neighboring ROI spanning the same lateral distance from the dendrite (dashed yellow box) is used to measure the fluorescent intensity of the adjacent background. (**c**) The average pixel intensity from the background box is multiplied with the number of pixels of the target ROI to generate the integrated background value, which is subtracted from the integrated pixel intensity of the target spine to calculate the actual fluorescence intensity of the target spine. Graph shows the time course of the estimated volume (normalized to the average of the two baseline images before HFU) of the stimulated spine (yellow) compared to the unstimulated neighbors (green). $\Delta V_{persistent}$ is the stable increase observed 30 min after HFU

One limitation to consider when studying simultaneous functional and structural plasticity under whole-cell recording conditions is the loss of the ability to induce synaptic plasticity soon after the start of the recording (~5 min) due to the washout (into the recording pipette) of intracellular signaling molecules critical for the induction of plasticity [62, 63]. The rate with which washout influences plasticity induction depends on the recording electrode resistance and the distance of the stimulated spine synapse from the electrode. Washout critically limits the time available for a stable baseline recording before induction of structural and functional plasticity, and in a CA1 pyramidal cell can occur within ~5 min after achieving whole-cell configuration for synapses within 100 μm from the soma. To avoid this washout of intracellular signaling molecules, perforated patch-clamp recordings from neurons transfected with fluorescent proteins can be used [46].

3.7 Quantification of Fluorescence Intensities in Dendritic Spines

Because the size of dendritic spines is often smaller than the resolution limit of the two-photon microscope, spine size and morphology cannot be determined by measuring the apparent spine head diameter. Instead, we use the integrated fluorescence intensity from the spine head as an approximate measure of spine size. Transfection with genetically encoded monomeric and freely diffusible cytoplasmic fluorescent proteins like EGFP highlights cellular morphology and also can be used to assess the relative volume of dendritic spines. Assuming a relatively homogeneous distribution of the fluorescent cell fill, spine brightness should be proportional to the spine volume. Indeed, comparison of fluorescence intensity measurements and subsequent reconstruction with serial section electron microscopy supports that spine brightness is a fairly accurate method for estimating spine volumes [64]. Spine size analysis can be performed using most image analysis software, including ImageJ/Fiji.

Spine brightness or the estimated spine volume is measured from the stimulated target spine and from those neighboring spines that are isolated from other spines and distinguishable from the parent dendrite (laterally projecting from the dendrite in the xy plane) throughout the whole time-lapse imaging series. Due to the lower resolution on the z-axis, spines projecting in the z plane are difficult to distinguish and therefore are not analyzed. Estimated spine volume is measured from background-subtracted green (EGFP) fluorescence images using the integrated pixel intensity of a boxed ROI surrounding the spine head (Fig. 6). The ROIs are drawn in the single z-stack slice in which the analyzed spine is the brightest and the summed pixel intensities (integrated fluorescence intensity) are calculated. To account for the higher background fluorescence in the immediate vicinity of the dendrite, the adjacent background fluorescence is subtracted. Because neighboring spines or other bypassing neurites often prevent placement of the same-sized box adjacent to the target spine, the average pixel intensity of a thinner box spanning the same lateral distance from the dendrite as the target spine box is calculated, multiplied with the pixel number of the target spine box, and subtracted (Fig. 6b, c). The background-subtracted fluorescence intensity is proportional to the spine volume and can be compared to baseline before plasticity induction or to that of unstimulated neighbors (Fig. 6c).

When a new uncaging-induced spine structural plasticity protocol is established, the measured structural changes should be compared to a mock stimulation, which is carried out under identical conditions in the absence of MNI-glutamate [33, 34]. In situations where glutamate spill over is a concern, a laterally shifted stimulation, which releases the same amount of glutamate at a similar distance away from the dendrite without an intervening spine, is an important control [32, 59].

The sample size required to achieve statistical significance depends on the response size and variability that results from the specific structural or functional plasticity paradigm. We average all spines from one cell and then calculate statistics across cells. At minimum, three independent slice culture dissections or acute slice preparations from rodents from both sexes are utilized for each measurement. Image analysis is performed blind to the experimental condition.

Step-by-Step Instructions:

1. Open the images from the time-lapse series and make sure the target spine is clearly identifiable and not obscured by any other structures throughout the whole time-lapse series.

2. Starting with the first time point, identify the image slice in the z-stack in which the target spine is the brightest.

3. Draw a rectangular box around the target spine. Make sure the size of the box can accommodate the changes in spine morphology observed throughout the time-lapse series (*see* **Notes 6 and 7**). Measure the *integrated* fluorescence intensity from this box.

4. Draw a thin box in the neighboring region to the spine box that spans equal distance from the parent dendrite. Due to the high spine densities in mature hippocampal slices, the measurement of a thin background box (as shown in Fig. 6b) is usually preferable, because it will not overlap with neighboring spines or dendritic structures. Measure the *average* fluorescence intensity in this box.

5. To calculate the actual fluorescence intensity of the target spine, background subtraction must be performed. Multiply the average pixel intensity for the thinner background box with the pixel number of the target spine box to calculate the background value, which is now normalized for the difference in size of the background box and can be subtracted from the integrated pixel intensity of the target spine box to give the fluorescence intensity of the target spine (Fig. 6c).

6. Copy the same boxes over to the appropriate image slice in the z-stack from the next time point and repeat the measurements for the remainder of the time-lapse series. For each time point, make sure that the values are measured in the z-stack slice where the target spine is the brightest.

4   Notes

1. Although MNI-glutamate does not spontaneously hydrolyze or photolyze easily, for exposures longer than a few minutes it is wise to keep the main room lights off or to cover the lights with a yellow filter (Rosco #10 filter). These precautions are particu-

larly important for the red-shifted RuBi-glutamate, which is more light-sensitive than MNI-glutamate. For RuBi-glutamate usage, computer screens should be covered with red filters (Rosco #27 filter), absorbing the blue and green wavelengths [20].

2. To minimize the amount of MNI-glutamate used per experiment, ACSF volume is kept low (5–7 ml) and the bath is recirculated using a peristaltic pump (Watson Marlow Sci Q400). Recirculating ACSF is aerated with 95% O_2/5% CO_2, which is first bubbled through a conical tube containing ultrapure water. This additional step allows the humidification of the 95% O_2/5% CO_2 gas mixture before entering the ACSF solution, thus, reducing evaporation and minimizing changes in ACSF osmolarity.

3. Use only intact hippocampal slices with clearly visible dentate gyrus (DG), CA3 and CA1 regions, which are thinned down to 4–8 layers of cells with smooth visible cell somata. Exclude any unhealthy slice cultures and cells. Healthy transfected CA1 pyramidal cells should display fluorescence throughout their dendritic arbor (in case of small, freely diffusible fluorescence molecules) and show no signs of degeneration, with a typical spine density and no signs of dendritic blebbing. Make sure overall spine density does not significantly change during the time-lapse imaging session, because widespread spine elimination or excessive formation of filopodia-like structures can be a sign of compromised cell health.

4. When studying mechanisms of spine growth, choose small spines for studies of spine enlargement and stabilization, as large spines have been reported to show only a transient increase in spine size [7]. For studies on spine shrinkage and elimination, use spines that have been present already for at least one baseline time point during the time-lapse imaging series prior to uncaging stimulation in order to reduce the percentage of transient spines that spontaneously shrink and eliminate in the absence of stimulation.

5. Signaling molecules from the stimulated spine can spread into the parent dendrite over 5–10 μm and therefore, unless the goal is to study cross-talk, it is best to avoid stimulation of a second spine on the same dendritic segment [46, 65, 66]. Note that it also has been reported that the stimulation of three or more spines on dispersed dendritic branches can be sufficient to activate nuclear Erk and thus modify cellular transcription rates [67].

6. When choosing the target spine ROI for analysis of structural plasticity, keep in mind that the spine will grow, so be sure to draw the ROI large enough to include all of the spine even on the final image. Also, pay attention that the ROI is aligned to the spine in all time points, even in the case that the image is drifting.

7. It is not possible to make accurate measurements of stubby spines, which exhibit no detectable spine neck. Without an obvious separation from the dendrite due to the spine neck, the distinction of the spine from the dendrite is arbitrary and fluorescence signals from spine and dendrite will be mixed and inseparable.

Acknowledgments

This research was supported by NIH grants R01 NS062736 and U01 NS103571 from the National Institute of Neurological Disorders and Stroke.

References

1. Harris KM, Kater SB (1994) Dendritic spines: cellular specializations imparting both stability and flexibility to synaptic function. Annu Rev Neurosci 17:341–371

2. Yuste R, Majewska A, Holthoff K (2000) From form to function: calcium compartmentalization in dendritic spines. Nat Neurosci 3:653–659

3. Hayashi-Takagi A, Yagishita S, Nakamura M et al (2015) Labelling and optical erasure of synaptic memory traces in the motor cortex. Nature 525:333–338

4. Segal M (2017) Dendritic spines: morphological building blocks of memory. Neurobiol Learn Mem 138:3–9

5. Stein IS, Zito K (2018) Dendritic spine elimination: molecular mechanisms and implications. Neuroscientist. https://doi.org/10.1177/1073858418769644

6. Matsuzaki M, Ellis-Davies GC, Nemoto T et al (2001) Dendritic spine geometry is critical for AMPA receptor expression in hippocampal CA1 pyramidal neurons. Nat Neurosci 4:1086–1092

7. Matsuzaki M, Honkura N, Ellis-Davies GC et al (2004) Structural basis of long-term potentiation in single dendritic spines. Nature 429:761–766

8. Okamoto K, Nagai T, Miyawaki A et al (2004) Rapid and persistent modulation of actin dynamics regulates postsynaptic reorganization underlying bidirectional plasticity. Nat Neurosci 7:1104–1112

9. Zhou Q, Homma KJ, Poo MM (2004) Shrinkage of dendritic spines associated with long-term depression of hippocampal synapses. Neuron 44:749–757

10. Penzes P, Cahill ME, Jones KA et al (2011) Dendritic spine pathology in neuropsychiatric disorders. Nat Neurosci 14:285–293

11. Kaplan JH, Forbush B 3rd, Hoffman JF (1978) Rapid photolytic release of adenosine 5'-triphosphate from a protected analogue: utilization by the Na:K pump of human red blood cell ghosts. Biochemistry 17:1929–1935

12. Engels J, Schlaeger EJ (1977) Synthesis, structure, and reactivity of adenosine cyclic 3',5'-phosphate benzyl triesters. J Med Chem 20:907–911

13. Callaway EM, Katz LC (1993) Photostimulation using caged glutamate reveals functional circuitry in living brain slices. Proc Natl Acad Sci U S A 90:7661–7665

14. Wieboldt R, Gee KR, Niu L et al (1994) Photolabile precursors of glutamate: synthesis, photochemical properties, and activation of glutamate receptors on a microsecond time scale. Proc Natl Acad Sci U S A 91:8752–8756

15. Wilcox M, Viola RW, Johnson KW et al (1990) Synthesis of photolabile precursors of amino acid neurotransmitters. J Org Chem 55:1585–1589

16. Aujard I, Benbrahim C, Gouget M et al (2006) O-nitrobenzyl photolabile protecting groups with red-shifted absorption: syntheses and uncaging cross-sections for one- and two-photon excitation. Chemistry 12:6865–6879

17. Ellis-Davies GC, Matsuzaki M, Paukert M et al (2007) 4-Carboxymethoxy-5,7-dinitroindolinyl-Glu: an improved caged glutamate for expeditious ultraviolet and two-photon photolysis in brain slices. J Neurosci 27:6601–6604

18. Chiovini B, Turi GF, Katona G et al (2014) Dendritic spikes induce ripples in parvalbumin interneurons during hippocampal sharp waves. Neuron 82:908–924

19. Fedoryak OD, Sul JY, Haydon PG et al (2005) Synthesis of a caged glutamate for efficient one- and two-photon photorelease on living cells. Chem Commun (Camb) 29:3664–3666

20. Fino E, Araya R, Peterka DS et al (2009) RuBi-glutamate: two-photon and visible-light photoactivation of neurons and dendritic spines. Front Neural Circuits. https://doi.org/10.3389/neuro.04.002.2009

21. Olson JP, Kwon HB, Takasaki KT et al (2013) Optically selective two-photon uncaging of glutamate at 900 nm. J Am Chem Soc 135:5954–5957

22. Gug S, Charon S, Specht A et al (2008) Photolabile glutamate protecting group with high one- and two-photon uncaging efficiencies. Chembiochem 9:1303–1307

23. Specht A, Bolze F, Donato L et al (2012) The donor-acceptor biphenyl platform: a versatile chromophore for the engineering of highly efficient two-photon sensitive photoremovable protecting groups. Photochem Photobiol Sci 11:578–586

24. Canepari M, Nelson L, Papageorgiou G et al (2001) Photochemical and pharmacological evaluation of 7-nitroindolinyl-and 4-methoxy-7-nitroindolinyl-amino acids as novel, fast caged neurotransmitters. J Neurosci Meth 112:29–42

25. Palma-Cerda F, Auger C, Crawford DJ et al (2012) New caged neurotransmitter analogs selective for glutamate receptor sub-types based on methoxynitroindoline and nitrophenylethoxycarbonyl caging groups. Neuropharmacology 63:624–634

26. Ellis-Davies GC (2007) Caged compounds: photorelease technology for control of cellular chemistry and physiology. Nat Methods 4:619–628

27. Kantevari S, Passlick S, Kwon HB et al (2016) Development of anionically decorated caged neurotransmitters: in vitro comparison of 7-Nitroindolinyl- and 2-(p-Phenyl-o-nitrophenyl)propyl-based photochemical probes. Chembiochem 17:953–961

28. Matsuzaki M, Hayama T, Kasai H et al (2010) Two-photon uncaging of gamma-aminobutyric acid in intact brain tissue. Nat Chem Biol 6:255–257

29. Lutz C, Otis TS, DeSars V et al (2008) Holographic photolysis of caged neurotransmitters. Nat Methods 5:821–827

30. Nikolenko V, Watson BO, Araya R et al (2008) SLM microscopy: scanless two-photon imaging and photostimulation with spatial light modulators. Front Neural Circuits. https://doi.org/10.3389/neuro.04.005.2008

31. Kantevari S, Matsuzaki M, Kanemoto Y et al (2010) Two-color, two-photon uncaging of glutamate and GABA. Nat Methods 7:123–125

32. Zito K, Scheuss V, Knott G et al (2009) Rapid functional maturation of nascent dendritic spines. Neuron 61:247–258

33. Hill TC, Zito K (2013) LTP-induced long-term stabilization of individual nascent dendritic spines. J Neurosci 33:678–686

34. Oh WC, Hill TC, Zito K (2013) Synapse-specific and size-dependent mechanisms of spine structural plasticity accompanying synaptic weakening. Proc Natl Acad Sci U S A 110:E305–E312

35. Stein IS, Gray JA, Zito K (2015) Non-ionotropic NMDA receptor signaling drives activity-induced dendritic spine shrinkage. J Neurosci 35:12303–12308

36. Hamilton AM, Lambert JT, Parajuli LK et al (2017) A dual role for the RhoGEF Ephexin5 in regulation of dendritic spine outgrowth. Mol Cell Neurosci 80:66–74

37. Hamilton AM, Oh WC, Vega-Ramirez H et al (2012) Activity-dependent growth of new dendritic spines is regulated by the proteasome. Neuron 74:1023–1030

38. Kwon HB, Sabatini BL (2011) Glutamate induces de novo growth of functional spines in developing cortex. Nature 474:100–104

39. Pologruto TA, Sabatini BL, Svoboda K (2003) ScanImage: flexible software for operating laser scanning microscopes. Biomed Eng Online 2:13

40. Yasuda R, Nimchinsky EA, Scheuss V et al (2004) Imaging calcium concentration dynamics in small neuronal compartments. Sci STKE 2004:pl5

41. Stoppini L, Buchs PA, Muller D (1991) A simple method for organotypic cultures of nervous tissue. J Neurosci Meth 37:173–182

42. De Simoni A, Yu LM (2006) Preparation of organotypic hippocampal slice cultures: interface method. Nat Prot 1:1439–1445

43. Gogolla N, Galimberti I, DePaola V et al (2006) Preparation of organotypic hippocampal slice cultures for long-term live imaging. Nat Prot 1:1165–1171

44. Opitz-Araya X, Barria A (2011) Organotypic hippocampal slice cultures. J Vis Exp. https://doi.org/10.3791/2462

45. Woods G, Zito K (2008) Preparation of gene gun bullets and biolistic transfection of neurons in slice culture. J Vis Exp. https://doi.org/10.3791/675

46. Harvey CD, Svoboda K (2007) Locally dynamic synaptic learning rules in pyramidal neuron dendrites. Nature 450:1195–1200

47. Lein PJ, Barnhart CD, Pessah IN (2011) Acute hippocampal slice preparation and hippocampal slice cultures. Methods Mol Biol 758:115–134

48. Feng G, Mellor RH, Bernstein M et al (2000) Imaging neuronal subsets in transgenic mice expressing multiple spectral variants of GFP. Neuron 28:41–51

49. Sjulson L, Cassataro D, DasGupta S et al (2016) Cell-specific targeting of genetically encoded tools for neuroscience. Annu Rev Genet 50:571–594

50. Tabata H, Nakajima K (2001) Efficient in utero gene transfer system to the developing mouse brain using electroporation: visualization of neuronal migration in the developing cortex. Neuroscience 103:865–872

51. Bedbrook CN, Deverman BE, Gradinaru V (2018) Viral strategies for targeting the central and peripheral nervous systems. Annu Rev Neurosci 41:323–348

52. Eilers J, Konnerth A (2009) Dye loading with patch pipettes. Cold Spring Harb Protoc 2009:pdb prot5201

53. Nevian T, Helmchen F (2007) Calcium indicator loading of neurons using single-cell electroporation. Pflugers Arch 454:675–688

54. Svoboda K, Yasuda R (2006) Principles of two-photon excitation microscopy and its applications to neuroscience. Neuron 50:823–839

55. Suter BA, O'Connor T, Iyer V et al (2010) Ephus: multipurpose data acquisition software for neuroscience experiments. Front Neural Circuits. https://doi.org/10.3389/fncir.2010.00100

56. Raghavachari S, Lisman JE (2004) Properties of quantal transmission at CA1 synapses. J Neurophysiol 92:2456–2467

57. Bloodgood BL, Sabatini BL (2007) Nonlinear regulation of unitary synaptic signals by CaV(2.3) voltage-sensitive calcium channels located in dendritic spines. Neuron 53:249–260

58. Zhang YP, Holbro N, Oertner TG (2008) Optical induction of plasticity at single synapses reveals input-specific accumulation of αCaMKII. Proc Natl Acad Sci U S A 105:12039–12044

59. Oh WC, Parajuli LK, Zito K (2015) Heterosynaptic structural plasticity on local dendritic segments of hippocampal CA1 neurons. Cell Rep 10:162–169

60. Bosch M, Castro J, Saneyoshi T et al (2014) Structural and molecular remodeling of dendritic spine substructures during long-term potentiation. Neuron 82:444–459

61. Lee SJ, Escobedo-Lozoya Y, Szatmari EM et al (2009) Activation of CaMKII in single dendritic spines during long-term potentiation. Nature 458:299–304

62. Kato K, Clifford DB, Zorumski CF (1993) Long-term potentiation during whole-cell recording in rat hippocampal slices. Neuroscience 53:39–47

63. Malinow R, Tsien RW (1990) Presynaptic enhancement shown by whole-cell recordings of long-term potentiation in hippocampal slices. Nature 346:177–180

64. Holtmaat AJ, Trachtenberg JT, Wilbrecht L et al (2005) Transient and persistent dendritic spines in the neocortex in vivo. Neuron 45:279–291

65. Hedrick NG, Yasuda R (2017) Regulation of Rho GTPase proteins during spine structural plasticity for the control of local dendritic plasticity. Curr Opin Neurobiol 45:193–201

66. Murakoshi H, Wang H, Yasuda R (2011) Local, persistent activation of Rho GTPases during plasticity of single dendritic spines. Nature 472:100–104

67. Zhai S, Ark ED, Parra-Bueno P et al (2013) Long-distance integration of nuclear ERK signaling triggered by activation of a few dendritic spines. Science 342:1107–1111

Chapter 5

Two-Photon Fluorescence Imaging of Visually Evoked Glutamate Release Using iGluSnFR in the Mouse Visual System

Bart G. Borghuis

Abstract

Neurons within the central nervous system communicate by releasing neurotransmitter at synapses. While electrophysiological recording and calcium imaging allow quantitative measurements of neuronal activation pre- and postsynaptically, a recently developed tool, *iGluSnFR*, permits direct measurements of neurotransmitter release from one neuron onto another. By visualizing neurotransmitter release, iGluSnFR allows innovative, quantitative studies in at least three major areas. First, by resolving synaptic release locally on dendrites, iGluSnFR enables efficient studies of the spatial organization of synaptic input. Second, by reporting the transmitted signal, rather than the pre- or postsynaptic neuronal response, iGluSnFR can help identify specific contributions from presynaptic release vs. postsynaptic receptor mechanisms during synaptic plasticity. Third, by reporting the presence of neurotransmitter in the extracellular space, iGluSnFR permits new studies of the transporter mechanisms responsible for the removal of neurotransmitter following synaptic release and changes in these mechanisms during neurological disease. In this chapter we describe how to target iGluSnFR expression to select neuron populations and how to measure stimulus-evoked iGluSnFR responses using two-photon fluorescence (2P) microscopic imaging.

Key words iGluSnFR, Glutamate, Synaptic release, Neurotransmission, Mouse retina, Bipolar cell, Ganglion cell

1 Introduction

1.1 Functional Imaging in the Mammalian Nervous System

In the central nervous system of vertebrates, most excitatory signaling relies on synaptic release of the neurotransmitter glutamate. Glutamate is released through exocytotic vesicle fusion at synapses—anatomically and functionally specialized sites within a neuron's axon terminal. Following release, glutamate molecules diffuse from the presynaptic neuron across the narrow (~20 nm) synaptic cleft and bind to glutamate receptors expressed in the dendritic membrane of the postsynaptic neuron [1]. By activating the postsynaptic receptor, depending on receptor type, glutamate exerts its effect on the postsynaptic neuron either by causing a conformational change

Espen Hartveit (ed.), *Multiphoton Microscopy*, Neuromethods, vol. 148,
https://doi.org/10.1007/978-1-4939-9702-2_5, © Springer Science+Business Media, LLC, part of Springer Nature 2019

that leads to opening of the receptor non-selective cation channel (ionotropic glutamate receptors) or by activating a G-protein-coupled receptor signaling pathway that leads to opening or closure of ion channels expressed in the cell membrane (metabotropic glutamate receptors). In both cases, the signal that carries information from one neuron to another is glutamate. Therefore, direct visualization of synaptic glutamate release is a powerful approach for studies of information encoding and signal processing within a neural circuit. The fluorescent biosensor iGluSnFR, developed in 2013, allows such measurements [2].

iGluSnFR, acronym for *intensity-based Glutamate-Sensing Fluorescent Reporter*, is an engineered sensor protein that changes fluorescence intensity following binding of its principal ligand, glutamate. iGluSnFR is currently the most advanced member of a larger family of glutamate biosensors [2–7] and the only glutamate biosensor that has been successfully applied in neuronal tissue preparations both in vitro and in vivo [2].

iGluSnFR combines a periplasmic glutamate/aspartate-binding protein cloned from *E. coli* (Glt1) with circularly permuted green fluorescent protein (cpGFP; Fig. 1), a GFP variant with enhanced capacity for transducing protein domain interactions into a fluorescence signal [8]. In iGluSnFR, Glt1 and cpGFP are linked in such a manner that a conformational change in Glt1 caused by ligand binding shifts the cpGFP group from a quenched (dimly fluorescent) to an unquenched (brightly fluorescent) state. Like Glt1, iGluSnFR is expressed in the plasma membrane, with the ligand-binding domain and fluorescent group located extracellularly. By reporting changes in extracellular glutamate concentration locally,

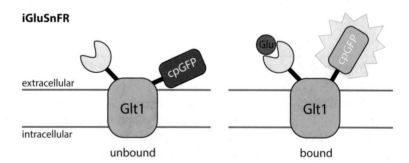

Fig. 1 iGluSnFR is an engineered fluorescent sensor protein. The molecular basis of iGluSnFR is Glt1, the periplasmic component of the ABC-transporter complex for glutamate and aspartate in *E. coli*. Linking circularly permuted GFP (cpGFP) close to the interdomain hinge region of Glt1 resulted in a fluorescent sensor with glutamate affinity in the single micro-molar range and a dynamic response range of approximately 200% ($\Delta F/F_0$) when expressed in neurons. In the unbound state (left), cpGFP is quenched and iGluSnFR fluorescence is low. Glutamate (Glu) binding (right) causes Glt1 to undergo a conformational change that unquenches cpGFP and strongly increases iGluSnFR fluorescence

and with physiologically relevant temporal resolution, iGluSnFR has enabled imaging-based studies of synaptic glutamate release in functionally intact neuronal tissues in worms, fish, and mouse, in vitro and in vivo [2, 9–15].

iGluSnFR is expressed by introducing cDNA that encodes it into neurons either transiently, for example, through viral transduction, or constitutively, through insertion into the model organism's genome. This genetic approach is similar to the application of other genetically encoded indicators including those developed for sensing calcium, such as the family of GCaMPs [16]. Genetic expression differs from approaches that use small molecule indicators like Oregon Green BAPTA-1 and Fluo-4 (both calcium indicators), which are typically injected or superfused for non-selective uptake, or loaded into single neurons using glass micropipettes. Genetically encoded sensor proteins can be expressed broadly to label many cell types within a tissue similar to small molecule indicators, but also selectively to label only a subset of cell types, for example, through Cre recombinase-dependent expression in Cre-expressing transgenic mouse lines [17]. In many experiments, cell-type specificity has important advantages over (non-selective) small molecule approaches. For most applications the choice of paradigm will depend on the goal of the experiment and available transgenic animal lines, gene promoters, and viral vectors.

1.2   iGluSnFR Expression: Presynaptic vs. Postsynaptic

Current versions of iGluSnFR are not selectively targeted to synapses. Instead, they express throughout the plasma membrane of the labeled cell. When selecting an experimental approach to visualize synaptic glutamate release using iGluSnFR it is important to realize that iGluSnFR can be used presynaptically as well as postsynaptically. Since the concentration of glutamate immediately following release into the synaptic cleft will be similar at the pre- and postsynaptic cell membrane, synaptic transmission can be monitored both from the axon terminal membrane of the presynaptic (glutamate releasing) neuron and the dendritic membrane of the postsynaptic neuron. Studies in retina further demonstrated that expression in Müller glia cells, whose processes closely associate with retinal synapses, permits sensitive measurements of neuronal glutamate signaling, even during low-contrast visual stimulation associated with low rates of synaptic vesicle release [14, 18].

iGluSnFR senses glutamate in the extracellular space, so that conceptually, the cells that express it can be thought of as a mere scaffold for locating iGluSnFR near to sites of synaptic release. It is important to realize that the iGluSnFR signal does not provide information about how the measured glutamate signal impacts the postsynaptic cell. This is a major difference between iGluSnFR imaging and, for example, calcium imaging where the fluorescence signal reflects the level of activation of the postsynaptic neuron.

1.3 Understanding the Response Properties of iGluSnFR as a Measurement Tool

The usefulness of any tool for measuring depends on three key properties: operating range, resolution, and linearity. This is true also for sensors designed to measure biological variables, such as iGluSnFR. *Operating range* is defined by the smallest and largest value of the biological variable that the sensor can report (Fig. 2a). Ideally, the operating range brackets the range of naturally occurring values. *Resolution* reflects the smallest change in value that the sensor can reliably distinguish (Fig. 2b). Resolution depends on the response gain, i.e., the unit change in signal per unit change in stimulus value, as well as the intrinsic variability (noise level) of the detected signal. *Linearity* pertains to the magnitude of the change in signal per unit change in stimulus value across the operating range (Fig. 2c). A purely linear sensor shows the same change in signal per unit change in value of the biological variable throughout its operating range. Linear scaling of the measured signal and biological variable makes interpretation of the recorded signal computationally straightforward. Linearity, therefore, is a highly desirable feature for any sensor.

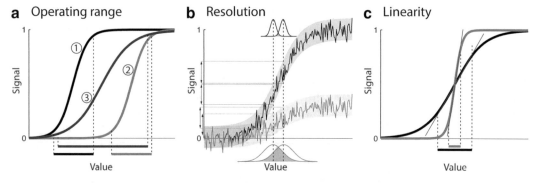

Fig. 2 Fundamental properties of a sensor. (**a**) *Operating range* reflects the smallest through largest value of the biological variable that the sensor can distinguish. Curves show response functions for three different sensors. The first has a narrow operating range and is sensitive to low values of the measured variable (1). The second also has a narrow operating range but is sensitive to higher values (2). The third has a broader operating range that spans the low and high end of the other two sensors (3). Approximate operating ranges of the three sensors are indicated by the horizontal lines (bottom). (**b**) *Resolution* is a measure of the smallest change in value of the biological variable that the sensor can reliably report. Resolution depends on the response amplitude as well as the noise level of the measured signal. With equal noise, a sensor with larger gain (black) will resolve smaller differences in value of the biological variable. This is illustrated by horizontal lines that show the noise margins for a given value of the measured variable. Vertical arrows delineating these margins show increased overlap at low gain. Overlap in the measured signal translates into uncertainty in the actual value of the measured biological variable, indicated by the amount of overlap (pink area) of the Gaussian curves shown (top and bottom). (**c**) *Linearity* pertains to the change in signal following a change in the value of the measured variable. Within a sensor's linear range, signal and variable value are proportional, e.g., a twofold change in value causes a twofold change in the signal. Curves show response functions of two sensors, one with a low (black) and one with a high response gain (gray). Reference lines (red) demonstrate that both sensors respond nearly linearly around the center of their response range. Horizontal lines (bottom) indicate the approximate linear range of each sensor

An additional property that is particularly important for studies of dynamic systems, such as neurons during live imaging, is temporal resolution. Based on neuronal integration times, monitoring circuit function at a physiologically and behaviorally relevant timescale in most systems requires a temporal resolution in the 1–50 ms range, which places important constraints on sensor kinetics.

The molecular properties of iGluSnFR have proven remarkably favorable for the study of glutamate signaling in neuronal tissue. This is in part due to protein engineering decisions made during its development and in part due to serendipity. To be useful, the operating range must match the transient change in glutamate concentration following synaptic release. According to the most detailed measurements, the concentration of glutamate within a synaptic vesicle in a mammalian CNS neuron is about 60 mM [19]. Release of a single synaptic vesicle into the synaptic cleft causes the concentration of extracellular glutamate at the synapse to rapidly rise from an estimated 25 nM [20] to about 1.1 mM (time constant τ_{rise} ~300 µs) before rapidly returning to baseline (τ_{decay} ~1 ms) through diffusion, reuptake, and binding to receptors and transporters [21, 22].

During iGluSnFR development, glutamate titration-based screening for candidates with favorable properties (high affinity, large fluorescence response amplitude, fast kinetics) by expression in HEK293 cells and in cultured neurons showed for the best candidate a glutamate equilibrium dissociation constant (K_D) of 107 µM [2]. Expression of this candidate in intact neuronal tissue shifted iGluSnFR's glutamate K_D favorably to ~5 µM, likely through interactions with other membrane proteins and biochemical conditions of the extracellular milieu [2]. Empirical data showed that this positioned iGluSnFR within the working range necessary for sensing glutamate under physiological conditions in intact circuits in various model systems [2]. Furthermore, recordings of glutamate release from bipolar cells in the in vitro mouse retina showed that visual stimulation across a 10–100% contrast range evoked fluorescence responses that scaled linearly with simultaneously recorded excitatory postsynaptic currents in retinal ganglion cells (Fig. 3), indicating that the sensor is approximately linear across a large part of the physiological response range. While the exact resolution limits remain to be determined, it appears that iGluSnFR can reliably report release of individual synaptic vesicles [23] and data obtained so far show that it is adequate for sensing low through high rates of synaptic glutamate release under physiological conditions in intact neuronal tissues (Figs. 3 and 5).

iGluSnFR can be expressed using several different methods. The preferred method for any given experiment will depend on the type of preparation (e.g., neuronal cell culture, ex vivo, or in vivo), the required level of cell type specificity (narrow or broad), and the

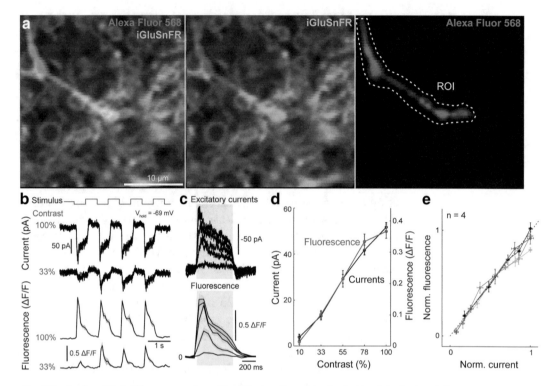

Fig. 3 Evaluating iGluSnFR response linearity with simultaneous electrophysiological and 2P recordings. (**a**) 2P images of the dendritic arbor of an iGluSnFR-expressing (middle) OFF-alpha type ganglion cell targeted for whole-cell recording. The cell was filled with a red fluorescent dye (Alexa Fluor 568; Thermo Fisher Scientific) through the patch pipette (right). iGluSnFR fluorescence intensity was averaged for the ROI indicated by the dashed outline (right). (**b**) Simultaneously recorded excitatory currents (top) and fluorescence responses (bottom) evoked with visual stimulation at two contrast levels (150 μm diameter spot centered at the soma; same cell as panel **a**). The fluorescence response is reported as fractional change in fluorescence intensity compared with the baseline intensity ($\Delta F/F$). Solid line and shaded region represent mean ± SEM of four repeats at each contrast level. (**c**) Average excitatory current (top, shown inverted for comparison) and fluorescence response (bottom) to five stimulus contrasts (10–100%). Curves show the average response during the dark phase of the stimulus (average of 12 stimulus periods). Pink region indicates time window used for quantification of the evoked response. (**d**) Contrast response functions based on the excitatory current (black) and fluorescence response (red) averaged over the duration of the response (pink region in panel **c**). Error bars indicate ± SEM across repeats ($n = 12$). (**e**) Scatter plot of excitatory current and fluorescence response amplitude from 10% to 100% contrast ($n = 4$ cells, indicated by different gray levels). Responses were normalized by fitting a line to the data for each cell and scaling currents and fluorescence responses to the current amplitude of the fit at 100% contrast. Error bars indicate ± SEM across trials. (Reproduced from Borghuis et al. [9], with permission from the Society for Neuroscience)

available genetic tools (Cre recombinase-expressing transgenic lines, celltype-specific gene promoters). Viral transduction has proven useful for many applications and, following intravitreal injection, in the retina results in dense labeling of ganglion cells and amacrine cells (Fig. 4a, b). This allows imaging of glutamate release from the retinal bipolar cells, an anatomically and functionally diverse group of excitatory neurons that signal visual

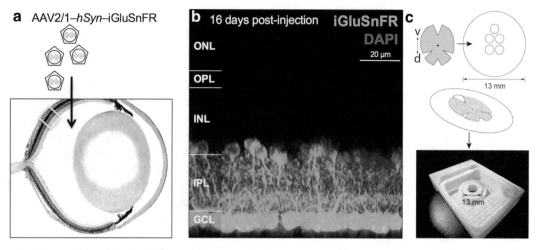

Fig. 4 Using intravitreal AAV injection to express a transgene in retinal ganglion cells and amacrine cells. (**a**) To transduce cells in the inner retina, a small volume (0.5–1.5 μl) of AAV virus in suspension (typically around 10^{13} IU/ml in 1× PBS, 5% sorbitol storage solution) is injected into the eye intravitreally. Driving expression with the human synapsin 1 promoter [40] targets transgene expression to neurons and avoids expression in retinal glia, including Müller glia cells. Yellow box approximates retinal area shown in panel (**b**). (**b**) Confocal image of a retinal slice 16 days after injection with AAV2/1-hSyn-iGluSnFR. Green, iGluSnFR fluorescence; blue, nuclear stain (DAPI). A typical incubation period of 2–3 weeks should give strong transgene expression in ganglion cell and amacrine cell populations. ONL outer nuclear layer, OPL outer plexiform layer, INL inner nuclear layer, IPL inner plexiform layer, GCL ganglion cell layer. (**c**) After incubation, the retina is harvested and dissected (top, pink; v, ventral, d, dorsal), mounted on a nitrocellulose filter disc with apertures for visual stimulation of the photoreceptors (see Materials Sect. 2.3) and mounted in a perfusion chamber (bottom). A ring with nylon wires placed over the retina and disc stabilizes the tissue for imaging and recording

information from the photoreceptors to the ganglion cells [18]. Bipolar cell response types can be distinguished based on anatomical location of the axon terminal and response time course measured with iGluSnFR [10]. Because dense labeling following viral transduction enables a wide variety of experiments to further our understanding of signal processing in parallel bipolar cell pathways, this is the approach that we will focus on in this chapter. Alternative approaches, including those aimed at obtaining sparse labeling in the retina, and imaging of glutamate output from ganglion cell axon terminals in the visual thalamus will be covered in Notes below.

2   Materials

2.1   Reagents

iGluSnFR is expressed in neuron populations within the retina through intravitreal injection with a viral vector containing the iGluSnFR gene. Adeno-associated viral vectors (AAVs) containing the iGluSnFR gene under control of various gene promoters have been deposited at Addgene (www.addgene.org), where they can be obtained for research purposes. AAVs can also be synthesized in-house using standard methods not covered in this chapter. For use in the retina, the goal is to obtain virus stock at a titer of ~$1 \cdot 10^{13}$ infectious units (IU; number of AAV particles with full genome content) per ml. A titer of less than $5 \cdot 10^{12}$ IU typically proves insufficient and leads to low-level expression in spatially restricted areas, only. A titer of more than $3 \cdot 10^{13}$ may cause an inflammatory response, indicated by full or partial coverage of the retinal surface with white blood cells, apparent overexpression of sensor protein, and cell death. Virus stocks should be stored at $-80\ °C$. However, aliquots thawed for use may be kept at $4\ °C$ for up to several months without substantial loss of efficacy. Freeze-thaw cycles cause loss of infectious virus particles and should be avoided.

2.2 Viral Injection

To transduce retinal neurons, a microliter volume of virus suspension is injected into the eye. There are two general approaches for injecting. The first uses a glass capillary advanced into the eye using a manual or motorized manipulator, connected to a pressure injector device such as a picospritzer. The second approach uses a metal needle mounted on a hand-held microliter syringe, either general-purpose (e.g., Hamilton Co.) or specialized for eye injections (e.g., www.BorghuisInstruments.com). See Methods for details of the eye injection procedure.

2.3 Retina Dissection

Retinas are dissected and maintained in pH-buffered Ames medium ($NaHCO_3$, 1.9 g/l + A1420; Sigma Aldrich)—a saline solution with added ingredients optimized for retina—and oxygenated through continuous bubbling with a mixture of 95% O_2 and 5% CO_2. To preserve light sensitivity, dissection is ideally performed using infrared illumination, aided by night vision scopes (Micro Night Vision monoculars; Night Vision Devices, USA) custom-fitted onto the eye pieces of a stereoscope with zoom-optics (Olympus SZ61), or in dim red light if necessary.

Tools required for isolating the retina from the eye are a #11 surgical scalpel blade to puncture the eye at the *ora serrata*, and fine forceps and spring scissors (Item numbers 11295-20; 15009-08; Fine Science Tools) to dissect and radially incise the retina.

The retina is separated from the pigment epithelium either mechanically by prying with spring scissors and forceps, or ideally, using a custom-built jet knife—a stream of dissecting medium expelled from a small-gauge needle (e.g., 30 gauge hypodermic needle; Becton Dickinson) using a miniature peristaltic pump, aimed at the junction between the retina and sclera. By avoiding physical contact between surgical tools and the photoreceptor outer segments, this approach minimizes potential damage to the photoreceptors and helps maintain visual sensitivity throughout the retinal preparation.

The isolated retina is transferred to a glass microscope slide, flattened, and adhered to a cellulose filter disc (HAWP01300; Millipore) for subsequent mounting and handling. To enable visual stimulation of the photoreceptors, each disc features four 1.2 mm diameter holes in a tight square pattern (Fig. 4c), cut out using a custom punch. This punch can be manufactured from an 18 gauge needle (Becton Dickinson) by cutting off the tip and sharpening the open end with an internal bevel. The tissue preparation is stabilized in a perfusion chamber using a ring with parallel, small diameter nylon strings (e.g., Frog Hair 8X, 0.003″/0.076 mm; Amazon.com) spaced at ~1 mm intervals, placed over the filter disc and retina (Fig. 4c).

2.4 Microscope System

2P imaging of light-evoked iGluSnFR responses in the retina in vitro requires an ultrashort pulse IR laser and a 2P microscope. Ultrashort pulse lasers emit photons not as a continuous stream but as pulses with extremely short duration (pulse width ~10^{-13} s) and high photon flux necessary for evoking two-photon fluorescence. The unattenuated emission beam of a pulse IR laser is extremely damaging to the tissues in the eye and when viewed directly will cause instant and irreversible eye damage. Appropriate safety measures must be taken while working with an exposed beam at all times.

Currently, the most common choice of laser light source for 2P imaging with GFP-based biosensors like iGluSnFR are titanium-sapphire lasers. The output wavelength of these lasers is tunable in the 740–1040 nm range; typical output power is in the single Watt range and for most systems decreases with increasing wavelength (e.g., 4.0 W @ 800 nm; 2.4 W @ 910 nm). In almost all cases it is necessary to limit laser power after the objective, incident on the biological sample. This is best done using a Pockels cell in the laser light path (e.g., Conoptics Model 350-80-LA-02 KD∗P Series E-O Modulator; Conoptics; Danbury, CT), which allows fast (µs) dynamic control of laser transmission. A low-cost alternative is to insert into the light path a tunable half-wave plate (e.g., AHWP05M-980, RSP1; Thorlabs, USA) and polarizing beam splitter (e.g., 05FC16PB.7; Newport, USA) to reject a fixed fraction of the laser power (rejected beam) into a laser beam block (LB2; Thorlabs).

When it comes to selecting a 2P microscope system, there are currently many options, both commercially available and custom-built. A common custom design comprises an Olympus BX51 microscope base and conventional (non-resonant) galvanometric scan mirrors (model 6215H; Cambridge Technology, Lexington, MA), controlled by ScanImage software (www.vidriotechnologies.com). A 60× 1.0 NA objective lens is commonly used for imaging individual dendrites, groups of dendrites, and retinal areas up to 50 × 50 µm. Larger retinal areas may be imaged with a 40× 0.8 NA objective.

To position the retina preparation under the microscope objective, a custom-designed, 3D-printed perfusion chamber containing the retina preparation (Fig. 4c) is fixed onto the microscope stage. A microscope coverslip functions as the bottom of the perfusion chamber and allows visual stimulation of the photoreceptors through the microscope condenser light path. Using a setup similar to that used in conventional slice recording, gravity-fed perfusion medium heated to physiological temperature (~35 °C) using an in-line solution heater (SH-27B; Harvard Apparatus) enters the chamber on one side and is removed by suction at the opposite side, using a low-gauge hypodermic needle set at the desired fluid level and attached to a vacuum line. A motorized x-y translating stage (e.g., XYR-AB-4040; Dover Motors) controlled by ScanImage software through a micromanipulator controller (e.g., Sutter MP-285; Sutter Instrument) is highly recommended. Computer-controlled stage motion strongly facilitates locating sites of useful iGluSnFR expression using 2P imaging and returning to these sites for 2P imaging of stimulus-evoked responses.

2.5 Visual Stimulation

Visual stimulation of the retinal photoreceptors during 2P imaging is a challenge because stimulus photons must be kept strictly separate from evoked fluorescence photons. There are two general approaches that help achieve this separation.

The first approach prevents stimulus light from entering the fluorescence detection light path by separating stimulus light and fluorescence emission spectrally. This approach uses a stimulus wavelength that is relatively far from the fluorescence emission wavelength and dichroic mirrors that maximally pass stimulus photons onto the tissue but block them from the emitted-photon detection pathway. An effective strategy uses stimulation light around 400 nm, which strongly activates predominantly UV opsin-expressing cone photoreceptors in the ventral mouse retina. A potential long-wavelength tail of the stimulus LED is blocked by passing the stimulus output through a short-pass filter with a cut-off wavelength between 420 and 450 nm. Collected light in the detector light path is passed through conventional band-pass dichroic filters to send fluorescence emission into the photomultiplier tubes (R3896, Hamamatsu; PS310, Stanford Research Systems). Using commercially available filters (e.g., www.semrock.com), this configuration allows visual stimulation of the retina at a mid-photopic light level (photon flux ~10^4 photons/μm^2/s, measured with a radiometer, e.g., UDT S470 Benchtop Optical Meter, Gamma Scientific) with negligible bleed-through of stimulus light into the fluorescence image. Insufficient separation or block of stimulus light will appear in the fluorescence image as shot noise at low stimulus light intensities and distinct striping at high stimulus light intensities. If stimulus bleed-through limits experiments then additional measures must be taken to resolve, for example by using more selective filters at the stimulus output and fluorescence input stage, or by reducing stimulus light intensity using a neutral density filter.

The second approach to prevent stimulus light from mixing with fluorescence emission separates visual stimulation and fluorescence detection temporally. This is achieved by stroboscopic presentation of the visual stimulus in counter-synch with fluorescence acquisition. This approach takes advantage of the particular way that laser scanning 2P microscopes build up the fluorescence image. In one-directional scan mode, the fluorescence image is acquired by scanning the laser beam across the field of view using the x-axis mirror to collect a single line of fluorescence emission. After acquiring a line, the beam is sent back to the opposite image margin (fly-back) and the orthogonal (y-axis) mirror is adjusted over an angle corresponding to one line-width before initiating the next line scan. The stroboscopic approach makes use of the fact that during the fly-back period no fluorescence emission light is collected. By turning on the stimulus light source briefly and exclusively during this period, the retina is illuminated but stimulus light does not corrupt the fluorescence image. Because the typical repetition rate of x-direction scanning (1–2 kHz) exceeds the temporal resolution of cone phototransduction by more than tenfold, the strobed visual stimulus appears continuous in time at the level of the photoreceptors and postsynaptic visual circuits. Stroboscopic stimulus presentation requires the scan hardware and the stimulus light source to be tightly synchronized, which is feasible but requires some level of expertise to implement depending on the specific systems used.

3   Methods

3.1   Expressing iGluSnFR Through Viral Transduction

Viral transduction is a powerful method for introducing transgenes into neuronal tissues. The approach uses a genetically modified viral vector as a vehicle to deliver genetic material into cells through binding to receptors expressed on the cell surface followed by endocytosis. In many cases these vectors are replication deficient and their harmful infectious components have been removed. Thus, they enable transgene expression without killing the host cell. Examples of commonly used vectors are adenovirus, adeno-associated virus (AAV), and lentivirus.

The vector of choice for many applications is AAV, a single-stranded DNA dependovirus belonging to the parvovirus family [24]. AAV exhibits low cytotoxicity, enabling incubation periods of weeks or months to obtain a high level of transgene expression. AAV is not currently known to cause disease in humans and in most cases can be handled at biosafety level 1 (BSL-1), facilitating production and use. AAV has a DNA packaging capacity of just under 5 kb. This is small compared with other vectors but sufficient for many applications due to the availability of compact, highly truncated gene promoter sequences capable of driving strong transgene expression. AAV vectors expressing iGluSnFR are commercially available (e.g., Vector Core, University of Pennsylvania) and also may be synthesized using established methods either in-house or at a molecular biology core facility, which are increasingly available at many universities.

The AAV vector family available for research purposes comprises various serotypes that are either naturally occurring or have been generated through targeted mutation of existing serotypes. Serotype impacts expression level and may bias transgene expression to different neuron populations. Empirically, within the retina, AAV2/1 (an AAV serotype 2 vector, "pseudotyped" by expressing an AAV serotype 1 capsid) gives broad expression in ganglion cells, amacrine cells, and Müller glia at useful expression levels [14, 18, 25]. AAV2/2, 2/5, 2/7, 2/8, and 2/9 also each give useful expression in retinal neurons and glia, and additional serotypes are being generated through targeted mutation and in vivo directed evolution approaches [26–28]. In addition to serotype, a major determining factor for which cells will express the transgene—and how strongly—is the gene promoter. In the retina, CAG, a synthetic promoter developed by fusing a cytomegalovirus (CMV) early enhancer element with the promoter, first exon, and first intron of the chicken beta-actin gene, gives strong expression in most retinal neuron classes as well as Müller glia [25]. To avoid expression in glia, selective expression in neurons can be obtained using the synapsin promoter, cloned from the human or mouse gene sequence driving expression of the synapsin protein. Ganglion cells, amacrine cells, horizontal cells, and photoreceptor cells (through sub-retinal injection) are readily targeted using the aforementioned AAV serotypes and gene promoters [25]. One retinal neuron class that has remained notoriously difficult to transduce—for reasons that are not well understood, are the bipolar cells, although some progress is being made [29–31].

3.2 Expressing iGluSnFR Through Intravitreal AAV Injection

For viral transduction of the retina, 0.7–1.5 µl of AAV-hSyn-iGluSnFR virus suspended in 0.01 M PBS (8.5 g NaCl, 1.4 g Na_2HPO_4, 0.2 g NaH_2PO_4 in 1 l dH_2O) + 5% sorbitol at a titer of $\sim 1.0 \cdot 10^{13}$ IU/ml is injected intravitreally into the eye of an anesthetized mouse (Fig. 4). For most applications, mice aged 6–10 weeks of age are best, but the same procedures work from young through old mice (p14–2y). When injecting with a metal needle seated on a Hamilton-type syringe, first puncture the conjunctiva and sclera at the level of the *ora serrata* using a 30 gauge hypodermic needle. Then insert the metal needle, its tip curved with a 3 mm radius, through this puncture hole. Advance approximately 2–3 mm into the vitreal space following the back-curvature of the lens and inject the loaded virus in suspension over the central retina. With practice, the injection procedure for two eyes can be performed in less than two minutes. The preferred method for anesthesia is isoflurane inhalation (in O_2, 2.25% to induce; 1.25% to maintain) because of the shorter recovery time (typically <2 mins) and reduced risk of hypothermia compared with other conventional methods, e.g., intraperitoneal injection with Ketamine/Xylazine.

After injection, animals are typically housed in an animal care facility under biohazard protocol for incubation. While complications associated with intravitreal eye injections in mice are extremely rare (incidence of <1:500), Institutional Animal Care and Use Committees at most institutions require animals to be monitored daily for signs of discomfort or distress following the procedure and throughout the incubation period. The typical incubation period for AAV2/1 with a construct driven under the CAG or synapsin promoter is 16–21 days. However, virus titer may vary from one batch to the next, and the rate of transgene expression may differ between neuron types. When starting a new series of experiments it is therefore highly recommended to include a pilot phase to determine the optimal incubation period for the available virus stock and neuron type, by harvesting tissue and assessing fluorescence expression at a series of incubation time points. For iGluSnFR, expression can be readily assessed by imaging of fixed tissue on a confocal microscope or by 2P imaging of live tissue, in vitro.

When optimized, the virus injection procedure and incubation should give retinal tissues with useful transgene expression reliably and reproducibly. Using established vectors, the typical success rate is around 80%, i.e., four out of five injected retinas will have adequate levels of transgene expression in a retinal area that is sufficiently large to be useful for imaging experiments.

3.3 Retina Preparation

Following incubation, transduced retinas are harvested. First, the mouse is euthanized and the eyes removed and transferred to oxygenated and pH-buffered Ames medium. Next, each eye is hemisected and the anterior half with lens removed and discarded. The posterior eyecup including the retina is radially incised to allow the dome-shaped tissue to be flattened as a whole-mount preparation (Fig. 4c). The incised retina is dissected out of the eyecup by separating it from the pigment epithelium, and by severing its connection with the optic nerve by means of a cut below the retina under the optic nerve head. The isolated retina is then transferred onto a cellulose filter disc with the optic nerve head in the center and retinal quadrants covering the puncture holes. The filter disc helps flatten and stabilize the tissue through its adhesive properties, and facilitates handling and mounting the tissue in a perfusion chamber for imaging experiments.

Spectral sensitivity of the mouse retina changes along the dorso-ventral axis [32]. For consistent results it is therefore essential to keep track of retinal orientation during dissection and mounting, and to keep orientation constant from one experiment to the next. This is achieved through radial cuts into the retina (approximately 1/3 of the radius) in an asymmetric pattern, e.g., Gerald Holtom's peace symbol.

To transfer the retina onto filter paper, the isolated retina is moved onto a glass slide with the ganglion cell-side down, and a cellulose filter disc is gently placed on top of the photoreceptor side. The filter will stick to the retina by adhesion. The filter paper is positioned centered on the tissue, with each hole covering one leaflet of the retina. A fifth hole in the filter paper, offset with respect to the square pattern (Fig. 4c), will be used to focus the stimulus at the level of the retina's photoreceptors. Next, the filter with retina attached is floated off the glass slide into to the dissecting bath and mounted in the perfusion chamber filled with oxygenated Ames to be mounted on a stage under the 2P microscope.

3.4 Preparation for Image Acquisition

On the imaging stage of the 2P microscope, the retina is continuously perfused with heated, oxygenated Ames medium (34–36 °C) at a flow rate of 4–6 ml/min. This can be achieved using a gravity-fed system with flow rate set by the bore of the perfusion tubing and height of the source reservoir with respect to the perfusion chamber. iGluSnFR expression may be assessed immediately by scanning for fluorescence-labeled axons converging on the optic nerve head using 2P imaging with the scan laser tuned to 910 nm and laser power level during imaging adjusted to ~10 mW after the objective. However, an equilibration period of 15–30 minutes is typically required for the tissue to physiologically stabilize and for light-evoked fluorescence responses to appear.

When the tissue has stabilized, dynamic changes in iGluSnFR fluorescence following spontaneous and/or scan laser-evoked synaptic release from bipolar cell axon terminals should be sufficiently large that they are readily detectable during live viewing, for example, using the "focus view" setting in the microscope control software. In most cases, onset of the scan laser will stimulate retinal photoreceptors sufficiently to evoke a detectable fluorescence response. This laser-evoked response can be used to test for visual responsiveness of a potential target area for experiments. However, at low laser power or following repeated laser scanning a stronger visual stimulus may be required, such as turning on and off a visual display by opening and closing a shutter in the light path, or by presenting light flashes using an LED or laser light source at a wavelength that activates the photoreceptors but is blocked from the microscope's photomultiplier tubes (see Materials, Sect. 2.5).

3.5 Image Acquisition

After locating an area with clearly defined anatomical structure at the level of the inner plexiform layer (intersecting neuronal arbors and dendritic varicosities, and fluorescence responses to test stimuli) the experimental data are collected. For most applications data collection consists of repeated trials of several seconds to tens of seconds of imaging paired with visual stimulation of the retina. Stimulus presentations are typically separated by intertrial-intervals of two to five seconds to allow visual responses to return to baseline following each trial. Typical rates of image acquisition are between 8 and 32 frames per second. When necessary, faster acquisition can be achieved using a line scan mode, where the y-mirror is stationary and fluorescence data are obtained from a single line across the field of view only. These data are obtained at the repetition rate of the x-scan mirror, typically 1–2 kHz, or ~ 7 kHz in resonant scan systems. The choice of scan mode depends on the process that is studied and may be substantially slower, for example, during *time-lapse* interval image acquisition to measure changes in iGluSnFR fluorescence over a time course of minutes or hours to study light-adaptive or circadian changes in baseline or evoked glutamate release [33–35].

3.6 Data Analysis

The outcome of most imaging experiments is fluorescence time series with three dimensions (x, y, time); data obtained in line scan mode will have two dimensions (x, time). The first step in analyzing these time series is to define one or more regions of interest (ROIs) within the image and extract a vector representing the time-varying fluorescence signal within each region of interest. This is readily achieved in software, either using custom-written scripts in, e.g., R (www.r-project.org) or Matlab (www.mathworks.com), or using image analysis software such as ImageJ with the Time Series Analyzer plugin (imagej.nih.gov). The resulting vectors represent the fluorescence signal in each ROI as a function of time. ROI-based analysis is useful because it increases the signal-to-noise ratio of the measurement compared with the individual pixels within each ROI. Dimensionality reduction also sharply reduces the size of the data set and makes data portable for analysis independent of the original image data. For example, the raw image data for a single 20 s acquisition of 512×128 pixels at 16 frames/s may require ~100 MB of storage space, whereas the combined size of fluorescence responses for ten ROIs extracted from these data is less than 10 kB.

Due to inevitable differences in expression level in the cell membrane, absolute fluorescence intensity will vary across experiments. To make measurements invariant to differences in expression level, the fluorescence response is typically expressed as a fold-change from baseline, mathematically expressed as the time-varying response $r(t) = \Delta F / F_0$. Here, F_0 is the measured fluorescence signal just prior to stimulus onset and $\Delta F = (F_t - F_0)$, i.e., the difference between the fluorescence signal measured at a given time point F_t and F_0. By divisively normalizing the fluorescence change to baseline fluorescence, measurements can be directly compared or averaged across experiments even if absolute fluorescence levels vary.

3.7 Interpreting the Fluorescence Signal

iGluSnFR senses glutamate in the extracellular space and signals a local increase in extracellular glutamate concentration with an increase in fluorescence. iGluSnFR expressed in neurons has a K_D of approximately 5 µM—more than two decades lower than the estimated peak concentration of glutamate in the synaptic cleft following synaptic release (~1.1 mM) [2, 21]. Thus, a single release event will cause nearly all available iGluSnFR molecules near the synapse to bind ligand and become activated. However, the glutamate concentration in the cleft following release of a single vesicle is high only for a very short period of time as it decays with a time constant of ~1 ms. Rapid glutamate clearance is necessary for transmitting information at high rates and is achieved through binding to and active reuptake by glutamate transporters expressed pre- and postsynaptically, in addition to passive diffusion.

The fluorescence onset and decay time constants of iGluSnFR are approximately 5 and 40 ms, respectively [2]. Therefore, iGluSnFR cannot resolve the (sub-millisecond) time course of glutamate changes in the cleft during synaptic transmission. But just as in calcium imaging, where the fluorescence response is a low-pass filtered version of the spike response, the iGluSnFR signal will reflect synaptic activity, and empirical data obtained in retina show that differences in fluorescence response amplitude tightly correlate with graded changes in the synaptic rate of release based on excitatory postsynaptic currents in the postsynaptic neuron (Fig. 3). Thus, iGluSnFR fluorescence provides a direct measure of glutamate release, but due to the temporal response limit it cannot resolve the time course of individual release events. As such, the iGluSnFR signal must be interpreted as a temporally integrated (low-pass filtered) measure of glutamate concentration dynamics in the extracellular space.

While iGluSnFR under-samples the temporal dynamics of glutamate in the synaptic cleft, its high affinity for glutamate suggests that it may be capable of reporting release of a single synaptic vesicle. Resolution of quantal release is implied in a published report [23] and consistent with unpublished observations in retina (BG Borghuis) where apparent single vesicle release events can be distinguished at low levels of spontaneous, nonsynchronous release (Fig. 5). Recordings in acute slices of the mouse lateral geniculate nucleus (LGN), containing iGluSnFR-expressing axon terminals of retinal ganglion cells transduced by intravitreal virus injection, showed transient fluorescence events consistent with spontaneous

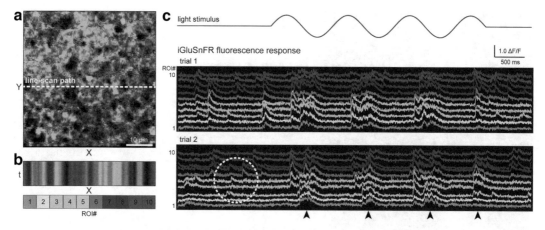

Fig. 5 iGluSnFR fluorescence imaging in line scan mode resolves glutamate release dynamics at high temporal resolution. (**a**) 2P image of the retinal inner plexiform layer at the level of the OFF bipolar cell axon terminals. Dashed line indicates the scan path selected for fluorescence acquisition in line scan mode. (**b**) X-t plot of the partial fluorescence signal obtained in line scan mode prior to visual stimulation ($t = 0$–30 ms). The scanned profile is divided into ten adjacent regions of interest (ROIs; colored squares, bottom). (**c**) Time course of the light stimulus (top; 100 µm diameter spot on a mid-photopic background centered on the scan area, contrast modulated at 1 Hz, 100% Michelson contrast). Fluorescence responses (bottom) during two consecutive stimulus presentations (ROIs shown in panel **b**). Fluorescence signals show common modulation across ROIs and stimulus presentations (arrowheads), consistent with OFF-type bipolar cell activation during light decrements. Traces also show fluorescence events local to single or directly adjacent ROIs on single trials (e.g., dashed circle), consistent with low-level spontaneous glutamate release events

activity of spiking neurons (Fig. 6). During visual stimulation, rates of release from graded potential synapses of bipolar cells are high and the iGluSnFR signal from individual release events overlap in time and sum to yield a compound signal (Figs. 3 and 5). At a low rate of synaptic vesicle release (e.g., spontaneous release), individual release events can be treated as discrete events, counted and expressed as a rate per unit time. At higher rates of release, the total response is best quantified as the amplitude of the fluorescence response.

Many experiments benefit from normalizing the fluorescence response to baseline fluorescence ($\Delta F/F$). However, some experiments may require comparing absolute fluorescence levels across cells and preparations. Assessing differences in absolute fluorescence is complicated by the fact that differences in expression level and bleaching of the fluorophore during prior imaging impact fluorescence intensity independent of experimental perturbations. To disambiguate, absolute iGluSnFR fluorescence intensities in some applications may be obtained by using a reference signal, such as an independent, co-expressed fluorescent label (e.g., PSD95-tagRFP; Fig. 7) for ratiometric comparison with iGluSnFR intensity across cells and preparations.

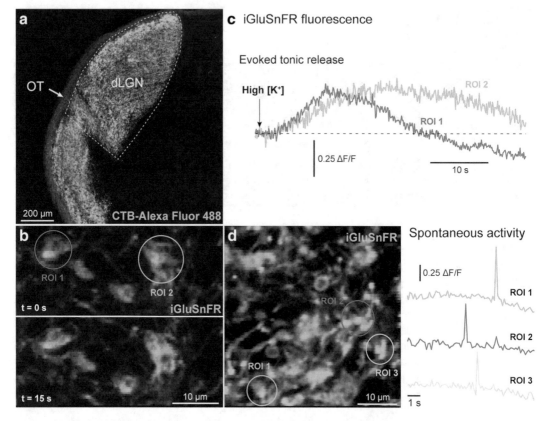

Fig. 6 Imaging synaptic release at central synapses with presynaptically expressed iGluSnFR. (**a**) Confocal fluorescence image of a coronal section of the right-side mouse thalamus at the level of the dorsal lateral geniculate nucleus (dLGN). Green, synaptic terminals from retinal ganglion cells labeled through intravitreal injection with a fluorescent neuronal tracer (Alexa Fluor 488-conjugated cholera toxin B). OT, optic tract. (**b**) 2P images of iGluSnFR-expressing ganglion cell axon terminals in an acute slice preparation containing mouse dLGN (approximate location indicated with magenta box in panel **a**). Ganglion cells were transduced by intra-vitreal injection with AAV-hSyn-iGluSnFR. Images show iGluSnFR fluorescence before (top) and 15 seconds after (bottom) onset of perfusion with 100 mM potassium chloride. (**c**) Fluorescence response for the ROIs indicated in panel (**b**). (**d**) 2P image (left) of a dLGN slice preparation with iGluSnFR expression in axon termi-nals of retinal ganglion cells. ROIs (colored circles) showed fast, localized fluorescence transients consistent with spontaneous release events (right)

4   Notes

4.1   Understanding Voxel Averaging

The fluorescence intensity assigned to each image pixel during 2P imaging represents fluorescence emission from a small volume—a *voxel*—centered on the focal point of the scan laser. Due to the nonlinear nature of 2P excitation, this volume is smaller by many-fold compared with single-photon fluorescence imaging. Nevertheless, the point-spread function for 2P imaging in neuro-nal tissue under optimized conditions is on the order of ~0.4 μm in x and y and about 1.7 μm in z (60× objective, 1.0 NA) [9]—large

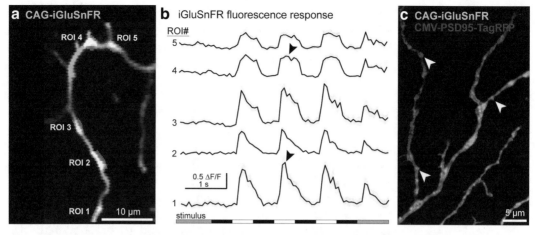

Fig. 7 Biolistic gene transfer permits iGluSnFR fluorescence imaging of sparsely labeled cell populations. (**a**) 2P image of a dendritic arbor of an iGluSnFR-expressing ON-alpha type ganglion cell in vitro (green). Image obtained in a whole-mount retina preparation 18 hours after biolistic transduction with cDNA encoding iGluSnFR under control of the CAG promoter. Magenta areas delineate ROIs used to quantify fluorescence responses during visual stimulation. (**b**) Visually evoked fluorescence responses for the ROIs indicated in panel (**a**) (contrast-reversing spot on a photopic background, 150 μm diameter, 1 Hz square wave, 100% Michelson contrast; stimulus time course is shown at bottom). Fluorescence responses were sustained in some regions (arrowhead, top) and transient in others (arrowhead, bottom), consistent with reported differences in bipolar cell response dynamics [10, 18]. The sparse expression pattern obtained with biolistic gene transfer is particularly useful to avoid voxel averaging of fluorescence signals from nearby labeled structures. (**c**) Confocal image of the dendritic arbor of an ON-alpha type ganglion cell co-labeled with cDNA encoding iGluSnFR (green) and a fluorescent marker for excitatory synapses (PSD95-TagRFP, red). Combining these plasmids helps guide fluorescence measurements to dendritic locations containing excitatory synapses

enough to combine fluorescence signals originating from iGluSnFR expressed on multiple overlapping dendrites into a single image pixel. Thus, voxel averaging precludes resolving fluorescence responses from directly adjacent structures in x and y, and particularly in the z dimension.

Voxel averaging of iGluSnFR responses from nearby structures can be avoided by labeling neuron populations sparsely. The rational is that when only a small fraction of dendrites within a volume expresses the sensor then it becomes feasible to find regions with iGluSnFR expression in dendrites isolated from nearby fluorescent structures by more than 1 μm in x, y, and z. There are currently two established methods for obtaining sparse iGluSnFR expression.

The first approach titrates the dose of AAV used in viral injections. Intravitreal injection with high-titer virus causes transgene expression in many ganglion cells and amacrine cells and results in dense labeling of the inner plexiform layer. Reducing titer will reduce the density of labeled cells. The retinal expression pattern may become patchy, with distributed hotspots that contain clusters of labeled cells, separated by non-labeled cells. Neurites extending from these hotspots may be sufficiently isolated to permit measuring their fluorescence response without interference from adjacent structures. A downside of this approach is that a lower viral dose will lower overall expression levels, which may pose a challenge in experiments. Low expression levels may be avoided by co-injecting a mixture of two viral vectors, mimicking an established non-viral (electroporation) method [36]. Here, the first vector expresses Cre-dependent (floxed) iGluSnFR under control of the CAG promoter to drive strong expression in Cre recombinase-expressing cells. The second vector expresses Cre-recombinase and for sparse labeling is administered at a much lower titer compared with the first (e.g., 1:10,000).

A second approach to obtain sparse label is biolistic transfection, where plasmid DNA is introduced into neurons using a gene gun [37]. For biolistic transfection, gold particles coated with DNA encoding the transgene are shot into a neuronal tissue or cell culture. An advantage of this method is that it obviates the additional step of virus synthesis and lifts constraints on compatibility between serotype and cell type of interest. More importantly, biolistic labeling gives highly discrete (all-or-none type) transgene expression: either a cell receives a DNA-labeled particle and expresses the construct, or it does not. Because each particle contains many copies of the DNA plasmid, a high level of transgene expression is typically reached within 24 hours of transfection. High expression levels give high signal-to-noise ratio of the iGluSnFR signal. This permits sensitive measurements and prolonged scan times relatively unchallenged by bleaching of the sensor protein. The biolistic approach requires tissue incubation and control experiments to assess visual sensitivity and general health of the retina. Preliminary studies showed light-evoked responses in retinal tissue after 18 hours of incubation (35 °C, oxygenated Ames medium with added antibiotic, e.g., Penicillin-Streptomycin; P0781, Sigma Aldrich) that were similar to responses recorded in an acute preparation of virally transduced tissue, indicating feasibility of the biolistic approach (Fig. 7; BG Borghuis, unpublished).

4.2 Fluorescence Imaging in a Light Sensitive Tissue

A challenge for 2P imaging in the retina—a tissue that is, itself, light sensitive—is that incident light from the IR scan laser will cause light responses in the rod and cone photoreceptors. This may be counterintuitive because light at wavelengths above 900 nm is typically invisible to the eye. This is readily explained, however, by the two-photon absorption cross section of mammalian rod and cone opsin and the nature of IR light emitted by ultrashort pulse lasers, which comprises extremely brief ($\sim$100 fs; 10^{-13} s) pulses of IR light with extremely high photon flux (see Materials, Sect. 2.4). The estimated photo-isomerization rate evoked by the IR laser in mouse cones at typical working power (10–15 mW after the objective) is $\sim$$10^3$ R*/cone/s [9, 25, 38, 39]. The impact of this laser-driven activation on retinal circuit function can be strongly reduced by adapting the retina to a photopic background light level of ~10^4 R∗/cone/s and presenting visual stimuli by modulating luminance intensity around this background. The IR activation of photoreceptors precludes most imaging studies of rod photoreceptor-mediated (scotopic) visual responses in whole-mount tissue. Laser-evoked visual activation can be avoided by imaging in a retinal slice preparation where the imaged region is spatially offset with respect to the photoreceptors (Fig. 8). A downside of this approach is that in the slice potentially important lateral circuit connections are compromised or lost.

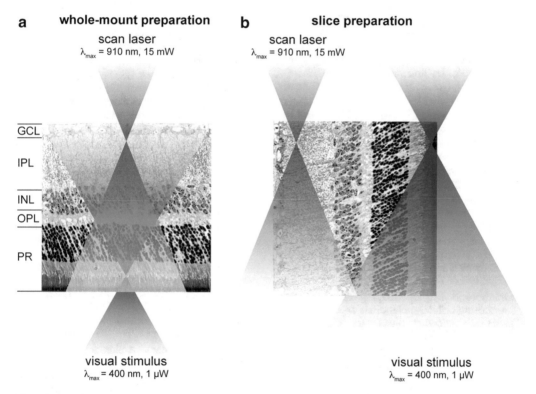

Fig. 8 Imaging in a retinal slice preparation minimizes photoreceptor activation by the scan laser. (**a**) Schematic of fluorescence imaging and visual stimulation in a retinal whole-mount preparation. Red, scan laser; blue, visual stimulus; gray-scale image, histological section of retina (PR photoreceptors, OPL outer plexiform layer, INL inner nuclear layer, IPL plexiform layer, GCL ganglion cell layer). Two-photon stimulation by the scan laser activates photoreceptors, even at wavelengths above 900 nm [25, 39]. This precludes studies of visual function at scotopic (rod photoreceptor-mediated) light levels in a whole-mount preparation, because the imaged region and laser-activated photoreceptor area overlap. (**b**) Laser activation of the photoreceptors can be avoided in a retinal slice preparation, where the imaged region and the photoreceptor population that provides input to it are spatially offset. This configuration also strongly reduces the potential for stimulus light to enter the fluorescence detection light path. An important downside of this approach is that in a slice preparation, lateral circuit connections are compromised, and this may negatively impact experimental outcomes

Conflict of Interest

Dr. Bart G. Borghuis is the owner of Borghuis Instruments, a company that sells specialized syringes for intra-ocular virus injections.

References

1. Clements JD (1996) Transmitter timecourse in the synaptic cleft: its role in central synaptic function. Trends Neurosci 19:163–171

2. Marvin JS, Borghuis BG, Tian L, Cichon J, Harnett MT, Akerboom J, Gordus A, Renninger SL, Chen TW, Bargmann CI, Orger MB, Schreiter ER, Demb JB, Gan WB, Hires SA, Looger LL (2013) An optimized fluorescent probe for visualizing glutamate neurotransmission. Nat Methods 10:162–170

3. Hires SA, Zhu Y, Tsien RY (2008) Optical measurement of synaptic glutamate spillover and reuptake by linker optimized glutamate-sensitive fluorescent reporters. Proc Natl Acad Sci U S A 105:4411–4416

4. Liang R, Broussard GJ, Tian L (2015) Imaging chemical neurotransmission with genetically encoded fluorescent sensors. ACS Chem Neurosci 6:84–93

5. Namiki S, Sakamoto H, Iinuma S, Iino M, Hirose K (2007) Optical glutamate sensor for spatiotemporal analysis of synaptic transmission. Eur J Neurosci 25:2249–2259

6. Okubo Y, Sekiya H, Namiki S, Sakamoto H, Iinuma S, Yamasaki M, Watanabe M, Hirose K, Iino M (2010) Imaging extrasynaptic glutamate dynamics in the brain. Proc Natl Acad Sci U S A 107:6526–6531

7. Okumoto S (2010) Imaging approach for monitoring cellular metabolites and ions using genetically encoded biosensors. Curr Opin Biotechnol 21:45–54

8. Nagai T, Sawano A, Park ES, Miyawaki A (2001) Circularly permuted green fluorescent proteins engineered to sense Ca^{2+}. Proc Natl Acad Sci U S A 98:3197–3202

9. Borghuis BG, Marvin JS, Looger LL, Demb JB (2013) Two-photon imaging of nonlinear glutamate release dynamics at bipolar cell synapses in the mouse retina. J Neurosci 33:10972–10985

10. Franke K, Berens P, Schubert T, Bethge M, Euler T, Baden T (2017) Inhibition decorrelates visual feature representations in the inner retina. Nature 542:439–444

11. Hikima T, Garcia-Munoz M, Arbuthnott GW (2016) Presynaptic D1 heteroreceptors and mGlu autoreceptors act at individual cortical release sites to modify glutamate release. Brain Res 1639:74–87

12. Parsons MP, Vanni MP, Woodard CL, Kang R, Murphy TH, Raymond LA (2016) Real-time imaging of glutamate clearance reveals normal striatal uptake in Huntington disease mouse models. Nat Commun 7:11251

13. Poleg-Polsky A, Diamond JS (2016) Retinal circuitry balances contrast tuning of excitation and inhibition to enable reliable computation of direction selectivity. J Neurosci 36:5861–5876

14. Rosa JM, Bos R, Sack GS, Fortuny C, Agarwal A, Bergles DE, Flannery JG, Feller MB (2015) Neuron-glia signaling in developing retina mediated by neurotransmitter spillover. Elife 14:4

15. Xie Y, Chan AW, McGirr A, Xue S, Xiao D, Zeng H, Murphy TH (2016) Resolution of high-frequency mesoscale intracortical maps using the genetically encoded glutamate sensor iGluSnFR. J Neurosci 36:1261–1272

16. Akerboom J, Chen TW, Wardill TJ, Tian L, Marvin JS, Mutlu S, Calderon NC, Esposti F, Borghuis BG, Sun XR, Gordus A, Orger MB, Portugues R, Engert F, Macklin JJ, Filosa A, Aggarwal A, Kerr RA, Takagi R, Kracun S, Shigetomi E, Khakh BS, Baier H, Lagnado L, Wang SS, Bargmann CI, Kimmel BE, Jayaraman V, Svoboda K, Kim DS, Schreiter ER, Looger LL (2012) Optimization of a GCaMP calcium indicator for neural activity imaging. J Neurosci 32:13819–13840

17. Madisen L, Garner AR, Shimaoka D, Chuong AS, Klapoetke NC, Li L, van der Bourg A, Niino Y, Egolf L, Monetti C, Gu H, Mills M, Cheng A, Tasic B, Nguyen TN, Sunkin SM, Benucci A, Nagy A, Miyawaki A, Helmchen F, Empson RM, Knopfel T, Boyden ES, Reid RC, Carandini M, Zeng H (2015) Transgenic mice for intersectional targeting of neural sensors and effectors with high specificity and performance. Neuron 85:942–958

18. Borghuis BG, Looger LL, Tomita S, Demb JB (2014) Kainate receptors mediate signaling in both transient and sustained OFF bipolar cell pathways in mouse retina. J Neurosci 34:6128–6139

19. Riveros N, Fiedler J, Lagos N, Munoz C, Orrego F (1986) Glutamate in rat brain cortex

synaptic vesicles: influence of the vesicle isolation procedure. Brain Res 386:405–408

20. Herman MA, Jahr CE (2007) Extracellular glutamate concentration in hippocampal slice. J Neurosci 27:9736–9741

21. Clements JD, Lester RA, Tong G, Jahr CE, Westbrook GL (1992) The time course of glutamate in the synaptic cleft. Science 258: 1498–1501

22. Diamond JS, Jahr CE (1997) Transporters buffer synaptically released glutamate on a submillisecond time scale. J Neurosci 17:4672–4687

23. Jensen TP, Zheng K, Tyurikova O, Reynolds JP, Rusakov DA (2017) Monitoring single-synapse glutamate release and presynaptic calcium concentration in organised brain tissue. Cell Calcium 64:102–108

24. Schultz BR, Chamberlain JS (2008) Recombinant adeno-associated virus transduction and integration. Mol Ther 16:1189–1199

25. Borghuis BG, Tian L, Xu Y, Nikonov SS, Vardi N, Zemelman BV, Looger LL (2011) Imaging light responses of targeted neuron populations in the rodent retina. J Neurosci 31: 2855–2867

26. Dalkara D, Byrne LC, Klimczak RR, Visel M, Yin L, Merigan WH, Flannery JG, Schaffer DV (2013) In vivo-directed evolution of a new adeno-associated virus for therapeutic outer retinal gene delivery from the vitreous. Sci Transl Med 5:189ra176

27. Flannery JG, Visel M (2013) Adeno-associated viral vectors for gene therapy of inherited retinal degenerations. Methods Mol Biol 935: 351–369

28. Kay CN, Ryals RC, Aslanidi GV, Min SH, Ruan Q, Sun J, Dyka FM, Kasuga D, Ayala AE, Van Vliet K, Agbandje-McKenna M, Hauswirth WW, Boye SL, Boye SE (2013) Targeting photoreceptors via intravitreal delivery using novel, capsid-mutated AAV vectors. PLoS One 8:e62097

29. Cronin T, Vandenberghe LH, Hantz P, Juttner J, Reimann A, Kacso AE, Huckfeldt RM, Busskamp V, Kohler H, Lagali PS, Roska B, Bennett J (2014) Efficient transduction and optogenetic stimulation of retinal bipolar cells by a synthetic adeno-associated virus capsid and promoter. EMBO Mol Med 6:1175–1190

30. Lu Q, Ganjawala TH, Ivanova E, Cheng JG, Troilo D, Pan ZH (2016) AAV-mediated transduction and targeting of retinal bipolar cells with improved mGluR6 promoters in rodents and primates. Gene Ther 23:680–689

31. Scalabrino ML, Boye SL, Fransen KM, Noel JM, Dyka FM, Min SH, Ruan Q, De Leeuw CN, Simpson EM, Gregg RG, McCall MA, Peachey NS, Boye SE (2015) Intravitreal delivery of a novel AAV vector targets ON bipolar cells and restores visual function in a mouse model of complete congenital stationary night blindness. Hum Mol Genet 24:6229–6239

32. Wang YV, Weick M, Demb JB (2011) Spectral and temporal sensitivity of cone-mediated responses in mouse retinal ganglion cells. J Neurosci 31:7670–7681

33. Choi SY, Sheng Z, Kramer RH (2005) Imaging light-modulated release of synaptic vesicles in the intact retina: retinal physiology at the dawn of the post-electrode era. Vis Res 45: 3487–3495

34. Ribelayga C, Cao Y, Mangel SC (2008) The circadian clock in the retina controls rod-cone coupling. Neuron 59:790–801

35. Zhang Z, Li H, Liu X, O'Brien J, Ribelayga CP (2015) Circadian clock control of connexin36 phosphorylation in retinal photoreceptors of the CBA/CaJ mouse strain. Vis Neurosci 32:E009

36. Chen JL, Villa KL, Cha JW, So PT, Kubota Y, Nedivi E (2012) Clustered dynamics of inhibitory synapses and dendritic spines in the adult neocortex. Neuron 74:361–373

37. Morgan JL, Kerschensteiner D (2011) Shooting DNA, dyes, or indicators into tissue slices using the gene gun. Cold Spring Harb Protoc 12:1512–1514

38. Baden T, Berens P, Bethge M, Euler T (2013) Spikes in mammalian bipolar cells support temporal layering of the inner retina. Curr Biol 23:48–52

39. Euler T, Hausselt SE, Margolis DJ, Breuninger T, Castell X, Detwiler PB, Denk W (2009) Eyecup scope–optical recordings of light stimulus-evoked fluorescence signals in the retina. Pflugers Arch 457:1393–1414

40. Kugler S, Kilic E, Bahr M (2003) Human synapsin 1 gene promoter confers highly neuron-specific long-term transgene expression from an adenoviral vector in the adult rat brain depending on the transduced area. Gene Ther 10:337–347

Chapter 6

Imaging Neuronal Signal Transduction Using Multiphoton FRET-FLIM

Paul R. Evans, Long Yan, and Ryohei Yasuda

Abstract

Synaptic plasticity, the ability of neurons to modulate the strength of specific inputs, is critical for neural circuits to adapt to experience throughout life. In excitatory pyramidal neurons, plasticity is induced by coincident neuronal activity and glutamate release at tiny postsynaptic protrusions called dendritic spines, which initiate the coordinated activity of hundreds of different proteins located in spines and throughout the neuron at distinct temporal phases. Thus, elucidating the spatiotemporal dynamics of individual signaling proteins is critical to refine our understanding of this process. The complex, polarized morphology of neurons can restrict protein activity to small cellular subcompartments, while other signals can spread over long distances, which poses unique challenges to monitoring protein dynamics. Fluorescence resonance energy transfer (FRET) is a useful photophysical phenomenon to visualize signaling in space and time within live cells by measuring the efficiency of energy transfer between two fluorescent proteins. Using two-photon fluorescence lifetime imaging microscopy (2pFLIM) to assay FRET-based signaling sensors permits chronic, high-resolution measurements of discrete neuronal signaling events, even in dense, light-scattering brain slices. Here, we describe the imaging setup required to perform 2pFLIM and highlight its application to decipher the orchestrated signaling underlying the structural plasticity of dendritic spines.

Key words Fluorescence lifetime imaging microscopy (FLIM), Two-photon microscopy, Fluorescence resonance energy transfer (FRET), Sensor biology, Dendritic spine, Neuronal signal transduction, Spine structural plasticity, Long-term potentiation (LTP)

1 Introduction

Neural circuits are composed of several cell types that are highly interconnected in specific patterns to maintain network function and stability. A hallmark of neurons is synaptic plasticity, the ability to rapidly adapt to experience and alter the strength of specific connections, which likely encodes aspects of learning and memory. Plasticity is a complex process that involves signaling occurring on various timescales (milliseconds to hours) that can be either compartmentalized or widespread [1, 2]. Spines are small, actin-rich protrusions emanating from dendritic processes of many glutamatergic neurons in the brain that serve as postsynaptic subcellular

Espen Hartveit (ed.), *Multiphoton Microscopy*, Neuromethods, vol. 148,
https://doi.org/10.1007/978-1-4939-9702-2_6, © Springer Science+Business Media, LLC, part of Springer Nature 2019

compartments to isolate electrical [3] and biochemical signals [4, 5]. Long-term potentiation (LTP) is a prominent form of Hebbian plasticity induced by coincident neuronal activity and glutamate release at specific spines that results in long-lasting enhanced sensitivity to glutamate and enlargement of the stimulated spine [6]. Thus, dendritic spines are an excellent model system to study protein dynamics during plasticity. In addition to compartmentalized signaling in stimulated spines that gives rise to synapse-specific LTP, signaling cross-talk with neighboring spines influences the local threshold for plasticity [7, 8], and long-range signaling integrated at the nucleus is necessary to initiate gene transcription to consolidate LTP [9–11].

A vast network of signaling molecules required for plasticity have been identified, but the spatial and temporal regulation of their signaling during plasticity are less well understood. Elucidating these aspects of protein activity during plasticity is crucial to construct a full picture of how individual signaling proteins mediate this process. To ascertain these signaling dynamics, one needs a method that collects ongoing measurements with sufficient temporal resolution to capture protein activity and spatial resolution to detect synaptic compartments.

Genetics and pharmacology are classic methods to define the necessary signaling proteins comprising a cellular process, but they provide little, if any, information about the localization or time-scale of signaling. Traditional biochemical techniques have been useful in identifying protein-protein interactions and posttranslational modifications required for LTP, but these approaches tend to measure global changes in protein states and are thus limited in their ability to resolve discrete signaling events. For example, Western blots of proteins from brain tissue lysates can probe the phosphorylation state of a protein, but this signal emerges from hundreds to thousands of homogenized cells at one time point, providing minimal insight into the spatiotemporal regulation of this event. Immunolabeling an endogenous protein with a specific antibody is often used to resolve subcellular localization of signaling proteins, but this method requires tissue fixation and limits signaling measurements to a single time point. Immunostaining also relies on a specific antibody to study a protein of interest, but even when equipped with sensitive antibodies, tissue fixation can alter morphology and destroy antibody epitopes obscuring the interpretation of these experiments.

Dynamic protein localization can be visualized in living tissue by performing time-lapse fluorescence microscopy of a genetically encoded fluorescent protein tag. The expression of genetically encoded material is widely used for the ease with which the cDNA vectors can be introduced into living cells and modified to examine the role of specific protein domains on signaling function. Traditionally, this approach requires overexpressing the fluorescently tagged protein, which can perturb the endogenous signaling pathways if not carefully controlled. With recent advances in genome editing gained through CRISPR/Cas9, it is now possible to tag the endogenous protein, eliminating the need for overexpression [12, 13].

Fluorescence resonance energy transfer (FRET) harnesses genetically encoded fluorescent proteins to provide sensitive readouts of cellular signaling [14]. FRET is the non-radiative energy transfer from an excited donor fluorophore to an acceptor fluorophore, which in turn emits a photon, when in close proximity (generally within nanometers) [15, 16]. When the two fluorophores are separated or misaligned, the excited donor fluorophore emits its own photons rather than transferring the energy to the acceptor. Because FRET efficiency is highly dependent on the orientation and distance between the two fluorophores, FRET-based sensors can report signaling activity by measuring conformational changes and protein interactions that alter FRET. Additional factors that influence FRET efficiency include the degree of overlap between the donor emission spectrum and the acceptor excitation spectrum, the angle between the donor and acceptor, and the quantum efficiency of the donor [15].

A variety FRET-based biosensors have been developed to detect molecular dynamics of a range of signaling proteins [14, 17]. For example, intramolecular conformational changes can be visualized by tagging two regions of the same protein with the donor and acceptor fluorophores. CaMKII activity has been visualized in hippocampal neurons and dendritic spines during plasticity using this sensor design because its conformation reports its kinase activity [18–20]. Some signaling proteins selectively engage binding partners in their active state; FRET sensors measure these activity-dependent interactions by tagging one protein with a donor fluorophore and the interacting partner with the acceptor. Signaling sensors have been developed based on this principle for small G proteins, which only bind certain effectors when in the active GTP-bound state, such as the FRas Ras activity sensors [8, 21, 22]. In these sensors, HRas is tagged with a donor fluorophore while the Ras Binding Domain (RBD) of Raf is flanked with acceptor fluorophores, and an increase in FRET signifies elevated Ras signaling. Another strategy for FRET sensor development is to concatenate binding domains from different proteins into a single polypeptide with linker regions and fluorophores that display FRET differences after a signaling event. For example, kinase activity sensors with this design include a phosphorylation motif for a specific kinase, a flexible linker, an interacting domain that is specific for the phosphorylated peptide, and donor/acceptor fluorophores fused to the ends of these domains. FRET can read out the balance of kinase and phosphatase activity as the phosphorylation-dependent binding of the motif with the interacting domain brings the donor and acceptor in close proximity producing an increase in FRET [23]. FRET sensors have been utilized to probe other aspects of signaling, including second messenger concentrations [17].

Following an excitation laser pulse to excite the donor fluorophore, relative FRET efficiency is often quantified as the ratio of acceptor to donor fluorescence emission intensities (i.e., ratiometric method), which can be separated by a dichroic mirror and wavelength filters [14, 16, 24]. An increase in FRET is reflected by a higher acceptor:donor intensity ratio because the donor energy is transferred to the acceptor and emitted as the acceptor fluorescence. However, it can be difficult to quantify FRET efficiency using intensity-based methods due to off-target, direct excitation of the acceptor by the laser pulse and spectral bleed-through of donor fluorescence into the acceptor channel. The fluorescence intensity ratio is also susceptible to local changes in donor and acceptor fluorophore concentrations, such that if a bimolecular sensor is used and one protein translocates after stimulation, it would falsely appear as a change in FRET [16, 24]. These technical limitations render ratiometric imaging to be useful only when local acceptor and donor fluorophore concentrations are fixed, e.g., by including both fluorophores in one polypeptide to report signaling. Moreover, the fluorescence ratio is also affected by the wavelength-dependent scattering of light, which poses issues for imaging intact synapses in light-scattering brain tissue. It is possible to correct the intensity ratio by using a second excitation wavelength to selectively measure acceptor intensity, but fluorescence lifetime imaging microscopy can avoid artifacts introduced by differences in acceptor/donor photostability.

Fluorescence lifetime imaging microscopy (FLIM) is another method to quantify FRET by measuring the decay of the fluorescence emitted over time by the donor fluorophore, which is shortened when the donor is undergoing FRET [16, 24, 25]. The decrease in fluorescence lifetime is proportional to FRET efficiency, and the exact percentage of donor undergoing FRET can be extracted by curve fitting the fluorescence lifetime decay curve. An attractive feature of FLIM is that fluorescence lifetime is an inherent property of the donor fluorophore that is independent of local fluorophore concentration and acceptor fluorescence, which avoids the stoichiometric limitations of ratiometric imaging. FLIM also permits chronic measurements of FRET over time generating a dynamic view of signaling throughout a cellular process.

Combining FLIM with two-photon laser scanning microscopy (2pLSM) enables imaging of individual synapses in light-scattering tissue [21, 26], such as the intact brain or brain slice preparations. The increased penetration depth in 2pLSM is due to the wavelength-dependent scattering of light, and the longer wavelengths used for two-photon excitation produce much less scatter than conventional one-photon excitation wavelengths. 2pLSM can acquire high spatial resolution images in tissue because the low probability of excitation produces little contaminating background photons from out-of-focus excitation. In contrast to dissociated cell culture, brain slice preparations offer a more physiologically relevant context to investigate synaptic signaling because neuronal morphology, local synaptic connections, and developmental history are preserved.

In the following sections, we will review the optical setup for 2pFLIM as well as the design, execution, and data analysis of 2pFLIM experiments to measure neuronal signal transduction. In particular, we will focus on paradigms to induce synaptic plasticity at individual dendritic spines in brain slice cultures and visualize the spatiotemporal signaling dynamics with FRET-based protein activity sensors. We will also review the previous and current applications of FRET-based sensors to elucidate the biochemical network underlying spine structural plasticity.

2 Materials/Methods

2.1 Two-Photon Laser Scanning Microscopy (2pLSM) Setup for Fluorescence Lifetime Imaging Microscopy

Two-photon laser scanning microscopy is widely used for its ability to image fluorescent samples with high sensitivity and resolution deep in light-scattering tissue. Time-correlated single photon counting (TCSPC) is the most common method to measure fluorescence lifetime decay. TCSPC measures the time lapsed between single photon detection and the subsequent excitation laser pulse, and when combined with 2pLSM, which already uses a pulsed laser, this method provides high sensitivity and highest temporal resolution of fluorescence lifetime. Since every photon detected contributes to the signal, TCSPC is ideal for imaging dendritic spines in brain slices where the fluorescence photons are limiting [21]. The following modifications (Fig. 1a) can equip a standard 2pLSM to be combined with TCSPC for 2pFLIM imaging [21, 27]:

1. Install a TCSPC imaging card (e.g., TimeHarp 260 two-channel PICO model PCIe card, PicoQuant; or SPC-160 PCIe card, Becker & Hickl).

2. Use a photomultiplier tube (PMT) with low transit time spread (<0.2 ns) and single photon counting sensitivity (e.g., H7422-40P, Hamamatsu).

3. Connect the built-in pulse output trigger of the 2p laser to the "SYNC" input of the TCSPC card. The laser pulse may also be detected by reflecting it to a fast photodiode with biased voltage circuit (e.g., DET10A2, Thorlabs) if there is no built-in photodiode output. A signal inverter (comes with Becker & Hickl cards) may be necessary.

4. Feed the detected photon signal from the PMT to the CFD (constant fraction discriminators) port of the TCSPC card. Note: a preamplifier is optional in this setup.

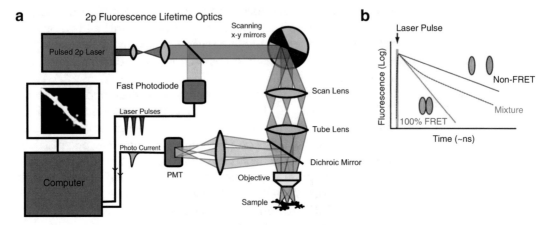

Fig. 1 Two-photon fluorescence lifetime (2pFLIM) setup and data acquisition. (**a**) Schematic of a representative microscope setup and signal detection for 2pFLIM. A pulsed laser is projected onto x-y scanning mirrors which direct the excitation beam through an optical path (shown here as a scan lens, tube lens, dichroic mirror, and objective) to focus on a fluorescent neuron. Note the fast photodiode preceding the scanning mirrors in the excitation laser path that allows the computer to detect timing of the laser pulses. The photons emitted by the sample are then collected by the objective lens and deflected onto a photoelectron multiplier tube (PMT) that converts single photons into electrons that can be read by the system as current. The computer boards are synced so the time lapsed between excitation laser pulses and detected photon current can be measured to enable time correlated single photon counting (TCSPC). Figure adapted from Yasuda et al. [21] with copyright permission. (**b**) Semi-log plot of fluorescence decay curves acquired with 2pFLIM. Following the excitation laser pulse in the absence of acceptor, a typical FLIM donor fluorophore (e.g., mEGFP) will display a monoexponential fluorescence lifetime. The donor fluorescence lifetime becomes shorter when it undergoes FRET when in close proximity to the acceptor (orange), and the decay fits with two time constants representing a mixture of the two populations of donor (FRET and non-FRET components). (Figure adapted from Yasuda [25] with permission)

5. Feed the pixel, line, and frame clocks into the TCSPC card (for TimeHarp 260 or SPC-160, it is a D-sub connector).

6. Obtain a fluorescence decay curve, averaged over ~1 s, and position the signal so that the entire curve is visible (the decay should begin at approximately 2 ns) by adjusting the cable lengths (either lengthen the SYNC cable or shorten the CFD cable) or delay setting of the hardware. Note that electrical signals travel approximately 30 cm/ns. For some TCSPC cards, the lifetime curve can be shifted from the software.

7. The decay onset must be recalibrated every time the microscope is moved. For this reason, a fixed 2pLSM with a movable stage is preferred to movable objective or movable microscope.

8. Source code for FLIM image acquisition software is available online   https://github.com/ryoheiyasuda/FLIMimage_Matlab_ScanImage and https://github.com/ryoheiyasuda/FLIMage_public.

2.2 Sensor Design for 2pFLIM

Optimal fluorophore pairs for FLIM imaging have different requirements than intensity-based FRET measurements [24]. FLIM only measures the fluorescence lifetime of the donor fluorophore; therefore, optimal FLIM donors like mEGFP and EYFP are bright and photostable. It is also desirable for a FLIM donor to display a monoexponential fluorescence lifetime decay because this feature enables FRET and non-FRET components of the lifetime decay to be separated by curve fitting (see Sect. 2.5) to yield the exact fraction of donor fluorophore undergoing FRET [24]. Ratiometric imaging most often uses the ECFP-EYFP FRET pair, but these fluorophores are not optimal for FLIM as the ECFP donor is dim and displays a multi-exponential decay [21, 28]. mRFP1 or mCherry1 are typical FRET acceptor fluorophores used with mEGFP because the spectral separation from the mEGFP donor fluorescence emission produces little contaminating acceptor signal in the donor channel at the 2p imaging wavelengths [21, 24, 29]. However, mRFP1 and mCherry1 have relatively low folding efficiency, and unfolded fluorophores do not undergo FRET. Therefore, the acceptor protein is normally tagged with two mRFP1 or mCherry1 fluorophores to enhance the signal-to-noise ratio [24]. A non-radiative variant of YFP (super resonance energy transfer accepter chromophore, or sREACh) is also used as an acceptor with mEGFP for 2pFLIM [27, 30], but this dim fluorophore does not readily produce fluorescence making it difficult to verify expression or quantify concentration. Therefore, the mEGFP-sREACh pair is best utilized in unimolecular sensors where mEGFP can be visualized for expression measurements. Recently, a new red-shifted FRET donor named monomeric cyan-excitable red fluorescent protein (mCyRFP1) was developed that enables 2pFLIM with its FRET acceptor mMaroon1 [31]. mCyRFP1 and mEGFP are excited at similar laser wavelengths but display spectrally separated fluorescence emission, which allows simultaneous dual-color FLIM measurements of FRET sensors expressed in the same cell. mCyRFP1 also displays a monoexponential decay so the fraction of donor undergoing FRET can be calculated [31].

When designing FRET sensors, it is critical that the signaling functions of the protein are not disrupted by tagging with the donor and/or acceptor fluorophores. Control experiments are required to verify that the desired function of the protein is not affected. The insertion site, linker sequence, and fluorophore selected may be adjusted to mitigate off-target effects. Similarly, protein engineering can be employed to improve the kinetics or signal-to-noise ratio produced by the sensor, such as with variants of the Ras activity sensor [8, 21, 22]. It is also important to consider effects of sensor expression on endogenous signaling when interpreting experimental findings. Ectopic expression of a protein activity sensor modulates that signaling pathway to some degree [14], and in general, biosensors with higher sensitivity tend to more strongly perturb endogenous biochemical cascades. Therefore, it is important to balance optimizing the sensor's signal-to-noise ratio with effects on native signaling components [24].

2.2.1 Sensors Developed for 2pFLIM (Table 1)

The FRET-based biosensors that have been developed or modified for 2pFLIM imaging of neuronal signal transduction are listed in Table 1.

2.3 Slice Culture Preparation for Imaging Sensors

2.3.1 Organotypic Slice Culture

Hippocampal organotypic slice cultures are an attractive model system for studying synaptic signaling because local synaptic connections remain intact, providing a more physiologically relevant context to investigate synapses than dissociated neuron cultures. Moreover, slice cultures are viable for weeks in culture and are amenable to transfection ex vivo to deliver genetically encoded material such as FRET-based biosensors.

Table 1
FRET-based biosensors for 2pFLIM of signal transduction

Sensor target	Signaling class	Sensor name	Sensor readout	References
CaMKII	Kinase	Green-Camuiα	Conformation change	[19, 20]
Ras	Small G protein	FRas	Protein interaction (GTP-dependent)	[8, 21, 22]
Rac1	Small G protein		Protein interaction (GTP-dependent)	[7]
RhoA	Small G protein		Protein interaction (GTP-dependent)	[31, 32]
Cdc42	Small G protein		Protein interaction (GTP-dependent)	[31, 32]
TrkB	Receptor tyrosine kinase		Protein interaction (BDNF-dependent)	[33]
PIP_3	Lipid	FLIMPA3	Intramolecular FRET (PIP_3-dependent)	[27, 34, 35]
ERK	Kinase	EKAR	Intramolecular FRET (phospho-dependent)	[9, 11, 36]
PKA	Kinase	AKAR	Intramolecular FRET (phospho-dependent)	[36–38]
PKA	Kinase		Protein interaction	[39]
SNARE complex	Membrane fusion		Protein interaction	[40]
Cofilin	Cytoskeleton		Protein interaction	[41]
Actin	Cytoskeleton		Protein interaction	[27]
NADH-NAD$^+$ (Redox)	Cell metabolism	Peredox	Conformation change	[42, 43]

Hippocampal slice cultures are prepared from postnatal day 6–8 mice or rats [44]. First, the animals are deeply anesthetized with isoflurane and rapidly decapitated with surgical scissors. The skin and skull are cut sagitally using a scalpel from between the eyes to the occipital bone while laterally folding back both sides of the skull. The brain is then removed by lifting the brain from its base with a spatula and transferring it to a petri dish containing chilled dissection medium kept over ice. The brain is bisected between the hemispheres using a spatula, and the striatum and diencephalon are removed to expose the hippocampi. A smoothly curved iris spatula is used to remove the hippocampi by unfolding the cortex and severing connections between the cortex and hippocampus.

The dissected hippocampi are laid flat with the long axis perpendicular to the blade and sliced coronally into 350 μm sections with a tissue chopper (Ted Pella, Inc.), transferred into a new dish containing dissection medium (1 mM $CaCl_2$, 5 mM $MgCl_2$, 10 mM glucose, 4 mM KCl, 26 mM $NaHCO_3$, and 248 mM sucrose, bubbled with 95% O_2 and 5% CO_2), and carefully separated by gently pipetting up and down. Hippocampus slices are placed on Millipore membrane cell culture inserts (0.4 μm pore size/30 mm diameter) in wells containing 1 ml of warmed tissue medium (8.4 g/ml HEPES-based minimum essential medium, 20% horse serum, 1 mM L-glutamine, 1 mM $CaCl_2$, 2 mM $MgSO_4$, 12.9 mM D-glucose, 5.2 mM $NaHCO_3$, 30 mM HEPES, 0.075% ascorbic acid, and 1 μg/ml insulin, pH adjusted to 7.2 for storage in 95% O_2/5% CO_2). Generally, three to four hippocampus slices are placed on each membrane insert. Once the slices have adhered to the membrane, carefully remove any residual dissection medium from the membrane using a 20 μl pipette (Gilson P20) to allow the slices to aerate. Note, avoid directly touching the slice with the pipette tip as this may damage the tissue. Organotypic slice cultures are kept in a cell culture incubator at 35 °C (for rat, 37 °C for mouse) in a humidified environment of 95% O_2/5% CO_2, and tissue medium is changed every 2–3 days. Generally, cultured hippocampus slices can be maintained under these conditions for ~3 weeks.

2.3.2 Biolistic Transfection

When imaging single dendritic spines under 2pFLIM, it is optimal to have few neurons transfected in each slice to acquire high contrast images. Ballistic gene transfer [45] is a convenient method to sparsely transfect neurons in cultured slices by shooting slices with gold bullets coated with cDNA encoding the desired FRET biosensor (or other protein of interest). The uptake of the gold particles at a neuronal nucleus is required for expression, which is a low probability event resulting in the desired sparse transfection.

Gold particles (Bio-Rad, ~1 μm diameter, 8–11 mg) are coated with plasmids containing sensor cDNA and dehydrated onto the interior of plastic tubing [46]. More than one cDNA can be coated onto the same gold beads by pre-mixing, e.g., when imaging a bimolecular FRET sensor or when co-expressing a sensor with a separate fluorescent protein as a volume marker. The tubing is loaded into a Helios gene gun (Bio-Rad), and pressurized air ejects the gold particles from the tubing, launching them into the culture well containing slice cultures. Organotypic hippocampus cultures are typically transfected after 7–10 days in vitro, and neurons are imaged once fluorescent protein expression is sufficient, generally 2–10 days following transfection. However, expression time prior to imaging varies according to the sensor used and plasmid properties, such as the promoter. For example, slice cultures transfected with a bright fluorophore, e.g., mEGFP, under the expression of a strong CAG promoter can be imaged as soon as 24 hours after transfection. Note that prolonged sensor expression may cause undesired effects on endogenous cellular signaling.

2.4 Two-Photon Glutamate Uncaging

2.4.1 Setup for Simultaneous 2p Imaging/Uncaging

Simultaneously imaging neuronal signaling during 2p glutamate uncaging requires two two-photon lasers tuned to different wavelengths. The laser wavelengths are determined by the excitation spectra of the donor fluorophore in the sensor and the two-photon cross section of the caged compound. For 2pFLIM imaging with mEGFP as the donor and uncaging MNI-glutamate, one Ti:Sapphire laser is tuned to 920 nm to excite the mEGFP donor for imaging while the second Ti:Sapphire laser is set to 720 nm to uncage MNI-glutamate [8]. The uncaging beam should be controlled by an independent shutter (e.g., LS 6 mm high-speed laser shutter with AlMgF$_2$-coated BeCu blades (Part # LS6ZM2) from Vincent Associates) from the imaging laser to avoid contaminating illumination resulting in off-target glutamate uncaging.

2.4.2 2p Glutamate Uncaging to Induce Spine Structural Plasticity

Glutamate is the main excitatory neurotransmitter in the mammalian central nervous system, and the majority of glutamatergic synapses terminate onto dendritic spines. In the hippocampus, synaptic plasticity has been extensively characterized at CA3 Schaffer collateral synapses onto CA1 neurons where long-term potentiation (LTP) of synaptic transmission is readily induced. LTP is a form of Hebbian plasticity that is induced by synchronous neuronal activity and glutamate release at specific synapses, which activates postsynaptic NMDARs on spines through which calcium (Ca^{2+}) enters and initiates biochemical cascades required for plasticity induction [47–49]. LTP is observed as a sustained strengthening of the glutamatergic synapse due to upregulated surface expression of AMPA-type glutamate receptors and is associated with enlargement of the stimulated spine, i.e., structural plasticity. LTP is classically induced by high-frequency electrical stimulation of Schaffer collaterals to evoke glutamate release from terminals, but these paradigms activate numerous synapses making it difficult to study signaling initiated at one synapse. Two-photon glutamate uncaging is a powerful method to selectively stimulate a single spine [6, 50] as the low probability of simultaneous absorption of two photons greatly restricts the volume in which glutamate is uncaged. Recently, caged glutamate variants with an appreciable two-photon cross section have been developed including MNI-glutamate [51], CDNI-glutamate [52], and Rubi-glutamate [53]. Here, we will focus on experimental details for inducing spine structural plasticity by MNI-glutamate uncaging on cultured hippocampus slices as performed by our group.

The activation of postsynaptic NMDA-type glutamate receptors is a key requirement for activity-dependent LTP induction at Schaffer collateral-CA1 synapses in the hippocampus. The magnesium block in the NMDA receptor allows it to act as a coincidence detector such that only synapses receiving glutamate stimulation during neuronal depolarization become potentiated. At negative resting membrane potential, the NMDAR channel pore is blocked by magnesium ions, but neuronal depolarization relieves the magnesium block and permits subsequent Ca^{2+} influx through NMDARs. Ca^{2+} then activates intracellular [2] and extracellular [33] signaling cascades that drive enlargement of the stimulated spine as well as the insertion of new AMPA-type glutamate receptors into the spine [54]. Glutamate uncaging on a single spine produces very little electrical depolarization at the soma; therefore, LTP produced by glutamate uncaging requires removal of the magnesium block by pairing with postsynaptic depolarization or by performing uncaging in the absence of extracellular magnesium [6]. We will describe experimental details for the latter scenario.

MNI-glutamate uncaging experiments to induce plasticity at a single spine are typically performed at room temperature on cultured hippocampus slices (DIV 12–21) with neurons transfected with FRET-based signaling sensors and/or a fluorescent protein to serve as a volume marker. Organotypic slices are perfused with artificial cerebrospinal fluid (ACSF) containing (in mM) 127 NaCl, 2.5 KCl, 25 $NaHCO_3$, 1.25 NaH_2PO_4, 4 $CaCl_2$, 25 glucose, 0.001 tetrodotoxin (Tocris), and 4 MNI-caged L-glutamate (Tocris), bubbled with 95% O_2 and 5% CO_2. Importantly, this solution contains 0 $MgCl_2$ and 4 mM $CaCl_2$, and the ACSF is also supplemented with 0.001 mM tetrodotoxin to block spontaneous activity and mitigate off-target antagonization of GABA receptors by MNI-glutamate [53]. Slice cultures are transferred to the perfusion chamber on the microscope by cutting the mesh insert surrounding the slice with a fine surgical blade and grasping the remaining mesh adhering to the slice with fine forceps. Organotypic slice cultures may alternatively be prepared by allowing slices to adhere to poly-L-lysine coated coverslips, rather than to mesh inserts, and stored in a rolling drum incubator [55]. To reduce focal drift during imaging, slices are secured in place underneath a harp fabricated from gold wire (e.g., 0.5 mm diameter gold wire from Alfa Aesar) and allowed to rest in the perfusion chamber for approximately 30 mins prior to imaging.

First, a secondary or tertiary dendritic segment bearing healthy spines is selected and baseline images are collected. The number of images and time period over which baseline images are acquired depends on the sensor. During baseline imaging, it is helpful to monitor sensor fluorescence lifetime online to ensure that protein activity is stable before uncaging. A train of 30–60 uncaging laser pulses at 720 nm (4–8 ms duration) are delivered at 0.5–1 Hz adjacent to the head of a dendritic spine of interest [6, 8, 32]. In the absence of extracellular magnesium, this protocol results in a robust "transient phase" increase in spine volume (~300–500% increase) for 1–3 minutes after uncaging that slightly decreases to a long-term "sustained phase" that is typically 150–200% of the original spine volume and persists for over an hour [7, 19, 20, 32, 33]. Fast-framing 2pFLIM images may be acquired during uncaging if the FRET-based sensor shows a rapid response to glutamate uncaging, e.g., CaMKII [19, 20], or a longer imaging period may be selected if the protein of interest displays a slower activity profile, e.g., Ras, ERK, or TrkB [8, 9, 21, 33, 36]. See Sect. 3 for representative publications highlighting FRET-based signaling sensors that show these different activity profiles. When using fast-framing 2pFLIM imaging to capture fast protein activity [20], one may need to reduce the number of pixels acquired in each image to increase temporal resolution; the loss of pixels may be offset by increasing the imaging zoom.

2.5 2pFLIM Image Analysis

2.5.1 Fitting Fluorescence Lifetime Curves

Following excitation with a brief laser pulse, fluorophores exhibit an exponential decay in fluorescence. Fluorescence decay curves acquired with time-domain FLIM measurements (such as TCSPC) are convolved with the Gaussian pulse response function (PRF) of the system that can be expressed as [21]:

$$H\left(t,t_0,\tau_D,\tau_G\right)=\frac{1}{2}\exp\left(\frac{\tau_G^2}{2\tau_D^2}-\frac{t-t_0}{\tau_D}\right)\mathrm{erfc}\left(\frac{\tau_G^2-\tau_D\left(t-t_0\right)}{\sqrt{2}\tau_D\tau_G}\right) \quad (1)$$

where H is the convolution of the Gaussian PRF with standard deviation τ_G, τ_D is the fluorescence lifetime of the free donor, t is the photon arrival time for each photon collected by the PMT, t_0 is the time offset, and erfc is the complementary error function. While preferred FLIM donor fluorophores have a monoexponential decay, some fluorescent proteins display multiexponential decay and accordingly will have multiple decay time constants [24, 28].

A major advantage of fluorescence lifetime imaging microscopy is the ability to quantify binding fraction, the proportion of donor undergoing FRET with the acceptor, when the donor fluorophore possesses a monoexponential decay. The binding fraction is calculated by fitting the fluorescence lifetime decay with a double exponential curve:

$$F(t) = F_0 \left[P_D \cdot H(t, t_0, \tau_D, \tau_G) + P_{AD} \cdot H(t, t_0, \tau_{AD}, \tau_G) \right] \quad (2)$$

where τ_D is the fluorescence lifetime of the free donor, τ_{AD} is the fluorescence lifetime of the donor bound to the acceptor (i.e., undergoing FRET), and F_0 is the peak fluorescence. P_D and P_{AD} are the fractions of free donor and the donor bound to the acceptor, respectively ($P_D + P_{AD} = 1$). τ_D can be fixed to the fluorescence lifetime of the free donor measured separately under the same imaging condition.

2.5.2 Displaying/ Generating Fluorescent Lifetime Images

Fluorescence lifetime imaging microscopy can be visualized by coding the image with color to represent fluorescence lifetime and brightness to represent fluorescence intensity [21, 56]. Fluorescence lifetime imaging microscopy are generated by calculating the mean fluorescence lifetime (τ_m) in each pixel as follows:

$$\tau_m = t - t_0 = \frac{\int dt \cdot t F(t)}{\int dt \cdot F(t)} - t_0 \quad (3)$$

where $F(t)$ is the fluorescence lifetime decay curve (21). The offset time (t_0) is obtained by fitting. Because the number of photons in each pixel are limited, fitting is performed using Eqs. 1 or 2 after integrating all the pixels of an image to improve convergence with the fluorescence lifetime.

When the free donor and donor bound to its acceptor coexist (Eq. 2), τ_m is given by

$$\tau_m = \frac{\int dt \cdot t F(t)}{\int dt \cdot F(t)} = \frac{P_D \tau_D^2 + P_{AD} \tau_{AD}^2}{P_D \tau_D + P_{AD} \tau_{AD}}. \quad (4)$$

Thus pixel-by-pixel binding fraction can be calculated as

$$P_{AD} = \frac{\tau_D (\tau_D - \tau_m)}{(\tau_D - \tau_{AD})(\tau_D + \tau_{AD} - \tau_m)} \quad (5)$$

3 Application Examples

2pFLIM of FRET-based biosensors has provided unprecedented insight into the spatiotemporal dynamics of neuronal signaling driven by LTP induction at single dendritic spines. By measuring the activity of individual proteins, the signaling network downstream of NMDAR-Ca^{2+} influx that mediates plasticity has come into view [1, 2]. Below we highlight two examples of proteins with distinct spatiotemporal dynamics during LTP imaged by 2pFLIM:

a conformational sensor to detect CaMKII activity [20] and a protein interaction sensor to readout the activity of the BDNF receptor TrkB [33].

3.1 CaMKII

The calcium/calmodulin-dependent kinase II (CaMKII) is a serine/threonine kinase with a central role in LTP that acts as one of the first signaling enzymes to decode Ca^{2+} transients in spines [57]. In its basal state, CaMKII displays a closed conformation in which its N and C termini are in close proximity. When Ca^{2+} enters the spine, it binds calmodulin (Ca^{2+}/CaM) which in turn binds CaMKII causing it to adopt an open conformation and become active. Subsequently, CaMKII autophosphorylation occurs at T286 and allows CaMKII to retain activity even after the unbinding of Ca^{2+}/CaM. However, the spatiotemporal dynamics of CaMKII activity during LTP induction and the precise role of T286 autophosphorylation in this process were debated [57].

These questions were resolved using a conformation-based CaMKII activity sensor green-Camuiα, in which both ends of CaMKIIα are tagged with fluorophores that undergo FRET when in close proximity [18–20]. Thus, CaMKII activation is observed as an increase in the donor fluorescence lifetime, representing a transition from the closed (inactive) to the open conformation (active) that decreases FRET (Fig. 2a). Using this sensor in combination with fast-framing 2pFLIM imaging during standard glutamate uncaging to induce LTP (30 pulses, 0.49 Hz), it was found that CaMKII activity closely follows spine Ca^{2+} elevations elicited by each uncaging pulse [19, 20]. The improved millisecond temporal resolution of 2pFLIM revealed a step-wise activation of CaMKII that plateaus at a stable lifetime change after a few uncaging pulses (Fig. 2b, c). These results show that CaMKII integrates Ca^{2+} signals in spines, which summate to a plateau. The increase in fluorescence lifetime decays back to baseline levels once uncaging is finished, within approximately 1 minute. Of note, CaMKII activity is almost completely restricted to the stimulated spine [19, 20].

To ascertain the role of autophosphorylation in the activation profile of CaMKII in spines, similar 2pFLIM experiments were performed with a phosphorylation-null (T286A) mutant sensor. The wild-type (WT) and T286A Camuiα sensors showed similar activation, i.e., fluorescence lifetime change, in response to a single glutamate uncaging pulse, but the T286A mutant activity decayed more quickly [20]. These experiments indicate that T286 autophosphorylation prolongs the activity of CaMKII. Next, the WT and T286A Camuiα sensors were compared while performing repetitive glutamate uncaging to induce LTP (Fig. 2c). Unlike the step-wise activation and integration of WT Camuiα, the T286A mutant was unable to integrate the uncaging-evoked Ca^{2+} signals, resulting in a greatly attenuated change in fluorescence lifetime

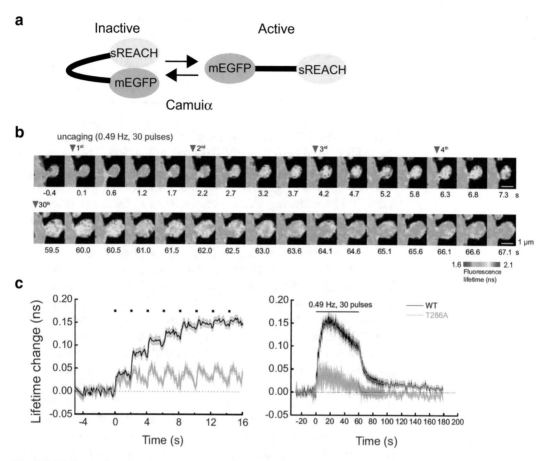

Fig. 2 CaMKII autophosphorylation underlies optimal signal integration during plasticity induction. (**a**) Schematic of the green-Camuiα CaMKII activity sensor. In its closed, inactive state, the termini of CaMKII are in close proximity and high FRET (shortened mEGFP lifetime) is observed. When CaMKII is active, it switches to an open conformation which results in a longer mEGFP lifetime as the fluorophores on either end of the sensor are separated. (**b**) Representative fluorescence lifetime images of CaMKII activation in dendritic spines acquired with millisecond temporal resolution during 0.49 Hz glutamate uncaging to induce LTP at a single spine. Warmer colors indicate longer fluorescence lifetime of Camuiα corresponding to the active, open conformation of CaMKII. (**c**) Average time courses of WT (black) and T286A (green, phospho-null mutant) Camuiα fluorescence lifetime change in dendritic spines during 0.49 Hz glutamate uncaging. Left panel is an expanded view of the data shown in the right panel. Black dots indicate glutamate uncaging pulses. (Figures adapted from Chang et al. [20] with copyright permission)

that was similar to results with a single uncaging pulse. Consistently, the deficit in signal integration to achieve full activity in the T286A sensor could be largely rescued with higher frequency stimulation. These results showed that CaMKII is a leaky integrator of synaptic stimulation, and further, that T286 autophosphorylation plays a critical role in slowing the decay of CaMKII activity to allow it to summate spine Ca^{2+} to achieve full activation and induce LTP [20].

3.2 BDNF-TrkB

Another signaling molecule with a critical role in synaptic plasticity is brain-derived neurotrophic factor (BDNF) and its receptor TrkB, but it was unknown if BDNF release and subsequent TrkB activation occurred during spine plasticity. To monitor BDNF-TrkB activity during LTP, a FRET-based sensor was generated based on the BDNF-dependent association of the TrkB receptor with phospholipase Cγ1 [33]. This sensor is composed of TrkB fused to mEGFP co-expressed with the SH2 domain of phospholipase Cγ1 fused with two mRFP1 copies (Fig. 3a). This sensor was expressed in CA1 pyramidal neurons in cultured hippocampus slices using ballistic gene transfer, and TrkB activity was monitored using 2pFLIM during glutamate uncaging to induce plasticity at a single spine [33]. Following glutamate uncaging pulses to induce LTP, TrkB receptor activity rapidly increased in the stimulated spine, which peaked within 1–2 minutes and persisted for at least an hour (Fig. 3b, c). Over the 20 minutes following uncaging, TrkB activity gradually increased in nearby dendritic segments and adjacent spines, indicating spreading of TrkB signaling. The rapid activation of TrkB following glutamate uncaging suggests that BDNF is poised for release in response to dendritic spine stimulation. Most surprising, this study found that the stimulated spine itself is an important source of BDNF driving TrkB activity. Glutamate uncaging experiments on neurons expressing BDNF fused to super-ecliptic pHluorin (BDNF-SEP), a pH-sensitive derivative of eGFP, to visualize exocytosed BDNF revealed that glutamate uncaging pulses evoke BDNF release from the stimulated spine itself (Fig. 3d). Together, these results indicate that glutamate released on dendritic spines initiates autocrine BDNF-TrkB signaling that is required for full activation of TrkB and intact spine plasticity [33].

These two example applications of 2pFLIM show the diversity of temporal scales and signaling molecules that can be measured with 2pFLIM. The Camuiα sensor revealed the rapid, stepwise activation of CaMKII during glutamate uncaging that decays soon after synaptic stimulation ends and is restricted to the stimulated spine [19, 20]. Uncaging experiments with the TrkB sensor uncovered that autocrine BDNF-TrkB signaling is activated by glutamate released on dendritic spines, and TrkB activity is sustained for

Fig. 3 (continued) Increased binding fraction indicates higher TrkB activity. Panel on the right is an expanded view of the time course shown to its left. (**d**) Glutamate release drives BDNF release from stimulated dendritic spines. *Left,* Representative two-photon fluorescence images of BDNF exocytosis visualized with a BDNF-SEP construct (green) during glutamate uncaging to induce spine plasticity. mCherry1 is co-expressed to visualize spine volume (red). *Right,* averaged time course of BDNF-SEP release shown on left in spines (green) and dendrites (blue). Black bars indicate glutamate uncaging pulses. Inset shows normalized changes in fluorescence intensity of a spine volume marker (mCherry1, $\Delta R/R_0$) during spine plasticity induction resulting in enlargement. (Figures adapted from Harward et al. [33] with copyright permission)

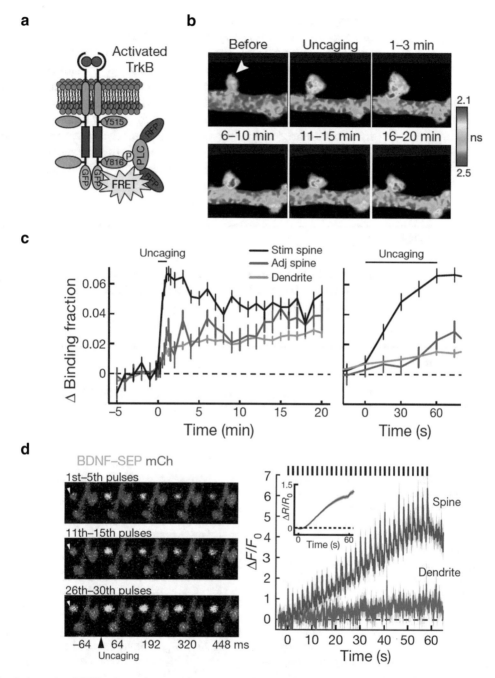

Fig. 3 Autocrine BDNF release from spines during LTP promotes sustained and widespread TrkB signaling. (**a**) Schematics of sensor designed to detect TrkB receptor activity. The TrkB receptor with mEGFP fused at its intracellular terminus serves as the donor, and the acceptor is composed of the SH2 domain of phospholipase Cγ1 double flanked with mRFP1. BDNF (purple circles) activates TrkB kinase activity thereby promoting interactions with the SH2 acceptor and resulting in decreased fluorescence lifetime. (**b**) Representative fluorescence lifetime images of TrkB signaling during glutamate uncaging to induce plasticity at a single spine. Warmer colors indicate shorter fluorescence lifetime corresponding to higher TrkB activity. (**c**) Average time courses of TrkB sensor binding fraction in stimulated spines, adjacent spines, and dendritic segments.

at least 1 hour after stimulation and spreads through the dendrite to neighboring spines [33]. Thus, 2pFLIM can be applied not only to study intracellular signaling through protein kinases like CaMKII, but is also useful in elucidating the dynamics of transmembrane receptors like TrkB. Although we have gained great insight from recent 2pFLIM studies, the few highly specific and sensitive FRET sensors is a limiting factor in our understanding of the coordinated signaling dynamics underlying spine plasticity. The future development of robust sensors targeting more signaling molecules will greatly aid our understanding of the process of synaptic plasticity as a whole.

Acknowledgments

We thank Dr. Lesley Colgan, Dr. Tal Laviv, and other members of the Yasuda lab for thoughtful discussion. We also thank Dr. Nathan Hedrick for contributing to the figures.

Conflict of Interest

Dr. Ryohei Yasuda is the founder of Florida Lifetime Imaging LLC., a company that helps people set up FLIM.

References

1. Murakoshi H, Yasuda R (2012) Postsynaptic signaling during plasticity of dendritic spines. Trends Neurosci 35:135–143

2. Nishiyama J, Yasuda R (2015) Biochemical computation for spine structural plasticity. Neuron 87:63–75

3. Tsay D, Yuste R (2004) On the electrical function of dendritic spines. Trends Neurosci 27:77–83

4. Bloodgood BL, Sabatini BL (2005) Neuronal activity regulates diffusion across the neck of dendritic spines. Science 310:866–869

5. Svoboda K, Tank DW, Denk W (1996) Direct measurement of coupling between dendritic spines and shafts. Science 272:716–719

6. Matsuzaki M, Honkura N, Ellis-Davies GCR et al (2004) Structural basis of long-term potentiation in single dendritic spines. Nature 429:761–766

7. Hedrick NG, Harward SC, Hall CE et al (2016) Rho GTPase complementation underlies BDNF-dependent homo- and heterosynaptic plasticity. Nature 538:104–108

8. Harvey CD, Yasuda R, Zhong H et al (2008) The spread of Ras activity triggered by activation of a single dendritic spine. Science 321:136–140

9. Zhai S, Ark ED, Parra-Bueno P et al (2013) Long-distance integration of nuclear ERK signaling triggered by activation of a few dendritic spines. Science 342:1107–1111

10. Impey S, Obrietan K, Wong ST et al (1998) Cross talk between ERK and PKA is required for Ca2+ stimulation of CREB-dependent transcription and ERK nuclear translocation. Neuron 21:869–883

11. Harvey CD, Ehrhardt AG, Cellurale C et al (2008) A genetically encoded fluorescent sensor of ERK activity. Proc Natl Acad Sci U S A 105:19264–19269

12. Suzuki K, Tsunekawa Y, Hernandez-Benitez R et al (2016) In vivo genome editing via CRISPR/Cas9 mediated homology-independent targeted integration. Nature 540:144–149

13. Mikuni T, Nishiyama J, Sun Y et al (2016) High-throughput, high-resolution mapping of protein localization in mammalian brain by in vivo genome editing. Cell 165:1803–1817

14. Miyawaki A (2003) Visualization of the spatial and temporal dynamics of intracellular signaling. Dev Cell 4:295–305

15. Förster T (1948) Zwischenmolekulare Energiewanderung und Fluoreszenz. Ann Phys 437:55–75

16. Lakowicz JR (1999) Principles of fluorescence spectroscopy. Kluwer Academic/Plenum, New York

17. Ueda Y, Kwok S, Hayashi Y (2013) Application of FRET probes in the analysis of neuronal plasticity. Front Neural Circuits 7:163

18. Takao K, Okamoto K-I, Nakagawa T et al (2005) Visualization of synaptic Ca^{2+}/calmodulin-dependent protein kinase II activity in living neurons. J Neurosci 25:3107–3112

19. Lee S-JR, Escobedo-Lozoya Y, Szatmari EM et al (2009) Activation of CaMKII in single dendritic spines during long-term potentiation. Nature 458:299–304

20. Chang J-Y, Parra-Bueno P, Laviv T et al (2017) CaMKII autophosphorylation is necessary for optimal integration of Ca^{2+} signals during LTP induction, but not maintenance. Neuron 94:800–808

21. Yasuda R, Harvey CD, Zhong H et al (2006) Supersensitive Ras activation in dendrites and spines revealed by two-photon fluorescence lifetime imaging. Nat Neurosci 9:283–291

22. Oliveira AF, Yasuda R (2013) An improved ras sensor for highly sensitive and quantitative FRET-FLIM imaging. PLoS One 8:e52874

23. Ni Q, Titov D, Zhang J (2006) Analyzing protein kinase dynamics in living cells with FRET reporters. Methods 40:279–286

24. Yasuda R (2006) Imaging spatiotemporal dynamics of neuronal signaling using fluorescence resonance energy transfer and fluorescence lifetime imaging microscopy. Curr Opin Neurobiol 16:551–561

25. Yasuda R (2012) Studying signal transduction in single dendritic spines. Cold Spring Harb Perspect Biol. https://doi.org/10.1101/cshperspect.a005611

26. Gratton E, Breusegem S, Sutin J et al (2003) Fluorescence lifetime imaging for the two-photon microscope: time-domain and frequency-domain methods. J Biomed Opt 8:381–390

27. Murakoshi H, Lee S-J, Yasuda R (2008) Highly sensitive and quantitative FRET–FLIM imaging in single dendritic spines using improved non-radiative YFP. Brain Cell Biol 36:31–42

28. Tramier M, Gautier I, Piolot T et al (2002) Picosecond-hetero-FRET microscopy to probe protein-protein interactions in live cells. Biophys J 83:3570–3577

29. Peter M, Ameer-Beg SM, Hughes MKY et al (2005) Multiphoton-FLIM quantification of the EGFP-mRFP1 FRET pair for localization of membrane receptor-kinase interactions. Biophys J 88:1224–1237

30. Ganesan S, Ameer-beg SM, Ng TTC et al (2006) A dark yellow fluorescent protein (YFP)-based Resonance Energy-Accepting Chromoprotein (REACh) for Forster resonance energy transfer with GFP. Proc Natl Acad Sci USA 103:4089–4094

31. Laviv T, Kim BB, Chu J et al (2016) Simultaneous dual-color fluorescence lifetime imaging with novel red-shifted fluorescent proteins. Nat Methods 13:989–992

32. Murakoshi H, Wang H, Yasuda R (2011) Local, persistent activation of Rho GTPases during plasticity of single dendritic spines. Nature 472:100–104

33. Harward SC, Hedrick NG, Hall CE et al (2016) Autocrine BDNF–TrkB signalling within a single dendritic spine. Nature 538:99–103

34. Sato M, Ueda Y, Takagi T et al (2003) Production of PtdInsP3 at endomembranes is triggered by receptor endocytosis. Nat Cell Biol 5:1016–1022

35. Ueda Y, Hayashi Y (2013) PIP_3 regulates spinule formation in dendritic spines during structural long-term potentiation. J Neurosci 33:11040–11047

36. Tang S, Yasuda R (2017) Imaging ERK and PKA Activation in single dendritic spines during structural plasticity. Neuron 93:1315–1324

37. Chen Y, Saulnier JL, Yellen G et al (2014) A PKA activity sensor for quantitative analysis of endogenous GPCR signaling via 2-photon FRET-FLIM imaging. Front Pharmacol 5:56

38. Bonnot A, Guiot E, Hepp R et al (2014) Single-fluorophore biosensors based on conformation-sensitive GFP variants. FASEB J 28:1375–1385

39. Tillo SE, Xiong W-H, Takahashi M et al (2017) Liberated PKA catalytic subunits associate with the membrane via myristoylation to preferentially phosphorylate membrane substrates. Cell Rep 19:617–629

40. Takahashi N, Sawada W, Noguchi J et al (2015) Two-photon fluorescence lifetime imaging of primed SNARE complexes in presynaptic terminals and β cells. Nat Commun 6:8531

41. Bosch M, Castro J, Saneyoshi T et al (2014) Structural and molecular remodeling of dendritic spine substructures during long-term potentiation. Neuron 82:444–459

42. Mongeon R, Venkatachalam V, Yellen G (2016) Cytosolic NADH-NAD + redox visualized in brain slices by two-photon fluorescence lifetime biosensor imaging. Antioxid Redox Signal 25:553–563

43. Díaz-García CM, Mongeon R, Lahmann C et al (2017) Neuronal stimulation triggers neuronal glycolysis and not lactate uptake. Cell Metab 26:361–374

44. Stoppini L, Buchs PA, Muller D (1991) A simple method for organotypic cultures of nervous tissue. J Neurosci Meth 37:173–182

45. McAllister AK (2000) Biolistic transfection of neurons. Sci STKE 2000:pl1

46. Woods G, Zito K (2008) Preparation of gene gun bullets and biolistic transfection of neurons in slice culture. J Vis Exp (12):675

47. Malenka RC, Kauer JA, Zucker RS et al (1988) Postsynaptic calcium is sufficient for potentiation of hippocampal synaptic transmission. Science 242:81–84

48. Malenka RC, Nicoll RA (1999) Long-term potentiation–a decade of progress? Science 285:1870–1874

49. Lynch G, Larson J, Kelso S et al (1983) Intracellular injections of EGTA block induction of hippocampal long-term potentiation. Nature 305:719–721

50. Matsuzaki M, Ellis-Davies GC, Nemoto T et al (2001) Dendritic spine geometry is critical for AMPA receptor expression in hippocampal CA1 pyramidal neurons. Nat Neurosci 4:1086–1092

51. Canepari M, Nelson L, Papageorgiou G et al (2001) Photochemical and pharmacological evaluation of 7-nitroindolinyl-and 4-methoxy-7-nitroindolinyl-amino acids as novel, fast caged neurotransmitters. J Neurosci Meth 112:29–42

52. Ellis-Davies GCR, Matsuzaki M, Paukert M et al (2007) 4-Carboxymethoxy-5,7-dinitroindolinyl-Glu: an improved caged glutamate for expeditious ultraviolet and two-photon photolysis in brain slices. J Neurosci 27:6601–6604

53. Fino E, Araya R, Peterka DS et al (2009) RuBi-Glutamate: two-photon and visible-light photoactivation of neurons and dendritic spines. Front Neural Circuits 3:2

54. Patterson MA, Szatmari EM, Yasuda R (2010) AMPA receptors are exocytosed in stimulated spines and adjacent dendrites in a Ras-ERK-dependent manner during long-term potentiation. Proc Natl Acad Sci USA 107:15951–15956

55. Gähwiler BH (1988) Organotypic cultures of neural tissue. Trends Neurosci 11:484–489

56. Hedrick N, Yasuda R (2014) Imaging signaling transduction in single dendritic spines. In: Nägerl U, Triller A (eds) Nanoscale imaging of synapses. Neuromethods, vol 84. Humana Press, New York

57. Lisman J, Yasuda R, Raghavachari S (2012) Mechanisms of CaMKII action in long-term potentiation. Nat Rev Neurosci 13:169–182

Chapter 7

Combining Multiphoton Excitation Microscopy with Fast Microiontophoresis to Investigate Neuronal Signaling

Espen Hartveit and Margaret Lin Veruki

Abstract

Multiphoton excitation (MPE) microscopy allows subcellular structural and functional imaging of neurons and can be combined with techniques for activating postsynaptic receptors at spatial and temporal scales that mimic normal synaptic transmission. Here, we describe procedures for combining MPE imaging of dye-filled neurons with fast microiontophoresis, by which neurotransmitter agonists can be applied from high-resistance micropipettes with subcellular resolution. With adequate compensation of the pipette capacitance, the effective time constant of the pipette is reduced, and this permits application of very brief pulses of receptor agonist (≤ 1 ms). The consequent high temporal and spatial resolution leads to the high specificity required for single-synapse investigations. This chapter includes detailed procedures for electrophysiological whole-cell recording, structural and functional (Ca^{2+}) MPE imaging of dye-filled neurons, targeting a microiontophoresis pipette to a specific subcellular compartment of a dye-filled neuron under visual control, and capacitance compensation of the microiontophoresis pipette, as well as examples of experimental results that can be obtained.

Key words Microiontophoresis, Ion channels, Ionotropic receptors, Receptor localization, Multiphoton excitation microscopy, Patch-clamp electrophysiology

1 Introduction

In conventional chemical synapses of the central nervous system (CNS), the organization and specialization of the pre- and postsynaptic structures convey high precision in both the spatial and temporal domains (for review, see [1]). When neurotransmitter molecules are released from presynaptic vesicles, they diffuse across the narrow synaptic cleft and bind to and activate heterogeneous populations of receptors, located in the membranes of postsynaptic neurons. The different types of receptors typically differ in functional properties such as affinity for the neurotransmitter, single-channel conductance, kinetics, selectivity and permeability for specific ions, developmental regulation, and influence on downstream signaling pathways [2, 3]. Ideally, the investigator would like to study and manipulate functional properties at both pre- and postsynaptic sites with

Espen Hartveit (ed.), *Multiphoton Microscopy*, Neuromethods, vol. 148,
https://doi.org/10.1007/978-1-4939-9702-2_7, © Springer Science+Business Media, LLC, part of Springer Nature 2019

subcellular (single-synapse) resolution in intact, functioning neural tissue. This includes mechanisms of presynaptic exocytosis, properties of pre- and postsynaptic receptors, and postsynaptic signaling and transduction mechanisms. The advent of 2-photon (2P) or multiphoton excitation (MPE) microscopy has enabled investigations at the required level of spatial and temporal resolution [4], with respect to both structural imaging of subcellular neuronal components and functional imaging of pre- and postsynaptic intracellular Ca^{2+} signals within intact neural tissue [5]. As such, MPE microscopy does not enable activation of postsynaptic receptors, but when combined with either 2P/MP uncaging of neurotransmitter [5] (see chapters by Tran-Van-Minh et al. and Stein et al., this volume) or fast microiontophoresis [6], such receptors can be activated at spatial and temporal scales that (approximately) mimic and are directly relevant for normal chemical synaptic transmission. Here, we describe the combination of MPE microscopy (for structural and functional imaging) and fast microiontophoresis.

Ca^{2+} plays important roles as a second messenger in a large number of intracellular signaling pathways [7], and changes in intracellular Ca^{2+} concentration can be triggered by a number of different mechanisms acting through receptors and ion channels located in the plasma membrane and intracellular organelles of neurons and other cells. Over the last 20 years, powerful imaging methods have been developed that allow the measurement of Ca^{2+} signals in subcellular neuronal compartments, including spines and varicosities, with high spatial and temporal resolution [5]. For such subcellular measurements, made from structures deep in intact, highly scattering tissue like the CNS, MPE microscopy is still the major technique of choice and has contributed to several major discoveries [8].

When imaging Ca^{2+} signals and dynamics in subcellular neuronal compartments, it should ideally be possible to apply stimuli with the same degree of spatial and temporal resolution as is possible for recording the Ca^{2+} signal itself. In some cases, action potentials triggered by extracellular stimulation of axons can evoke Ca^{2+} signals that lead to presynaptic release of neurotransmitter onto the postsynaptic target. With adequate placement of the stimulation electrode, it is possible to activate single axons [9]. Alternatively, extracellular stimulation can be used to depolarize non-spiking neurons to the threshold for releasing neurotransmitter [10]. In the latter case, it is difficult to ensure single-synapse stimulation and the specificity may be suboptimal. In some situations, it is possible to record intracellularly from a presynaptic neuron and use this to trigger release at specific synapses, but with whole-cell recording there is often a rapid rundown of release (especially with small neurons). In situations where presynaptic activation is not an option (either because of technical limitations or because desired pharmacological manipulations of postsynaptic mechanisms interfere with or compromise presynaptic release), it may be possible to directly activate postsynaptic receptors by agonist application. Ideally, however, the method of application should be designed to mimic the spatial and temporal aspects of normal synaptic activation. That is, the method should, as far as possible, enable activation of a single postsynaptic structure and the time course of activation of receptors should be similar to that experienced during presynaptic release. Fine-tipped glass micropipettes, e.g., the sharp microelectrodes used for intracellular recording, are sufficiently small and can be used for drug application within a sufficiently small region [11, 12], using either iontophoresis or pressure. The challenge, however, is to target the application to specific, visually identified subcellular structures and to ensure that the speed of application is sufficiently fast. Pressure application causes mechanical disturbances that can compromise Ca^{2+} imaging. In addition, if the pipette tip is not sufficiently small, leakage of the substance can make it problematic to position the tip close to the target. A larger distance between the pipette and the target compromises the speed and spatial restriction of application. A smaller tip size can make it difficult to use pressure ejection. Many of these problems can be avoided with microiontophoresis, where the agonist (or any other charged solute) can be ejected in a very small amount from a fine-tipped glass micropipette. With such pipettes and conventional microiontophoresis, leakage of substance can be further reduced to a minimum by applying a retaining current. However, the high pipette resistance in combination with the pipette capacitance acts as a low-pass filter that seriously compromises the speed of application. Prolonging the duration of the stimulus will compromise not only the temporal control, but also the spatial resolution and specificity of the application. One solution to these challenging problems has been to fully compensate the capacitance of the high-resistance glass micropipette, reducing the effective pipette time constant and thereby enabling application of agonist with a time course that, in the best cases, mimics the time course of synaptic release [13, 14]. This technique was initially applied to neurons in cell cultures where the conditions for high-resolution visualization are particularly favorable, but has recently been successfully applied to neurons in in vitro CNS slice preparations [6, 15, 16]. When the goal is to use fast microiontophoresis to apply agonists to study subcellular Ca^{2+} dynamics with MPE microscopy, concomitant structural imaging allows visual targeting of the specific subcellular compartments of interest. Structural imaging is also extremely useful for accurate positioning of a microiontophoresis pipette, also for studies where the primary goal is to investigate, e.g., the subcellular localization and function of neurotransmitter receptors by electrophysiological recording. Here, we provide a detailed description of experimental procedures for the combination of fast microiontophoresis, electrophysiological whole-cell recording, and structural and functional MPE microscopy applied to in vitro slice preparations from CNS tissue, together with examples of experimental results that can be obtained.

2   Materials and General Methods

2.1   Animal Anesthesia

All work with animals must be approved and performed in accordance with local (institutional) and national regulations. Deeply anesthetize the experimental animal (e.g., a rat or a mouse) with isoflurane (~2 to ~3%) in oxygen. Use 100% oxygen as opposed to ordinary air, as this strongly reduces the risk of laryngeal spasms and consequent hypoxia that compromises tissue quality. When an adequate depth of anesthesia has been reached, kill the animal by cervical dislocation or decapitation.

2.2   In vitro Slice Preparation

Depending on the region of the CNS from which the slice preparation will be made, different procedures for cutting the slices will have to be applied. The procedure described here is from our laboratory where the work is focused on the mammalian retina. After removing the eyes, rinse the eyes briefly in cold dissection solution (for details, see [16] and references therein). With the eyeball immersed in solution, locate it under a dissection microscope and use fine forceps and a scissor to remove extraocular connective and muscle tissue from the eyeball. Open the eyeball with a syringe needle, and use a small scissor to make an encircling cut slightly posterior to the ora serrata. Remove the front half corresponding to the cornea, lens, and the bulk of the vitreous body. Use fine forceps to thoroughly but gently remove any remaining pieces of vitreous. Gently dissect the retina from the rest of the eyeball. Divide the retinal eyecup into four quadrants and store them on small pieces of lens paper on top of a nylon mesh in an interface chamber with Ames solution (Sigma A1420) buffered (pH) by addition of $NaHCO_3$ and bubbling with gas containing 5% CO_2 (and 95% O_2). Cut retinal slices (at a thickness of 100–200 μm) by hand under a dissection microscope by using a curved scalpel blade. Alternatively, embed pieces of retina in low-temperature gelling agar (Sigma A0701) and cut slices from the agar block with a vibrating microtome (e.g., Leica VT1200). This kind of vibratome is the instrument of choice for cutting high-quality slices directly from trimmed tissue blocks from other regions of the CNS (e.g., hippocampus, thalamus, brain stem, etc. [17]), including slices suited for axon terminal and dendritic recordings [18, 19]. A single set of slices should not be used for more than 3–4 hours before being replaced by a new set.

2.3   Visualization with Dodt Gradient Contrast Videomicroscopy and MPE Microscopy

After preparing in vitro slices from a specific CNS region, place the slices in a perfusion or recording chamber (Fig. 1a) and secure them by gently placing a U-shaped "harp" (made from platinum-iridium wire) with thin nylon wires glued on one side. Place the chamber under the microscope to visualize the slices (Fig. 1b). For our experiments, we use a custom-modified "Movable Objective

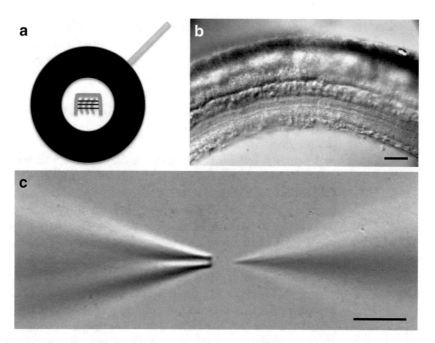

Fig. 1 Electrophysiological recording from and imaging of in vitro retinal slices. (**a**) Schematic figure of recording chamber (with central area corresponding to bottom covered with glass coverslip) with vertical retinal slices held in place by grid made by U-shaped profile (platinum-iridium wire) and nylon wires. (**b**) Infrared videomicrograph of a vertical retinal slice at low magnification. The layering of the retina can be clearly visualized, with the ganglion cell layer at bottom and the photoreceptor outer segments at top. (**c**) Infrared videomicrograph of a patch pipette for whole-cell electrophysiological recording (left) and a high-resistance pipette for microiontophoretic application (right). Note the relatively large opening of the patch pipette and that the opening of the high-resistance pipette cannot be resolved by the light microscope. Bars, 50 μm (**b**), 10 μm (**c**)

Microscope" (MOM) from Sutter Instrument (https://www.sutter.com) equipped with, e.g., a ×20 water immersion objective (XLUMPLFL, 0.95 NA; Olympus). All of the major microscope manufacturers offer MPE microscopes, typically modified confocal microscopes, that provide similar functionality. Some laboratories have designed their own MPE microscope (see chapter by Smith in this volume), and there are several published designs available, although the level of detail differs [20]). To adequately visualize the slices, it is necessary to employ some form of contrast enhancement. Infrared (IR) videomicroscopy works well with differential interference contrast (DIC; Nomarski), but the optical elements for DIC are difficult to combine with fluorescence microscopy without losing substantial signal intensity [21]. Instead, we use IR Dodt gradient contrast (IR-DGC) videomicroscopy [9, 22] with hardware from Luigs & Neumann (http://www.luigs-neumann.com) and a standard IR-sensitive analog CCD camera (e.g., VX55 from TILL Photonics). Powerful light-emitting diodes (LEDs) can be a convenient source for IR light. One example, with peak intensity around 780 nm, is the LED M780L2 from Thorlabs (https://www.thorlabs.com).

2.4 Patch Pipettes for Electrophysiological Recording

Patch pipettes should be made from thick-walled borosilicate glass, e.g., with outer diameter of 1.5 mm and inner diameter of 0.86 mm (Fig. 1c). There are several companies that supply high-quality glass for pulling pipettes (e.g., Sutter Instrument and WPI [https://www.wpiinc.com]). For pulling patch pipettes, it is necessary to use a two-stage puller (e.g., PC-100 from Narishige [http://uk.narishige-group.com], PIP 6 from HEKA Elektronik [http://www.heka.com], and P-97 or P-1000 from Sutter Instrument). For easier filling of the tip of the pipette with intracellular electrolyte solution, we recommend using filamented glass. Depending on the type of cell the recording will be made from, the open-tip resistance of the pipette (when filled with intracellular solution and positioned in the extracellular bath solution) can range from about 4 to 12 MΩ.

2.5 Intra- and Extracellular Solutions

The extracellular bath solution should be continuously bubbled with 95% O_2–5% CO_2 for oxygenation (which is required when working with slice preparations of vascularized CNS tissue to compensate for the increased diffusion distance after termination of capillary circulation) and pH control (CO_2/bicarbonate buffer system). A standard extracellular solution can have the following composition (in mM): 125 NaCl, 25 NaHCO$_3$, 2.5 KCl, 2.5 CaCl$_2$, 1 MgCl$_2$, 10 glucose, pH 7.4. If the experimental design requires heating above room temperature, appropriate instrumentation is required for heating the bath solutions and the recording chamber (e.g., Warner TC-324C; https://www.warneronline.com).

Depending on the experimental design, the recording pipettes can be filled with a solution of the following composition (mM): 125 potassium gluconate, 8 NaCl, 5 KCl, 10 Hepes, 0.2 EGTA, 4 magnesium ATP, and 0.4 disodium GTP. pH is adjusted to 7.3 with KOH. When required, EGTA can be replaced with the faster Ca^{2+} buffer BAPTA [23]. For structural MPE imaging, we add Alexa Fluor 488 hydrazide or 594 hydrazide (both as sodium salts; Thermo Fisher/Invitrogen A10436, A10438) to the pipette solution for visualization of the cellular morphology. Depending on the cell size (including the length of the relevant processes and the distance from the soma to the targeted structures), recommended concentrations are in the range of 40–100 µM. Higher concentrations can result in excessive spill or pooling of dye around the site of recording. It is also important to check the osmolality of the solution. Values in the range of 290–300 mOsm/kg are adequate for slices from rat CNS, but other species may require moderately different values.

For MPE microscopic Ca^{2+} imaging, the recording pipettes can be filled with the same solution, but EGTA (or BAPTA) is replaced with an appropriate Ca^{2+} indicator, e.g., Oregon Green 488 BAPTA-1 (OGB-1; 200 µM; Thermo Fisher/Invitrogen), or another indicator with different Ca^{2+} affinity [9]. When the intracellular solution contains dyes for structural and/or functional imaging, it is useful to prepare a stock solution at ×1.25 of the final concentration without fluorescent dyes, filtered to remove particles that can obstruct the recording pipettes (e.g., 0.22 μm syringe filters from Millipore; SLGV004SL) and stored at −20 °C until use. Before the experiment, the ×1.25 stock solution is diluted with water and/or fluorescent dye(s) dissolved in water to the appropriate concentration. A stock solution of fluorescent dye can be made by dissolving it in water (Sigma W3500) at a concentration of 1 mM and storing it as 100 μl aliquots at −20 °C until use. To conserve expensive dyes for structural and functional imaging, we make up a small volume (20 μl) of intracellular solution (at ×1 concentration) and fill each recording pipette with as little as 1–2 μl of solution. With such small total volumes, the pipettes can conveniently be filled by using long plastic tips (e.g., Eppendorf Microloader tips).

2.6 Electrophysiological Recording and Data Acquisition

Whole-cell recordings are performed with a high-quality patch-clamp amplifier (e.g., the EPC10 series from HEKA Elektronik and the dPatch from Sutter Instrument). Depending on the design of the amplifier, it can either be integrated with the other components required for amplifier control and data acquisition or be connected to an independent device for data acquisition with analog-to-digital and digital-to-analog signal conversion. Some systems come with integrated software (e.g., Patchmaster from HEKA Elektronik and SutterPatch from Sutter Instrument). Some laboratories write their own software for amplifier control and data acquisition (e.g., in the IGOR Pro environment; https://www.wavemetrics.com).

After establishing a GΩ-seal, currents caused by the recording electrode capacitance (fast capacitive current) should be measured and neutralized by the amplifier, either manually or automatically. After breaking into the cell and establishing a whole-cell recording, currents caused by the cell membrane capacitance (slow capacitive current) should also be measured and (fully or partially) neutralized by the amplifier, either manually or automatically.

For ordinary electrophysiological responses, it is usually adequate to low-pass filter (analog or digital Bessel filters) the signals with a cut-off (corner) frequency (f_c; −3 dB) in the range between 1 and 4 kHz. The sampling interval should ideally be the inverse of 5× f_c, at most the inverse of 2× f_c (to satisfy the Nyquist criterion for digital sampling). If it is desirable to also sample the very brief stimulus signals used for microiontophoretic application (see below) for documentation purposes, it is necessary to use a correspondingly short sampling interval, e.g., 10 μs.

2.7 MPE Microscope

In our system, we switch from IR-DGC videomicroscopy to MPE microscopy by changing the optical paths of the MOM via movable mirrors. In the MOM, scanning is performed by galvanometric scanners (MicroMax 673XX dual axis; Cambridge Technology, https://www.cambridgetechnology.com), equipped with either 3 or 6 mm mirrors (6210H; Cambridge Technology). The smaller mirrors are preferred for faster imaging, whereas the larger mirrors are preferred for higher resolution (as the expansion of the laser beam can be increased). To maximize the spatial resolution, the laser beam should be expanded (before hitting the scanning mirrors) such that it overfills the back aperture of the microscope objective [24]. The intensity of the laser is typically so high that the corresponding loss of light never poses a problem. Our setup is equipped with a computer-controlled, mode-locked, ultrafast-pulsed Ti:sapphire laser with tunable wavelength (690–1040 nm; Mai Tai HP Deep See; Spectra-Physics; https://www.spectra-physics.com). To minimize the total exposure time of the preparation during image acquisition, the laser light should be controlled by a high-speed laser shutter (e.g., LS6ZM2 from Vincent Associates; https://www.uniblitz.com). To attenuate the intensity of the laser, and thus control the exposure of the preparation to the laser light, it is useful to include an electro-optic modulator (EOM; "Pockels cell") in the optical pathway. A specific advantage of an EOM is that it has an extremely fast response time (≤ 1 μs rise/fall time). We use the "350-80LA with BK option" from ConOptics (https://www.conoptics.com). An EOM is driven by an external amplifier that should be controlled by the MPE acquisition software/hardware system. We use the 302RM amplifier from ConOptics.

The emitted epifluorescence light for MPE imaging and the forward-scattered laser light for IR-laser scanning gradient contrast (IR-LSGC) microscopy (see below) are detected by separate photomultiplier tubes (PMTs). Our system is equipped with multialkali PMTs (R6357; Hamamatsu; https://www.hamamatsu.com). There are other alternatives for PMTs, including the more sensitive GaAsP-based detectors (e.g., H10770PA-40; Hamamatsu). The analog signals from the PMTs are digitized by a fast acquisition board (e.g., NI-6110E from National Instruments; http://www.ni.com).

2.8 Computer Software for MPE Microscopy

MPE microscopy and image acquisition is controlled by dedicated computer software. Commercial MPE microscopy systems typically come with custom software sold by the microscope manufacturer. Alternatively, there are software programs, typically originating from academic research environments, that have been developed as generic programs with sufficient flexibility that they can be adapted to different microscope systems with a range of hardware components. The most commonly used programs are ScanImage ([25]; http://scanimage.vidriotechnologies.com; https://vidriotechnologies.com), HelioScan ([26]; http://helioscan.github.io/HelioScan), and MPScope ([27]; https://neurophysics.ucsd.edu/software.php). Our system is controlled by

ScanImage software (version 3.8 and higher; developed under MATLAB; MathWorks, https://www.mathworks.com). This program digitizes the image data at 12- or 16-bit resolution and stores the files as TIFF files.

2.9 Pipettes for Microiontophoresis in in Vitro Slice Preparations

For high-resolution microiontophoresis [13–16] with spatially and temporally restricted activation of, e.g., glutamate receptors on subcellular neuronal structures, it is necessary to use fine-tipped micropipettes with high resistance. This is necessary to match the dimensions of the relevant structures (e.g., small varicosities and thin dendritic processes), to minimize tissue damage, and to reduce the passive diffusion (leakage) of agonist (and dye; see below) from the pipette tips. Pipettes with resistances in the range between 90 and 120 MΩ perform well and can be pulled in two stages on a horizontal puller (in our case, the Flaming Brown P-97 pipette puller; Sutter Instrument).

Traditionally, the way to prepare micropipettes with such high resistances is to use thin-walled glass designed for sharp microelectrodes and intracellular recording (e.g., outer diameter, 1.0 mm; inner diameter, 0.78 mm). Such pipettes can work well when applied to cells in tissue culture [13, 14]. However, in our experience, using a slice preparation, such pipettes only work well for targeting visually identified structures located a few μm below the surface of a slice. For deeper structures, the tips of these pipettes are too flexible and tend to bend and thus deviate from their trajectory inside the tissue before reaching the target. To produce pipettes with stiffer tips, we instead use the same thick-walled glass used for patch pipette recording (filamented borosilicate glass; outer diameter, 1.5 mm; inner diameter, 0.86 mm). With such glass, we are able to prepare pipettes with long thin tips that do not bend when inserted into the neural tissue and have sufficiently high resistance that leakage of drug can be controlled with acceptable retaining currents (Fig. 1c, and see below). It is important that the pipettes are pulled such that the distance from the tip to the wider shaft (i.e., the taper) is sufficiently long to avoid compression of the tissue when inserted into the slices. Shorter tapers will increase the relative instability between the pipette and the tissue and lead to slow spatial drift ("relaxation") between tissue and pipette after positioning the pipette close to a given subcellular structure. The extent to which this is a problem is likely to vary with the mechanical properties of the tissue, related to the specific region of the CNS from which the in vitro slices are prepared. We have designed our pipettes with inspiration and guidance from the Sutter Instrument Pipette Cookbook (https://www.sutter.com/PDFs/pipette_cookbook.pdf; see Table 1 for specific parameters). The parameters ("velocity" and "time") that we regularly tweak to obtain desired resistances are provided as ranges.

Some authors have suggested that iontophoresis micropipettes can be re-used between experiments, especially desirable with pipettes that perform particularly well. To enable such re-use, it is important to store the micropipette in a closed container with a humidified atmosphere to prevent drying out and clogging at the tip. For certain drugs, it is also necessary to protect the pipette solution from light. In our experience, we have not found it worthwhile to re-use pipettes between experiment days, but we generally use the same pipette during the course of an experiment day.

2.10 Pipette Solutions for Microiontophoresis

Depending on the receptors to be activated by microiontophoresis, the pipettes can be filled with L-glutamate, L-aspartate, NMDA, GABA, etc. The concentration of the drug must be sufficiently high (both in absolute and relative terms) such that it will carry a substantial part of the total current applied during iontophoresis [11] but sufficiently low such that leakage from the tip of the micropipette does not become excessive. For the relevant drug molecules to be able to carry current, the pH of the solution must be such that the molecules exist in an ionized form. For glutamate, we have good experience with a concentration of 150 mM (in water) and pH adjusted to 7.0 (with NaOH). Additional examples can be found in [15].

For a typical micropipette with a resistance of 90–120 MΩ, the spatial resolution of light microscopy is insufficient to visualize and resolve the true size of the pipette tip (Fig. 1c; [28]). For visualization of the pipette tip during MPE microscopic imaging, a fluorescent dye (e.g., Alexa 594) must be added. This is also very useful for tuning the retaining current, verifying the ejection, and targeting the pipette tip to a specific subcellular structure of a neuron during MPE microscopy. It makes sense to use the same dye for structural imaging of both the neuronal processes and the microiontophoresis pipette. In our experience, a pipette concentration of 40 µM for Alexa 594 is a good balance between adequate visualization and potential leakage that might interfere with structural imaging of neuronal processes. For experiments that do not combine electrophysiological recording with imaging using a Ca^{2+} indicator with green fluorescence, Alexa 488 could be used instead of Alexa 594, but in our experience, Alexa 594 yields brighter fluorescence for a given concentration of dye.

2.11 Micro-iontophoresis Amplifier for Fast Drug Application

When the goal is to apply very brief pulses of a drug (e.g., <1 ms), it is necessary to use an amplifier for current injection that enables electronic compensation of the capacitance of the microiontophoresis pipette. With no or insufficient compensation, a brief current pulse will only charge the pipette capacitance instead of ejecting drug. Increasing the duration of the current pulse will eject drug but leads to reduced spatial and temporal resolution of the application. To our knowledge, there is currently only one commercially

Table 1
Instrument parameters for pulling high-resistance (90–120 MΩ) microiontophoresis pipettes suitable for in vitro tissue slices using the Flaming/Brown Micropipette Puller P-97 (Sutter Instrument)

Heat	Pull	Velocity	Time
325	0	30	200–250
325	55	85–90	200–250

The parameters were tuned for a heating filament of the "trough" type, and the "pressure" parameter was set to 200. Each line corresponds to one step of a two-step pull

available system that allows for electronic compensation of the pipette capacitance, made by NPI Electronic (http://www.npielectronic.de). This company offers several variants of a basic design and our experience is with the MVCS-C-02M system that allows application speeds that can simulate brief synaptic events. In the high-voltage version (up to ±225 V), it allows application of currents up to ±2.25 µA (into a 100 MΩ resistance), with positive or negative polarity set according to the charge of the molecule to be ejected. The MVCS-C-02M system allows capacitance compensation in the range 0–30 pF and maximum retaining currents of ±100 nA.

2.12 Capacitance Compensation and Tuning of Pipettes for Fast Microiontophoresis

As stated above, ejecting a substance by iontophoresis using current pulses in the submillisecond range (0.5–1 ms) depends crucially on the ability to adequately compensate the capacitance of the pipette. It is also necessary to adequately compensate the capacitance before the pipette resistance can be measured correctly. Because the effective capacitance of the micropipette will be related to the extent of immersion in the fluid of the recording chamber, the fluid level should be kept as low as possible. Capacitance compensation and resistance measurement should be first performed with the micropipette in the bath, but before insertion into the tissue, and should be verified several times throughout an experiment. Compensation of the capacitance of a microiontophoresis pipette is illustrated in Fig. 2, with application of repeated current pulses of ±10 nA (5 ms duration). The illustrated capacitance settings correspond to an under-compensated pipette, a correctly and accurately compensated pipette, and an overcompensated pipette, respectively. If the fluid level of the recording chamber does not change during the experiment, it is not expected that the capacitance will change. We have not attempted to reduce the pipette capacitance by coating it (e.g., with silicone elastomer, dental wax, or Parafilm) as this is likely to interfere with the movement of the pipette in the neural tissue and cause instability of its position relative to the tissue. When the capacitance has been correctly compensated, the resistance can be directly calculated from the

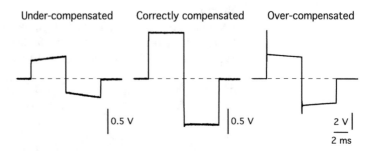

Under-compensated   Correctly compensated   Over-compensated

0.5 V   0.5 V   2 V
                              2 ms

Fig. 2 Tuning and compensation of the capacitance of a high-resistance pipette for microiontophoresis. The tuning was performed after filling the pipette with electrolyte and immersion into the recording chamber (as in Fig. 1) by applying current pulses of ±10 nA (5 ms duration). The capacitance setting of the microiontophoresis amplifier was adjusted to illustrate an under-compensated pipette (left), a correctly compensated pipette (middle), and an over-compensated pipette (right). The resistance of the pipette was approximately 95 MΩ

amplitude of the voltage deflection in response to the applied current, using Ohm's law. It can be useful to check the resistance of the pipette multiple times during the recording to ensure the tip is not clogged.

In addition to compensating the capacitance of the microiontophoresis pipette, we have also found it necessary to perform a procedure that "primes" the pipette before it is inserted into the neural tissue. For application of glutamate (as an example), this involves repeatedly passing current pulses of negative polarity (e.g., ≤−200 nA for 200 ms; 2–3 Hz) for 3–6 minutes to eject and optimally fill the tip of the pipette with glutamate (and fluorescent dye). The result of this priming is that the pipette tip is optimally visualized and that subsequent application of current pulses (of the correct polarity) will evoke precise and efficient ejection. This procedure should be performed while the tip of the pipette is in the bath solution, well above the tissue surface, and can be monitored by continuous MPE microscopy while focusing on the tip of the pipette. With the iontophoresis solution employed here, Alexa 594 is ejected by the same current polarity as glutamate (i.e., with negative current). We do not fully understand the mechanism underlying the "priming" phenomenon, but it has been suggested that current pulses of alternating polarity can serve to remove small air bubbles inside the pipette tip [29].

After successful priming of a pipette, it is relatively straightforward to adjust the retaining current (using the polarity opposite to that used for ejection) such that there is neither visible leakage of dye nor depletion of dye from the tip of the pipette. This procedure is performed with MPE microscopic imaging (see below) at high zoom while one attempts to focus as closely as possible at the expected position of the tip of the micropipette. When the retaining current is optimally adjusted, the MPE microscopic image of the pipette tip will correspond to the longest possible tip, with an optimal balance between the retaining current being too low (observed as an apparent shortening of the tip caused by leakage and consequent dilution of dye from the tip) and the retaining current being too high (observed as an apparent shortening of the tip caused by retraction of dye from the tip. For experiments with the glutamate-filled pipettes described here, we typically apply a retaining current in the range of +10 to +20 nA. If necessary, this current can be adjusted once the pipette has been positioned close to the targeted subcellular structure. If electrophysiological recording is performed simultaneously from the targeted neuron, leakage of glutamate can potentially be detected as an increase of current (or voltage) noise.

2.13 MPE Microscopy for Structural and Functional Imaging

In general, both Alexa Fluor 594 (20–60 μM) and Alexa Fluor 488 (50–60 μM) can be used for structural MPE imaging with good results. The wavelength of the laser is optimally tuned to 775 nm for Alexa 488 and to 810 nm for Alexa 594. We use the latter wavelength for simultaneous imaging with Alexa 488 and Alexa 594 (e.g., for simultaneous, dual recording and imaging of two neurons filled with different dyes) and also for cells filled with both Alexa 594 and the green fluorescent Ca^{2+} indicator OGB-1.

For acquisition, we usually sample two fluorescence channels, red fluorescence from Alexa 594 added to the solutions in the recording and microiontophoresis pipettes for structural imaging and green fluorescence (e.g., from OGB-1 added to the recording pipette) for functional imaging. In addition, we sample a third channel for IR-LSGC microscopy. The latter channel samples images using the forward-scattered laser light after it passes the substage condenser and a Dodt gradient contrast tube in reverse [9].

For structural imaging, we generally acquire image stacks as a series of optical slices (512×512 or 1024×1024 pixels; typically with averaging of two frames for each slice) at constant focal plane intervals (e.g., 0.4 μm). For imaging intracellular Ca^{2+} dynamics, we sample fluorescence from relevant subcellular structures, either as high-speed line scans (e.g., with 64 pixels/line and temporal resolution of ~960 Hz) or as lower-speed frame scans (e.g., 16×16 pixels at ~25 Hz or 32×32 pixels at ~14 Hz). To synchronize Ca^{2+} imaging and electrophysiological recording, we use a digital output from the patch-clamp amplifier to trigger image acquisition (with ScanImage).

2.14 Structural Imaging for Targeted Microiontophoresis

The neurons and structures of interest can be targeted based on their location in the tissue, details of the cellular morphology, and/or expression of specific fluorescent proteins (see chapter by Chen and Wei in this volume). In the examples used here, cell bodies of amacrine and bipolar cells in the mammalian retina can be identified based on their location in the inner nuclear layer and the specific morphology of cell bodies and proximal processes (Fig. 3). AII amacrine cells are targeted based on the position of a cell body at the border of the inner nuclear layer and the inner plexiform layer with a thick primary dendrite that descends into the inner plexiform layer (Fig. 3a). Bipolar cells are targeted based on their location in the outer and middle parts of the inner nuclear layer (Fig. 3b). A17 amacrine cells are targeted based on the location of a dome-shaped cell body (with the base toward the inner plexiform layer) at the border of the inner nuclear layer and inner plexiform layer (Fig. 3c).

Structural imaging with MPE microscopy is essential for identifying and targeting the subcellular structures of interest, both for microiontophoretic application combined with electrophysiological recording and for the addition of functional Ca^{2+} imaging. MPE microscopy is eminently suited to identify complex neuronal structures at high resolution with a low risk of phototoxicity that allows for long-lasting imaging. During the initial phase of a recording experiment, the limiting factor is the diffusion of the dye used for structural imaging. Diffusion from the pipette into the soma is rapid, but sufficient diffusion from the soma into the (distal) dendritic tree depends on the size and structure of the cell (including the length and thickness of the processes), and can take considerable time. For small neurons like AII amacrine cells (Fig. 3d) and bipolar cells (Fig. 3e, f) it is not necessary to wait more than 10–15 minutes after establishing the whole-cell configuration before commencing either microiontophoretic targeting and/or functional imaging. For neurons with large dendritic trees and thinner processes (e.g., A17 amacrine cells), it is typically necessary to wait at least 30 minutes such that the structural dye and/or the Ca^{2+} indicator has reached the desired subcellular neuronal compartments at a sufficient concentration (Fig. 3g). During this period there is ample time to use short periods of MPE imaging to verify the identity and condition of the neuron and to add pharmacological blockers and verify their action (by electrophysiological recording).

2.15 Targeting the Microiontophoresis Pipette to Specific Subcellular Structures

When using fast microiontophoresis to apply a drug to a specific and visually identified subcellular structure, the goal is to position the tip of the micropipette as closely as possible to the neuronal structure to make sure, to the largest extent possible, that any measured response originates from the targeted structure. Because the tip of a micropipette with a resistance in the range of 90–120 MΩ

Fig. 3 (continued) whole-cell recording, fluorescent dye diffuses from the patch pipette into the cell (Alexa 594: (d, f, and g; Lucifer yellow: e) and enables MPE microscopy to visualize the complete cellular morphology, as illustrated with maximum intensity projections for an AII amacrine cell (d), a rod bipolar cell (e), a cone bipolar cell (f), and an A17 amacrine cell (g). Bars, 10 μm (a–c), 10 μm (d), 10 μm (e), 10 μm (f), 20 μm (g). (Panel in (g) reproduced from Castilho et al. [16])

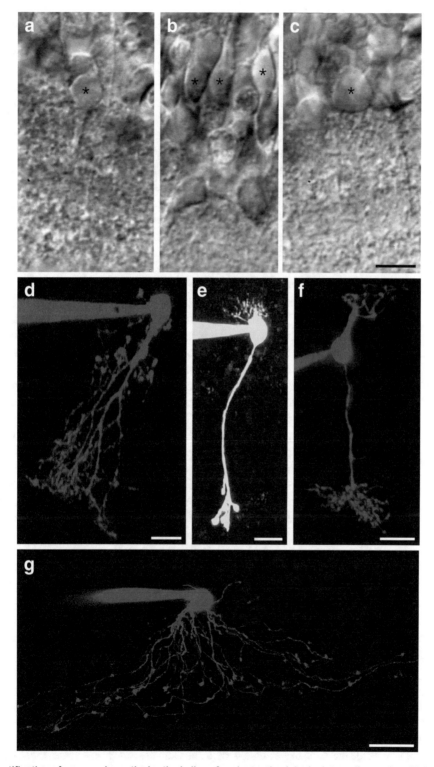

Fig. 3 Identification of neurons in vertical retinal slices for electrophysiological recording and multiphoton excitation (MPE) imaging. (**a–c**) Infrared videomicrographs of retinal slices with identified cell bodies (marked by asterisks) of an AII amacrine cell (**a**), bipolar cells (**b**), and an A17 amacrine cell (**c**). After establishing a

is extremely small, this can be quite challenging (see Fig. 1c). We have never measured the diameter of the tip of our pipettes used for microiontophoresis, but from general experience, pipettes with this resistance typically have diameters in the range of 0.05 µm when imaged by scanning electron microscopy [28]. Because this is below the limit of resolution of light microscopy, it means that we cannot know the true position of the pipette tip. Filling the pipette with a fluorescent dye and imaging the tip with MPE microscopy is a potential solution to this problem. When the pipette is filled all the way to the tip with fluorescent dye at a concentration sufficiently high to give a detectable signal, it is in principle possible to identify the location (but not the diameter) of the pipette tip with MPE microscopy.

For targeting the tip of the microiontophoresis pipette to a specific subcellular structure below the surface of the slice, we monitor both the dye-filled cell and the dye-filled pipette tip with MPE microscopy. For precise targeting, we have found it useful to follow a fixed procedure in which distances are estimated as accurately as possible. The main difficulty with a less stringent procedure is that once the micropipette is inside the slice, it is essentially impossible to move the pipette sideways to compensate for an inaccurate initial position. The following procedure (Fig. 4) depends on the MPE microscope being equipped with a motorized focus drive to accurately measure the distance between focal planes with µm resolution. It also depends on having access to a motorized three-axis micromanipulator where the x-axis corresponds to the long axis of the microiontophoresis pipette, the y-axis is horizontal and oriented 90° to the x- and z-axes, and the z-axis is vertical and oriented 90° to the y-axis (e.g., the Mini25 from Luigs & Neumann). In our setup, the angle between the long axis of the pipette (x-axis) and the horizontal plane is 20°. This shallow angle is determined by the use of a water immersion objective for visualization of the slice preparation and the need to position the pipette between the slice and the objective. The objective used in our setup (×20 XLUMPLFL; 0.95 NA; Olympus) is particularly wide, but similar objectives have been designed with a more pointed front that might allow a steeper angle of the pipette to be used.

Fig. 4 (continued) c (indicated by the red color). In position 3, the pipette has been moved downward (along the z-axis) a distance equal to $a + b$. In position 4, the pipette has been moved inward again along its long axis corresponding (almost) to the distance c. (c) Screenshot from the ScanImage imaging software [25] corresponding to position 4 in (b) with MPE microscopy of the subcellular target (left; a dendritic varicosity of an A17 amacrine cell) and the distal part of the microiontophoresis pipette (right). For better display of fluorescent structures, the MPE image has been oversaturated. Here and later, the vertical line across the varicosity indicates the spatial extent of the line scan for Ca^{2+} imaging. (d) As in (c), but an example of suboptimal targeting of a subcellular structure (A17 dendritic varicosity) with a microiontophoresis pipette. The frame around the varicosity indicates the spatial extent of the frame scan for Ca^{2+} imaging

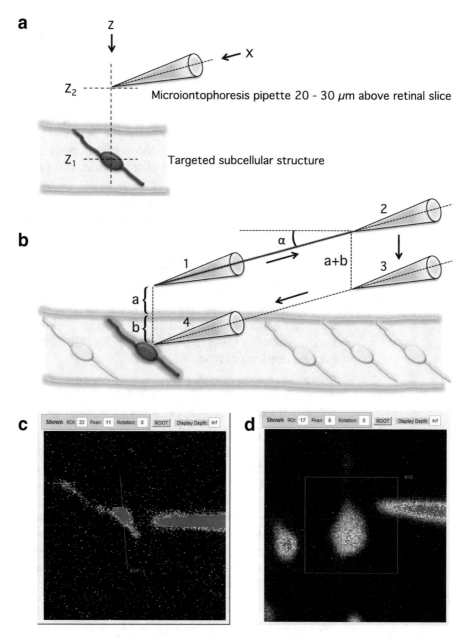

Fig. 4 Visually guided targeting of subcellular neuronal compartments of dye-filled neurons in in vitro slices using dye-filled microiontophoresis pipettes during MPE microscopic imaging. Schematic diagrams (**a**, **b**) show important stages of stepwise procedure. In this example, the neuron to be targeted has been filled with fluorescent dye via diffusion from a patch pipette during whole-cell recording. (**a**) First, with the help of MPE microscopy, the tip of the microiontophoresis pipette (at focal plane z_2; upper broken horizontal line) is aligned (x, y) with the center of the subcellular compartment (at focal plane z_1; lower broken horizontal line) inside the slice preparation imaged along the vertical z-axis (broken vertical line). In both (**a**) and (**b**), the slice is seen from the side (along the y-axis). (**b**) Stepwise procedure for targeting the subcellular structure located a distance *b* below the surface of the tissue. In position 1, the tip of the pipette is aligned vertically above the target and is located a distance *a* above the surface of the tissue. The angle α between the pipette's long axis and the horizontal plane is 20°. In position 2, the pipette has been moved outward along its long axis by a distance

With continuous visualization by MPE microscopy, proceed as follows:

1. Using a displayed crosshair (xy) as a reference in the imaging software, position the targeted subcellular structure at the center of the crosshair.

2. Position the tip of the microiontophoresis pipette above the slice (20–30 μm) such that it touches the center of the crosshair (xy) and thereby is superimposed vertically with the targeted subcellular structure (Fig. 4a; corresponding to position 1 in Fig. 4b).

3. By changing the focal plane between the target structure (focal plane z_1 in Fig. 4a) and the tip of the pipette (focal plane z_2 in Fig. 4a), measure this distance as accurately as possible. This distance is indicated in Fig. 4b as $a + b$, where a is the distance between the pipette tip and the slice surface and b is the distance between the slice surface and the target structure.

4. Move the pipette a distance c (marked red in Fig. 4b) outwards from the center of the field of view along its long axis (the x-axis), corresponding to position 2 in Fig. 4b. The pipette will now be outside the field of view.

5. Move the pipette downwards the same distance $a + b$ along the z-axis, corresponding to position 3 in Fig. 4b. This is done "blindly", with the pipette outside the field of view.

6. Move the pipette inwards the same distance c that it was moved outwards (in step 3 above) along its long axis (x-axis), corresponding to position 4 in Fig. 4b. For optimal control, the digital zoom of the MPE microscopic image should be decreased during the initial part of the movement to bring the pipette into the field of view and then gradually increased as the tip of the pipette approaches the targeted structure. In this way, the tip of the pipette can be detected as it enters the slice (or shortly thereafter) and visually tracked (with increasing digital zoom) on its way to the targeted structure.

In our experience, when performed with high-resolution micromanipulators, this procedure is successful in accurately positioning the pipette tip without requiring any re-positioning (Fig. 4c). In some cases, the final position will deviate along the y-axis ("north/south" as displayed in the imaging software) and/or along the z-axis (vertical; Fig. 4d). In such cases, the spatial offset (along the y-axis and/or z-axis) can be estimated before moving the pipette out of the tissue. Then the pipette can be re-positioned to compensate for the offset and moved back in again along the x-axis. The crucial part of the procedure is the retraction distance c that the pipette is moved outwards in step 4 above. If c is too short, the tip of the pipette will hit the surface of the slice as it is moved downwards in step 5. A simple geometrical analysis indicates that for the conditions used here, c should be (at least) ~3 times the distance ab. In this case, the pipette tip will be a distance a above the tissue surface in position 3, with the distance c equal to $ab/\sin \alpha$, or approximately $3ab$ for $\alpha = 20°$, as in our case. Accordingly, a multiplication factor of 3–3.5 should be sufficient to ensure that the tip of the pipette does not bump into the slice when moved downwards to position 3 (Fig. 4b). If the slice surface is very uneven, a larger factor can be used for calculating the retraction distance c. Depending on the experiment, it can be a good idea to check the capacitance compensation of the microiontophoresis pipette again after it has been moved into the tissue and is 20–30 μm away from the target structure.

2.16 Electrophysiological and Intracellular Ca^{2+} Responses Evoked by Fast Microiontophoretic Application of Agonist

The charge of the drug molecules in the solution of the microiontophoresis pipette will determine whether they can be ejected by applying a positive or a negative current (see [11, 12, 15] for other examples). The examples discussed here use application of glutamate to subcellular structures of AII and A17 amacrine cells in the rat retina.

After the tip of the microiontophoresis pipette has been positioned close (1–2 μm) to the structure of interest under visual control using MPE microscopy (Fig. 4), glutamate can be ejected by application of negative current pulses (typically −200 to −600 nA). When the capacitance of the microiontophoretic pipette has been adequately compensated (see Sect. 2.12), even very brief current pulses can evoke responses that can be detected in whole-cell recordings. Figure 5 illustrates a response evoked in an A17 amacrine cell (Fig. 3g) by a brief current pulse (−500 nA, 1 ms) when the microiontophoretic pipette was positioned close to a dendritic varicosity (Fig. 5a) where these cells receive synaptic input from glutamatergic rod bipolar cells [30]. Application of glutamate evoked a transient inward current (Fig. 5b) immediately after the iontophoretic current pulse. The rise times observed for such responses evoked in voltage-clamp recordings will typically be slower than the fastest rise times observed for spontaneous excitatory postsynaptic currents (spEPSCs). Because spEPSCs essentially can arise anywhere in the dendritic tree, it will only rarely be possible to directly compare their rise times with those of responses evoked by microiontophoresis at the same location. In addition, it is unlikely that the temporal concentration profile of agonist applied iontophoretically in intact neural tissue can completely mimic that of neurotransmitter released in synapses. Even if the speed of ejection from the tip of the micropipette could match that of release from synaptic vesicles fusing with the plasma membrane of the presynaptic neuron, the diffusion distance to the receptors in the postsynaptic density will inevitably be larger than the corresponding distance in the synaptic cleft.

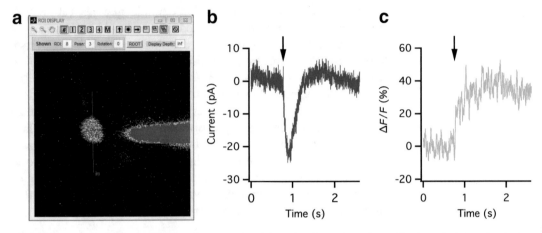

Fig. 5 Current and Ca²⁺ response evoked by subcellular microiontophoretic application of glutamate during whole-cell voltage-clamp recording and MPE microscopy of a neuron in a rat retinal slice. (**a**) Screenshot from ScanImage [25] with MPE microscopy of the subcellular target (left), a dendritic varicosity of an A17 amacrine cell (filled with 40 μM Alexa 594 and 200 μM of the Ca²⁺ indicator dye Oregon Green 488 BAPTA-1 [OGB-1] via the patch pipette), and the distal part of the microiontophoresis pipette (right; filled with 150 mM glutamate and 50 μM Alexa 594). (**b**) Inward current (recorded at a holding potential of −60 mV) evoked by microionto-phoretic application of glutamate (−500 nA, 1 ms; 150 mM in pipette). Here, and in (**c**), the time of the ionto-phoretic current pulse is marked by the arrow above the trace. (**c**) Ca²⁺ signal from varicosity in (**a**) recorded simultaneously with current in (**b**). Ca²⁺ signal measured as relative change in fluorescence (Δ*F*/*F*) of OGB-1

When structural MPE imaging, microiontophoretic drug application, and electrophysiological recording (voltage clamp or current clamp) are combined with functional MPE imaging of intracellular Ca²⁺, it is important to allow sufficient time for the Ca²⁺ indicator dye to diffuse into the target structure and reach a relatively stable concentration, thereby reducing a potential source of response variability. The increase in intracellular Ca²⁺ can be measured by using either line scans or frame scans, depending on the application and the temporal resolution required. Line scans sacrifice spatial information in favor of achieving higher temporal resolution (compared to frame scans). However, the advantage of higher temporal resolution must be considered in relation to the higher sensitivity to small changes in the position of the relevant subcellular structure caused by drift and instability. For applications where the goal is to quantify changes in the Ca²⁺ response over time in response to various experimental manipulations, frame scanning is preferred over line scanning.

As described earlier, line scans and frame scans of the target structure(s) must be precisely synchronized with the current stimulus applied to the microiontophoresis pipette and the electrophysiological recording. In our laboratory, we have found it most convenient to initiate data acquisition from the electrophysiological setup and use digital and/or analog output signals to trigger image acquisition and application of the microiontophoretic current. It can be a considerable challenge to keep track of the correlation between epochal image data and electrophysiological data during off-line analysis. For systems where the two types of data are saved in separate computer files, off-line analysis critically depends on recording the exact times of acquisition for both sets of data. It is common to use separate computers for hardware control and data acquisition for imaging and electrophysiology, and it is important to ensure that the date and time of the two computers are synchronized. Some programs for MPE imaging allow correlated electrophysiological data to be saved in a separate channel of the imaging data file [27].

Figure 5c illustrates a Ca^{2+} response obtained by MPE imaging performed in parallel with the electrophysiological recording (Fig. 5b) during microiontophoretic glutamate application to the A17 dendritic varicosity. For this A17 amacrine, Ca^{2+} imaging started ~30 minutes after establishing the whole-cell configuration to allow for maximal indicator loading at distal varicosities.

Ca^{2+} signals can be analyzed using standard procedures, with measurement of background fluorescence (F_b) as the average signal from a rectangular area close to the region of interest and baseline fluorescence (F_0) by averaging the signal in the region of interest during an interval before stimulus onset. For a given signal (F), the relative change in fluorescence related to a change in Ca^{2+} is then calculated as [9]:

$$\frac{\Delta F}{F_0} = \frac{F - F_0}{F_0 - F_b}$$

This equation is often referred to simply as $\Delta F/F$.

If we assume that the observed responses are generated by ejection of drug and activation of receptors in the membrane of the postsynaptic cell (in contrast to indirect effects mediated, e.g., by the electrical stimulation activating synaptic release of neurotransmitter from neighboring neurons; see Sect. 2.18), that the ejected amount of drug does not saturate the receptors, and that the increase of intracellular Ca^{2+} does not saturate the indicator dye, it should be possible to increase both the electrophysiological and Ca^{2+} responses by ejecting a larger amount of drug. This can be achieved by increasing either the amplitude and/or the duration of the current used for ejection. For the example illustrated in Fig. 6, the ejection current was varied from 0 to −500 nA in steps of −100 nA while the duration was kept constant at 1 ms. The evoked electrophysiological (Fig. 6a) and Ca^{2+} responses (Fig. 6b) increased in parallel with increasing stimulus intensity (Fig. 6c). The ability to establish tentative dose-response relationships like these suggests that the system is within its dynamic range and does not saturate for the range of stimuli applied. A more detailed examination is difficult, however, as the drug concentration in the vicinity of the

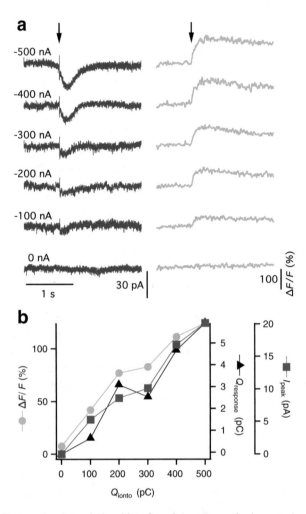

Fig. 6 Dose-response relationships for glutamate-evoked current and Ca²⁺ responses in a dendritic varicosity of an A17 amacrine cell during whole-cell voltage-clamp recording (holding potential −70 mV) and MPE microscopy (as in Fig. 5). Each response was evoked by microiontophoretic application of glutamate (150 mM in pipette, 1 ms pulse duration) with increasing iontophoretic current pulse amplitude (from 0 to −500 nA, steps of −100 nA). (**a**) Current (magenta) and Ca²⁺ (green) responses evoked by iontophoretic pulses of glutamate (arrows). (**b**) Peak amplitude of Ca²⁺ responses ($\Delta F/F$; green circles), peak amplitude of current responses (I_{peak}; magenta squares), and integral of current responses ($Q_{response}$; black triangles) as a function of iontophoretic stimulus charge (Q_{ionto}; product of iontophoretic current pulse amplitude and duration). Note increasing responses with increasing stimulus strength. Ca²⁺ signal measured as relative change in fluorescence ($\Delta F/F$) of OGB-1

receptors is essentially unknown. It is a consistent observation that electrophysiological responses such as the ones illustrated here can be maintained for longer periods of time than the corresponding intracellular Ca²⁺ responses. This could be related to a gradually developing phototoxicity that is unavoidable with functional imaging and/or the susceptibility of intracellular signal transduction mechanisms (involved in generating Ca^{2+} responses) to rundown.

2.17 Estimating the Spatial Profile of Glutamate After Micro-iontophoretic Ejection

One important reason for employing very brief pulses during microiontophoretic drug application is to limit the spatial extent of receptor activation. One way to investigate the spatial extent of activation is to examine how the evoked response changes as a function of the position of the microiontophoretic pipette as the pipette is gradually retracted from the targeted subcellular compartment. Figure 7 illustrates the results from an experiment where we targeted a specific dendritic varicosity of an A17 amacrine cell (Fig. 7a) and measured the voltage-clamp current evoked by brief iontophoretic current pulses while the pipette was slowly retracted in steps of 0.5 μm. When the pipette tip was close to the varicosity, a brief current pulse (−100 nA, 0.5 ms) evoked a transient inward current with peak amplitude of ~15 pA (Fig. 7b, top). As we retracted the pipette along its long axis (x), we measured the peak inward current of the response evoked at each position (stimulus interval ≥20 s). When the pipette tip was further from the varicosity, the peak amplitude of the response declined and the rise time

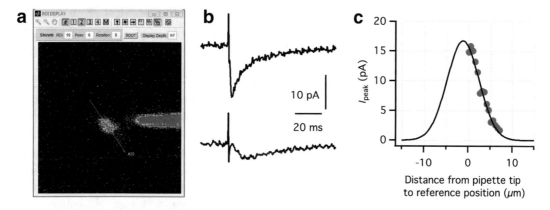

Fig. 7 Estimating the spatial profile of glutamate applied by brief current pulses from a high-resistance microiontophoresis pipette in a rat retinal slice. (**a**) Screenshot from ScanImage [25] with MPE microscopy of the subcellular target (left), a dendritic varicosity of an A17 amacrine cell (filled with 40 μM Alexa 594 and 200 μM of the Ca^{2+} indicator dye OGB-1 via the patch pipette), and the distal part of the microiontophoresis pipette (right; filled with 150 mM glutamate and 50 μM Alexa 594). Voltage-clamp recording with a holding potential of −80 mV. (**b**) Current responses evoked by brief iontophoretic pulses of glutamate (−100 nA, 0.5 ms) with the pipette tip close to the dendritic varicosity (top; single average of responses obtained with the pipette tip 0, 0.5, 1, and 1.5 μm from the reference position) or further away from the dendritic varicosity (bottom; single average of responses obtained with the pipette tip 4, 4.5, 5, and 5.5 μm from the reference position). The time of glutamate application indicated by the stimulus artifacts. (**c**) Peak response amplitude as a function of distance between the pipette tip and the reference position ("0 μm") during stepwise retraction of the pipette from the target structure. The reference position corresponds to the first position tested, with the minimum distance between the pipette tip and the target. The response at each position is the average of three repetitions. The data points have been fitted with a Gaussian function, $I(x) = I_{peak} \times \exp\left[-\left(\dfrac{x - x_0}{w}\right)^2\right]$, where I_{peak} is the peak amplitude, x_0 is the value at the center, and w is the width

of the response increased (Fig. 7b, bottom), consistent with a longer time required for diffusion of glutamate from the tip of the pipette to the location of the receptors. The data points for the peak response amplitude versus pipette position were well fitted by a Gaussian function (Fig. 7c), with best-fit values for width (SD × $\sqrt{2}$) and peak of 5.1 μm and 17 pA, respectively. From the Gaussian function, we also estimated the full width at half-maximum as 8.5 μm (Fig. 7c). From these measurements, we can conclude that fast microiontophoresis can apply drug with a high degree of spatial precision and confinement.

2.18 Control Experiments for Microiontophoresis

It is important to control for indirect effects evoked by applying current through a microiontophoresis pipette, i.e., to understand if an observed response (electrophysiological and/or intracellular Ca^{2+}) is due to iontophoretic ejection of the neuroactive substance or to direct electrical stimulation. Several control experiments can be performed, as suggested below:

1. When first setting up this technique in the laboratory, it can be useful to perform experiments where the microiontophoresis pipette contains only salt solution without active drug. Applying iontophoretic current of both positive and negative polarity can be used to test for direct effects of the electrical stimulation.

2. Reversing the polarity of the iontophoretic current can also be useful to verify a direct effect when applying a neuroactive substance. For example, if application of glutamate with negative current evokes a response, the stimulus can then be inverted (i.e., to a positive current) while keeping the other parameters (amplitude, duration) constant. If the response is still observed, it suggests that it is not due to iontophoretic ejection of glutamate.

3. If a response (electrophysiological, intracellular Ca^{2+}) is due to iontophoretic ejection of a neuroactive substance, as opposed to being directly evoked by the electrical stimulation itself, it should be possible to block the response by applying pharmacological blockers. If the response is reduced (or disappears) in the presence of a blocker, it is a good idea to examine if recovery can be observed after washout of the blocker.

4. When experiments combine electrophysiological recording and Ca^{2+} imaging and the goal is to investigate which ion channels are involved in mediating an intracellular Ca^{2+} response, one will typically control the membrane potential by recording with the voltage-clamp technique, thereby avoiding indirect effects mediated by postsynaptic depolarization and activation of voltage-gated Ca^{2+} channels. When recording from branched neurons, however, it can be difficult to achieve a satisfactory level of voltage control (space clamp) and avoid a local escape from the voltage-clamp potential in the dendritic tree. If one

suspects that activation of voltage-gated Ca^{2+} channels is responsible for, or significantly contributes to, an observed response, this can be examined by repeating the test after hyperpolarizing the neuron relative to the original holding potential. Even with imperfect space clamp control, hyperpolarization will decrease the likelihood of reaching the level of depolarization required for opening of voltage-gated Ca^{2+} channels and with sufficient hyperpolarization, a response mediated by voltage-gated Ca^{2+} channels will be reduced or blocked. In contrast, if a Ca^{2+} response is mediated by influx through ligand-gated channels, voltage clamping the cell at a more hyperpolarized potential should only increase the driving force for Ca^{2+} influx and no reduction of the response should be observed.

5. It is also possible that a response (electrophysiological or intracellular Ca^{2+}) can be evoked indirectly by action potentials generated in presynaptic axons by electrical stimulation from the iontophoresis pipette. Such responses should display an all-or-none behavior with a distinct threshold. By varying the stimulus intensity (amplitude and/or duration), it should be possible to reveal if this is the case or not.

3 Notes and Troubleshooting

Here we provide a list of potential problems and pitfalls that can occur during experiments that combine MPE structural and functional imaging, microiontophoretic drug application, and electrophysiological recording, together with suggestions for troubleshooting.

1. Problem: Attempted targeting of a specific subcellular structure with the microiontophoretic pipette is unsuccessful.

   Possible reason: The pipette touches the tissue when it is moved downwards after retraction (position 3 in Fig. 4b).

   Solution: For accurate targeting of the microiontophoretic pipette, it is necessary to measure the angle of the headstage of the microiontophoresis amplifier relative to the horizontal plane. Some micromanipulators come with a scale where the angle can be read directly, but the resolution may not be adequate. A very useful alternative for accurate measurement is to use a smartphone with a "Clinometer" app. The magnitude of the angle and the vertical distance between the target structure and the tip of the pipette (when located above the target; position 1 in Fig. 4b) determine the distance that the pipette should be retracted (position 2 in Fig. 4b) such that it does not move into the preparation when moving down to position 3 (Fig. 4b).

2. Problem: No electrophysiological or Ca^{2+} response is observed during microiontophoretic drug application.

Possible reason: The tip of the microiontophoresis pipette is too far from the targeted structure.

Solution: Move pipette closer (under visual control, i.e., during MPE imaging).

Possible reason: The tip of the microiontophoresis pipette is clogged.

Solution: Check the resistance of the pipette (following adequate capacitance compensation) and see if it has changed relative to the start of the recording. It is sometimes possible to reduce the resistance of a clogged pipette by repeated application of large amplitude current pulses, but make sure to move the pipette tip away from critical target structures if attempting this. Afterwards, check the capacitance compensation and measure the resistance again. If it is not possible to reduce the resistance sufficiently, the pipette must be replaced.

Possible reason: The neuron (or the targeted subcellular structure) does not express receptors that can be activated by the applied drug.

Solution: Attempt an alternative and less demanding experiment to verify that the type of neuron investigated expresses ligand-gated ion channels that are activated by the applied drug. Apply the drug by pressure ejection from a pipette with a larger opening such that the drug reaches a larger area of the dendritic tree or soma region of the cell.

3. Problem: During functional imaging of intracellular Ca^{2+}, no response can be evoked by microiontophoretic activation of Ca^{2+}-permeable receptors presumed to be expressed by the targeted neuron (e.g., NMDA receptors or Ca^{2+}-permeable AMPA receptors).

Possible reason: There are many potential explanations why the expected response is not observed, but it is important to first verify that *any* Ca^{2+} response can be observed.

Solution: A simple way to verify that the intracellular solutions, including the fluorescent Ca^{2+} indicator, and the microscope hardware/software imaging system are working as expected, is to apply depolarizing voltage steps and image from a neuron and subcellular compartment known to express voltage-gated Ca^{2+} channels.

4. Problem: Over the course of an experiment, there is excessive drift and spatial separation between the tip of the microelectrode and the target structure.

Possible reason: It is impossible to completely avoid mechanical drift, but one explanation is that the shape of the micropipette is suboptimal, e.g., the taper is too short. If such a pipette is used, it will push the tissue and this can often be followed by subsequent drift of the tip beyond the target structure after the absolute position of the pipette has been fixed.

Solution: It is important to use micropipettes with sufficiently long and slender tips for the iontophoretic application such that the pipette does not push the tissue too much as it is inserted into the slice. If possible, change the parameters of the pipette puller. To reduce the likelihood of drift during an experiment, only use high-quality micromanipulators and avoid abrupt changes of temperature at or near the setup. Such changes can be related to transient cooling or heating systems. If possible, try to shield the setup from such influences.

4 Outlook and Future Perspectives

Drug application by fast microiontophoresis has been used to study synaptic integration by applying agonists at different locations in the dendritic tree of hippocampal pyramidal neurons, including studying the interaction between glutamatergic excitation and GABAergic inhibition [6, 15]. In our laboratory, we have used the technique to investigate changes of the Ca^{2+} permeability of AMPA-type glutamate receptors in retinal amacrine cells [16]. The main advantage compared to conventional microiontophoresis is the compensation of the pipette capacitance. This enables the use of very brief pulses that allow for high spatial and temporal resolution of the drug application. When microiontophoresis is performed with high-resistance pipettes with long, slender tips, the mechanical damage inflicted on the neural tissue does not seem to have any serious consequences compared to other techniques that use electrodes for recording and stimulation. Compared to pressure application from fine-tipped micropipettes, microiontophoretic application involves no mechanical movement and is therefore compatible with high-resolution Ca^{2+} imaging of subcellular compartments where movement artifacts must be avoided.

The cost of adding equipment for fast microiontophoresis to a setup already equipped for electrophysiological recording and MPE imaging is modest. Microiontophoresis however, cannot compete with the flexibility offered by (more expensive) 1-photon (1P) and multiphoton (MP) microscopic uncaging of neuroactive substances. Of these, 1P uncaging using UV light has a higher uncaging efficacy, but suffers from higher risk of phototoxicity and lower axial resolution, because of the wavelengths used and because uncaging takes place in a larger volume (compared to MP uncaging). MP uncaging has a much improved axial resolution but is considerably more expensive to establish as it requires the addition of a second pulsed IR-laser to the setup (see chapters by Stein et al. and Tran-Van-Minh et al. in this volume).

Acknowledgments

This research was supported by The Research Council of Norway (NFR 182743, 189662, 214216 to EH; NFR 213776, 261914 to MLV).

References

1. Silver RA, MacAskill AF, Farrant M (2016) Neurotransmitter-gated ion channels in dendrites. In: Stuart G, Spruston N, Häusser M (eds) Dendrites, 3rd edn. Oxford University Press, New York, pp 217–257

2. Kew JNC, Davies CH (eds) (2010) Ion channels. From structure to function. Oxford University Press, New York

3. Zheng J, Trudeau MC (2015) Handbook of ion channels. CRC Press, Boca Raton

4. Denk W, Strickler JH, Webb WW (1990) Two-photon laser scanning fluorescence microscopy. Science 248:73–76

5. Higley MJ, Sabatini BL (2012) Calcium signaling in dendritic spines. Cold Spring Harb Perspect Biol 4:a005686

6. Müller C, Beck H, Coulter D, Remy S (2012) Inhibitory control of linear and supralinear dendritic excitation in CA1 pyramidal neurons. Neuron 75:851–864

7. Bootman MD, Berridge MJ, Putney JW, Roderick HL (eds) (2012) Calcium signaling. Cold Spring Harbor Laboratory Press, Cold Spring Harbor

8. Nguyen Q-T, Clay GO, Nishimura N, Schaffer CB, Schroeder LF, Tsai PS, Kleinfeld D (2008) Pioneering applications of two-photon microscopy to mammalian neurophysiology. In: Masters BR, So PTC (eds) Handbook of biomedical nonlinear optical microscopy. Oxford University Press, New York, pp 715–734

9. Yasuda R, Nimchinsky EA, Scheuss V, Pologruto TA, Oertner TG, Sabatini BL, Svoboda K (2004) Imaging calcium concentration dynamics in small neuronal compartments. Sci STKE 2004(219):pl5

10. Grimes WN, Li W, Chávez AE, Diamond JS (2009) BK channels modulate pre- and postsynaptic signaling at reciprocal synapses in retina. Nat Neurosci 12:585–592

11. Stone TW (1985) Microiontophoresis and pressure ejection. IBRO handbook series: Methods in the neurosciences. General ed: Smith AD. Wiley, Chichester

12. Lalley PM (1999) Microiontophoresis and pressure ejection. In: Windhorst U, Johansson H (eds) Modern techniques in neuroscience research. Springer-Verlag, Berlin, pp 193–212

13. Liu G, Choi S, Tsien RW (1999) Variability of neurotransmitter concentration and non-saturation of postsynaptic AMPA receptors at synapses in hippocampal cultures and slices. Neuron 22:395–409

14. Murnick JG, Dubé G, Krupa B, Liu G (2002) High-resolution iontophoresis for single-synapse stimulation. J Neurosci Meth 116:65–75

15. Müller C, Remy S (2013) Fast micro-iontophoresis of glutamate and GABA: a useful tool to investigate synaptic integration. J Vis Exp (77). https://doi.org/10.3791/50701

16. Castilho Á, Ambrósio AF, Hartveit E, Veruki ML (2015) Disruption of a neural microcircuit in the rod pathway of the mammalian retina by diabetes mellitus. J Neurosci 35:3344–3355

17. Geiger JRP, Bischofberger J, Vida I, Fröbe U, Pfitzinger S, Weber HJ, Haverkampf K, Jonas P (2002) Patch-clamp recording in brain slices with improved slicer technology. Pflügers Arch 443:491–501

18. Bischofberger J, Engel D, Li L, Geiger JRP, Jonas P (2006) Patch-clamp recording from mossy fiber terminals in hippocampal slices. Nat Prot 1:2075–2081

19. Davie JT, Kole MHP, Letzkus JJ, Rancz EA, Spruston N, Stuart GJ, Häusser M (2006) Dendritic patch-clamp recording. Nat Prot 1:1235–1247

20. Tsai PS, Kleinfeld D (2009) In vivo two-photon laser scanning microscopy with concurrent plasma-mediated ablation: principles and hardware realization. In: Frostig RD (ed) In vivo optical imaging of brain function, 2nd edn. CRC Press, Boca Raton, pp 59–115

21. Mainen ZF, Maletic-Savatic M, Shi SH, Hayashi Y, Malinow R, Svoboda K (1999) Two-photon imaging in living brain slices. Methods 18:231–239

22. Dodt H-U, Frick A, Kampe K, Zieglgänsberger W (1998) NMDA and AMPA receptors on neocortical neurons are differentially distributed. Eur J Neurosci 10:3351–3357

23. Bers DM, Patton CW, Nuccitelli R (2010) A practical guide to the preparation of Ca²⁺ buffers. In: Whitaker M (ed) Calcium in living cells. Methods in cell biology, vol 99. Wilson

L, Matsudaira P (series eds). Academic Press, Burlington, pp 1–26

24. Euler T, Hausselt SE, Margolis DJ, Breuninger T, Castell X, Detwiler PB, Denk W (2009) Eyecup scope–optical recordings of light stimulus-evoked fluorescence signals in the retina. Pflügers Arch 457:1393–1414

25. Pologruto TA, Sabatini BL, Svoboda K (2003) ScanImage: flexible software for operating laser scanning microscopes. Biomed Eng Online 2:13

26. Langer D, van 't Hoff M, Keller AJ, Nagaraja C, Pfäffli OA, Göldi M, Kasper H, Helmchen F (2013) HelioScan: a software framework for controlling in vivo microscopy setups with high hardware flexibility, functional diversity and extendibility. J Neurosci Meth 215:38–52

27. Nguyen Q-T, Driscoll J, Dolnick EM, Kleinfeld D (2009) MPScope 2.0: a computer system for two-photon laser scanning microscopy with concurrent plasma-mediated ablation and electrophysiology. In: Frostig RD (ed) In vivo optical imaging of brain function, 2nd edn. CRC Press, Boca Raton, pp 117–142

28. Brown KT, Flaming DG (1986) Advanced micropipette techniques for cell physiology. IBRO handbook series: Methods in the neurosciences. General ed: Smith AD. Wiley, Chichester

29. Dutta-Moscato J (2007) Microiontophoresis as a technique to investigate spike timing dependent plasticity. MSc thesis, University of Pittsburgh

30. Nelson R, Kolb H (1985) A17: a broad-field amacrine cell in the rod system of the cat retina. J Neurophysiol 54:592–614

Chapter 8

Multiphoton Excitation Microscopy for the Reconstruction and Analysis of Single Neuron Morphology

Espen Hartveit, Bas-Jan Zandt, and Margaret Lin Veruki

Abstract

Neurons are the main cellular components of the circuits of the central nervous system (CNS). The dendritic and axonal morphology of individual neurons display marked variability between neurons in different regions of the CNS, and there is evidence that the morphology of a neuron has a strong impact on its function. For studies of structure-function relationships of specific types of neurons, it is important to visualize and quantify the complete neuronal morphology. In addition, realistic and detailed morphological reconstruction is essential for developing compartmental models that can be used for studying neuronal computation and signal processing. Here we describe in detail how multiphoton excitation (MPE) microscopy of dye-filled neurons can be used for visualization and imaging of neuronal morphology, followed by a workflow with digital deconvolution and manual or semiautomatic morphological reconstruction. The specific advantages of MPE structural imaging are low phototoxicity, the ease with which it can be combined with parallel physiological measurements from the same neurons, and the elimination of tissue post-processing and fixation-related artifacts. Because manual morphological reconstruction can be very time-consuming, this chapter also includes a detailed, step-by-step description of a workflow for semiautomatic morphological reconstruction (using freely available software developed in our laboratory), exemplified by reconstruction of a retinal amacrine cell (AII).

Key words Computational neuroanatomy, Morphology, Neuronal reconstruction, Multiphoton excitation microscopy, 3D microscopy, Dendrites, Morphometry, Retina

1 Introduction

1.1 From Waldeyer and Cajal to MPE Microscopy

In what later became known as the neuron doctrine, Wilhelm Waldeyer proposed in a review published in 1891 that "the nerve cell is the anatomical, physiological, metabolic, and genetic unit of the nervous system" [1] (quoted and translated in [2]). Despite the wealth of discoveries that have accumulated over the more than 100 years that have since passed, the central tenet of this theory has stood the test of time: individual nerve cells, or neurons, are still considered to be a relevant level for investigating the function of the central nervous system (CNS) [3–5]. Neurons are the main cellular components of the CNS circuits that give rise to a rich variety of neural functions,

Espen Hartveit (ed.), *Multiphoton Microscopy*, Neuromethods, vol. 148,
https://doi.org/10.1007/978-1-4939-9702-2_8, © Springer Science+Business Media, LLC, part of Springer Nature 2019

including sensation and perception, motor action, and cognition. Despite enormous variability in the details of their morphology, neurons are characterized by the presence of multiple branching processes with specialized morphology. With the development of Golgi's technique for silver impregnation of individual neurons [6], and Cajal's technical refinements and subsequent application to essentially all regions of the CNS of several different species, this morphological variability was clearly described and documented [7, 8]. Despite delayed recognition of the importance of dendritic morphology for the functional properties of a neuron [9, 10], there is now strong evidence that the specific morphology of a neuron has a dramatic impact on its function [11]. Although the dendritic morphology of a given type of neuron can potentially be constrained by factors not directly related to signal processing and computation, e.g., development and innervation selectivity, it is generally accepted that, at least in principle, the computational and signal processing properties of a neuron are determined by its synaptic inputs, its three-dimensional (3D) dendritic morphology, and the properties and location of the different ion channels expressed in the cell membrane.

Because neurons are the fundamental building blocks of the nervous system, it is important to visualize, characterize, and quantitatively describe the morphology of single neurons. This workflow is essential for studies that aim to understand the structure-function relationships of specific types of neurons and how the morphological variability can support a correspondingly rich diversity with respect to differential signal processing [4, 12]. The morphology of single neurons can be visualized with several different techniques (for an extensive review, including technical descriptions, see [13]). For some purposes, digital images will suffice as documentation, but for quantitative studies, digital reconstruction of the 3D morphology is an essential step [14–16]. A simple variant of digital reconstruction is focused on creating a representation of the skeleton of the tree structure in 3D which can be adequate for studies that primarily aim to characterize, e.g., changes in branching pattern during development [17]. In a more advanced variant of digital reconstruction, however, a realistic and detailed morphological reconstruction is generated and used for compartmental modeling of neuronal computation and signal processing (reviewed by [14, 18]).

Classically, reconstructing the morphology of single neurons meant performing a manual reconstruction by drawing, either directly from the light microscope (using special optical arrangements) or from the projected images of photographs. A disadvantage of both approaches is that they essentially lack any ability to represent 3D information in a way that makes the reconstruction useful for quantitative analysis. This situation changed dramatically with the development of increasingly sophisticated methods for reconstruction that combine light microscopic imaging and computer-aided neuronal tracing software [18–21]. Digital reconstruction can be performed simultaneously with light microscopic imaging or subsequent to and independently of imaging, such that the reconstruction is performed from a previously stored digital image stack. Irrespective of the exact method, such manual reconstruction can be extremely time-consuming, and it can be challenging to maintain a consistent approach either throughout an extended reconstruction session or from one session to the next.

On this background, two major challenges of light-microscopic reconstruction of single neuron morphology can be summarized as follows: First, which method should be used for visualization of the neuron? Second, which method should be used for reconstruction of the neuron? As will be reviewed in the remainder of this section, the answers to these questions are interrelated and depend strongly on the purpose of the reconstruction. We will also review important aspects of using MPE microscopy of dye-filled neurons for visualization and describe workflows for both manual and semi-automatic morphological reconstruction.

1.2 Visualization

With light-microscopic imaging, there are several different alternatives available for visualization of single neuron morphology. A major drawback of the classical Golgi method, irrespective of problems related to inconsistency, variable success rate, and reproducibility, is that it cannot easily be extended to a workflow that combines visualization with correlated physiological recordings under conditions where the same cell is used for both physiological and morphological analysis. With the advent of intracellular recording with microelectrodes, it became possible to fill cells with water-soluble substances that could be visualized in the light microscope and used for display and morphological reconstruction. For high-quality visualization, however, there are essentially only two types of substances available: tracers (like biocytin [22] and Neurobiotin [23]) and fluorescent dyes. Fluorescent dyes can also be used for injecting neurons in fixed tissue slices (using sharp microelectrodes), and this can generate excellent morphology (e.g., [24]), but this approach cannot be extended to encompass correlated physiological measurements from the same neurons. Filling neurons with biocytin or Neurobiotin, using either sharp microelectrodes or patch pipettes, requires the use of live tissue (in vitro or in vivo). Subsequent visualization based on development of an insoluble reaction product has been used in a number of morphological, as well as correlated morphological-physiological investigations (e.g., [25], for review see [14]). Alternatively, biocytin and Neurobiotin can be reacted with avidin linked to a fluorescent dye. However, both visualization methods require tissue fixation which can compromise and distort exact morphological reconstruction [14].

When it is a goal to avoid tissue fixation altogether, a useful alternative is to fill neurons in live tissue with a fluorescent dye, using either sharp microelectrodes or patch pipettes. Such dye-filled neurons can be imaged with wide-field fluorescence microscopy, but this technique lacks optical sectioning and there is a substantial risk of phototoxicity. Images obtained in this way can often be remarkably improved with digital deconvolution (see below), but generally not to the extent that they are adequate for detailed quantitative morphological reconstructions, as needed, e.g., for compartmental modeling. In contrast, confocal laser-scanning microscopy provides optical sectioning and high resolution, but is not ideal when imaging fluorescent neurons in live tissue because of the high risk of phototoxicity (e.g., [26]). Confocal microscopy is ideal when imaging fluorescent neurons after tissue fixation, but, as indicated above, fixation introduces a considerable risk of tissue shrinkage and distortion that can lead to artifacts.

Most of the disadvantages discussed above can be strongly reduced (even eliminated) by imaging fluorescently filled neurons with MPE microscopy. This technique provides optical sectioning and the ability to image, with high resolution, deep into intact tissue [27]. MPE microscopy has few disadvantages (perhaps the only major one being the high cost of the required pulsed infrared [IR] laser), and for practical purposes, the resolution is essentially as good as that of confocal microscopy for the same conditions of imaging [28]. Because of the strongly reduced risk of phototoxicity compared to wide-field and confocal fluorescence microscopy, MPE microscopy is uniquely suited for imaging neurons in live tissue in combination with measurements of electrophysiological responses from the same cells [29–31]. This approach also eliminates the need to post-process tissue (which can save many hours of extra work), fixation-related artifacts, and the risk of tissue distortion that often take place when mounting and embedding tissue between microscope slides and cover slips. One limitation that is difficult to avoid results from the compromise between the motivation to image using an objective with the highest possible numerical aperture (NA) for maximum resolution and the need for a working distance that enables the positioning of micropipettes for recording from and/or filling the neuron. Whereas confocal imaging generally makes use of objectives with NA in the range of 1.3–1.4, the short working distance of these objectives prevents their use for imaging neurons with attached micropipettes. Water-immersion objectives that can be used for MPE microscopic imaging of dye-filled neurons have, at most, NA values in the range of 0.95–1.05. To partially compensate for this shortcoming, we have implemented the use of post-acquisition deconvolution in our workflow for morphological reconstruction (described in detail below). Although intracellular electrophysiological recordings with sharp microelectrodes have contributed substantially to combined structure-function investigations of single neurons, the current method of choice for such investigations is whole-cell recording with the patch-clamp technique [32, 33].

1.3   Digital Morphological Reconstruction: Manual or Automatic?

Because manual morphological reconstruction can be difficult and time-consuming, there has been a strong motivation to develop computerized morphological reconstruction procedures that require minimal user involvement (for recent reviews, see [34, 35]). Ideally, such automatic procedures would, in addition to saving time and increasing output, have the additional benefits of reduced subjectivity (i.e., errors resulting from variations in interpretation between different operators) and reduced susceptibility to operator fatigue. A number of different procedures for automatic reconstruction are available, both as freely available software from academic research laboratories (e.g., ORION [36], TREES [https://www.treestoolbox.org] [37, 38], Neuromantic [https://www.reading.ac.uk/neuromantic/body_index.php] [39], and neuTube [https://www.neutracing.com] [40]) and as commercial software for semiautomated and fully automated single neuron reconstruction (e.g., Neurolucida and Neurolucida 360 from MBF Bioscience [https://www.mbfbioscience.com], Imaris FilamentTracer from Bitplane [http://www.bitplane.com]). The extent to which the algorithms underlying the different procedures are freely available as source code varies, but for some of the academic software, the full source code is available (e.g., the TREEs toolbox running in the MATLAB environment; [38]). This chapter includes a detailed, step-by-step description of a workflow for semiautomatic reconstruction using software recently developed in our laboratory [41]. The description is focused on reconstruction of a retinal amacrine cell (AII), noting that this is just one example among several, but our software has performed reasonably well for this type of neuron. Our motivation for this development resulted from the specific challenge of reconstructing a densely branching neuron, but we expect the approach to be generally useful for reconstruction of neurons with widely varying morphologies.

2   Materials and General Methods

For animal anesthesia, in vitro slice preparation, visualization with IR Dodt gradient contrast videomicroscopy, intra- and extracellular solutions, electrophysiological recording, data acquisition, MPE structural (and functional) imaging, and IR-laser scanning gradient contrast (IR-LSGC) imaging, see the chapter "Combining Multiphoton Excitation Microscopy with Fast Microiontophoresis to Investigate Neuronal Signaling" by Hartveit and Veruki (this volume).

When electrophysiological responses are recorded for compartmental modeling, it is useful to correct membrane potentials for liquid junction potentials [42]. Such potentials can be measured experimentally, although it is difficult to do so without introducing considerable error. Instead, it is recommended to estimate the liquid junction potential by theoretical calculation, using, e.g., the software program JPCalcW/JPCalcWin. Some data acquisition programs can correct the holding potentials for liquid junction potentials online (e.g., Patchmaster from HEKA Elektronik) which can be very convenient.

2.1 Acquisition of an Image Stack

An image stack is acquired as a series of optical slices (xy) at fixed focal plane intervals from a given region. Depending on the size of the dendritic arborization, and potentially even the size of the axonal field of the imaged neuron, it might be sufficient to acquire a single stack. Alternatively, tiling acquisition can be used when more than one stack is required to cover the cell and all its processes. When this is necessary, the individual image stacks should be sampled with sufficient overlap such that they can be assembled into a single composite stack before reconstruction (see below).

To obtain image stacks that can be adequately processed with deconvolution (see Sect. 2.3), images must be sampled at a rate close to the ideal Nyquist rate, both in the x-, y-, and z-direction. This sets an upper limit for the voxel ("3D pixel") size along each dimension and constrains the size of the image as a function of the digital zoom, the number of pixels (along x and y), and the number of slices (along z). The Nyquist sampling distance in the lateral direction is calculated as:

$$\Delta_x = \Delta_y = \frac{\lambda_{ex}}{4kn \times \sin\alpha}$$

and for the axial direction, the Nyquist sampling distance is calculated as:

$$\Delta_z = \frac{\lambda_{ex}}{2kn \times (1 - \cos\alpha)}$$

where n is the lens medium refractive index, k is the number of excitation photons (photon count; set to 2 for MPE microscopy), λ_{ex} is the wavelength of the excitation light, and α is the half-aperture angle of the objective (reviewed by [43]). The web site for Huygens deconvolution software (https://www.svi.nl/NyquistCalculator) has a simple browser-based application to directly perform these calculations. As an example, for $\lambda_{ex} = 810$ nm and a water immersion ($n = 1.338$) objective with NA = 0.95 (NA = $n \times \sin\alpha$), we get a Nyquist sampling distance in the lateral direction of ~107 nm and in the vertical direction of ~512 nm.

After calibrating the microscope, it is then possible to calculate the voxel size (xyz) for any given combination of digital zoom and the number of pixels sampled in the x- and y-directions for a given image. A useful approach is to adjust the digital zoom until the desired structure (e.g., dendritic arborization) almost fills the imaged region and then increase the number of pixels for each image (e.g., to $1024[x] \times 1024[y]$) until Nyquist rate sampling is satisfied. The vertical interval between the slices of the image stack should be set directly in the image acquisition software. For acquisition of the complete arborization of an AII amacrine cell [44], we would typically sample image stacks with each slice set to 1024×1024 pixels and adjust the digital zoom to obtain an xy pixel size of 80–90 nm. Constrained by the available intervals of the vertical focus mechanism of the microscope, we would use a focal plane interval of 400 nm.

At each focal plane, a number of images can be averaged within each channel to improve the signal-to-noise ratio (SNR). It is tempting to increase the number of averages to improve the image quality, but there are trade-offs. Increased averaging does not improve the SNR linearly, it increases the risk of phototoxicity, and it increases the time required to sample a complete stack. When image stacks are subsequently processed by deconvolution (see below), we have found it optimal to average two frames for each slice.

For neurons with larger dendritic trees, the requirement of Nyquist sampling may not be compatible with acquiring the complete morphology in a single stack, and multiple stacks must be acquired in a tiling pattern. Combining the individual stacks to a single composite stack can be done with an "image montage" stitching function (e.g., in Neurolucida) that creates a 3D image montage by manually arranging the set of stacks to optimize x, y, and z alignment. An alternative computer program ("Volume integration and Alignment System"; VIAS) is freely available and designed to integrate multiple stacks into a single volumetric dataset (http://research.mssm.edu/cnic/tools-vias.html).

2.2 Image Processing Prior to Deconvolution

ScanImage [45], the MPE microscopy acquisition software used in our laboratory (see chapter "Combining Multiphoton Excitation Microscopy with Fast Microiontophoresis to Investigate Neuronal Signaling" by Hartveit and Veruki, this volume), stores the different channels of an image stack in an interleaved format. Thus, it is necessary to process the image stacks before importing the data to the deconvolution software whenever more than one image channel has been acquired. For this, we use a simple routine developed in the IGOR Pro environment (64-bit) that takes a multichannel image file, de-interleaves the data based on acquisition channels, and saves each channel as an individual file. To compensate for drift and mechanical instabilities that can occur during the image acquisition (due to, e.g., perfusion of the recording chamber), we use the Object Stabilizer module of Huygens to align images along the z-axis. The IGOR Pro *ImageRegistration* operation, as implemented in the SARFIA *RegisterStack* routines [46] can also be used to align the slices of an image stack. A specific advantage of the *ImageRegistration* operation is that it can align the image stack in one of the channels (e.g., the IR-LSGC channel) and then use the generated registration parameters to align the image stack of the other channels (e.g., fluorescence) according to the same parameters. This can be a useful feature when aligning image stacks of neurons with few processes.

2.3 Deconvolution of Fluorescence Image Stacks

Post-acquisition digital deconvolution is a very powerful technique for enhancing image quality [26, 47] and can be used with great advantage to increase the SNR and decrease the axial and lateral blurring [48] of image stacks before they are used for digital morphological reconstruction. Essentially, deconvolution removes noise and reassigns out-of-focus light, using a theoretically calculated or experimentally determined point spread function (PSF).

There are several options for deconvolution software, both freely available academic software as well as commercial software. With respect to commercial software, our experience is with the powerful and flexible Huygens package from Scientific Volume Imaging. The Huygens software comes in two different base versions, Essential and Professional (both 64-bit). Whereas they differ in some of the added functionality (and therefore the cost), the core of the deconvolution operation is identical between the two versions. The deconvolution procedure depends on specifying a PSF for the imaging system, but can use either a PSF that is experimentally determined for a specific setup and imaging condition or a PSF that is calculated theoretically. A number of different algorithms are available for deconvolution within the Huygens software. For deconvolving MPE microscopic images, we have obtained the best results using the Classic Maximum Likelihood Estimation (CMLE) algorithm.

Before starting deconvolution of an image or image stack, the Huygens software requires user input of several parameters related to the microscope and imaging. Some parameters can be left with their default values calculated from data available in the image stack. The most critical user-specified parameter is the SNR. In theory the SNR value can be deduced from the image intensities and photon shot noise. However, we have adopted a more heuristic strategy to obtain an optimal setting, as is advised by the Huygens software. Increasing the SNR value used during deconvolution increases the sharpness of the restoration result, but when set higher than an optimal value, it leads to enhanced noise and spatial fragmentation of the structures in the images.

Although it can be time-consuming, it is well worth the effort to run a series of repeated deconvolutions with several different values of the SNR on the same image stack in order to estimate an optimal SNR (while keeping all other parameters and settings constant). In the course of a study with morphological reconstruction of dye-filled AII amacrine cells from the rat retina [44], we investigated this process in detail. An example of the results that can be obtained with this procedure is illustrated in Fig. 1, for an arbitrary region with a number of tightly packed dendritic processes imaged in a single focal plane of an AII amacrine image stack. The raw (unprocessed) image data are shown in Fig. 1a, and b–g show the results for the same region after deconvolution with different values for the SNR (set to 1, 5, 10, 20, 40, and 80, respectively). Note that with increasing value for SNR, the sharpness of the deconvolved images increases and noise is removed (corresponding to removing out-of-focus light), but when the SNR for deconvolution is increased beyond an optimal value, the deconvolved images begin to display structural fragmentation of the dendritic processes. This is a sign that the SNR values are too high and the resulting images are over-deconvolved. For a more detailed analysis, we plotted linear intensity profiles for fluorescence at a series of locations across several dendritic processes. An example is illustrated in Fig. 1, where the position of the intensity profile corresponds to the line displayed in Fig. 1a. The population of linear intensity profiles for all the deconvolution results (Fig. 1b–g), together with that for the raw data, is illustrated in Fig. 1h. When the SNR for deconvolution is increased, it progressively increases the peak value of the intensity profile, and the profiles remain smooth until the optimum SNR value has been reached. However, when the SNR is increased beyond the optimum (approximately 20 for the example illustrated in Fig. 1), the intensity profile displays increased noise, reflecting the spatial (morphological) fragmentation in the corresponding deconvolved images. This procedure depends on the judgment of and input from the operator. To ensure that the optimally deconvolved image stack is selected for morphological reconstruction, the procedure described here with visual inspection and generation of linear intensity profiles should be applied to several different regions (and processes) of the same stack.

3   Manual 3D Morphological Reconstruction

3.1   General Aspects

In our laboratory, we routinely use the computer programs Neurolucida [19] and Neurolucida 360 for manual, computer-aided, quantitative morphological reconstruction of fluorescently labeled neurons. During reconstruction, the fluorescent processes

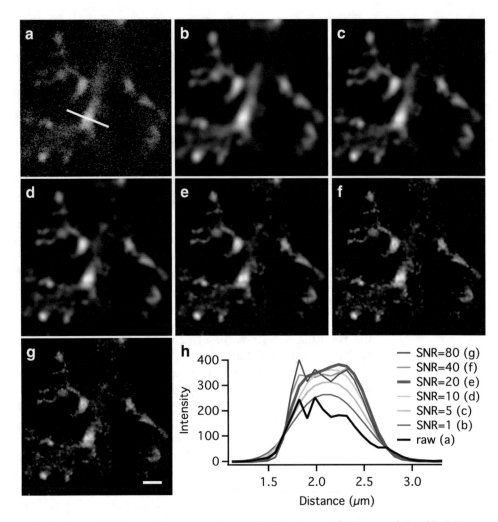

Fig. 1 Digital deconvolution of multiphoton excitation (MPE) microscopic images of dye-filled AII amacrine cells. (**a**) Subregion of individual image slice (average of two individual frames) with details of arboreal dendrites of an AII amacrine cell. Straight line (length 4.4 μm) across process used to create intensity profiles displayed in (**h**). Note how image is affected by noise and blurring. (**b–g**) Same image as in (**a**) after deconvolution with different settings for the signal-to-noise ratio (SNR) in the deconvolution software, as indicated in (**h**). Note how deconvolution reduces noise and blurring and how increasing the SNR progressively improves the images, but eventually leads to spatial fragmentation, most pronounced in (**f**) and (**g**). (**h**) Spatial intensity profiles of raw image (**a**) and deconvolved images (**b–g**) for different values of SNR during deconvolution. Note noisy profile from raw image, reaching a peak intensity of approximately 200 (thick black line) and how increasing the SNR (**b–d**; colored lines) increases the peak intensity from approximately 250 to 350–400. For SNR of 20 (**e**), the intensity profile reaches an overall maximum while still remaining relatively smooth (thick red line). For SNRs of 40 (**f**) and 80 (**g**), the profiles become noisy, corresponding to the spatial fragmentation seen in the images (**f** and **g**). Brightness, contrast and gamma settings were identical for all panels (**a–g**). Bar, 2 μm (**a–g**). (Adapted from Zandt et al. 2017 [44])

in the different slices of an image stack are tracked by scrolling up and down through the stack, and the computer mouse is used to delineate a given process and visually determine its diameter. Further details can be found in the on-line documentation (https://learn.mbfbioscience.com). 3D reconstruction of the soma can be done in two different ways. One can identify the image slice corresponding to the largest outline ("maximum projection") and trace the circumference with a single contour. The volume of the soma is then estimated by spatial integration of this contour (through rotation) during the subsequent analysis. Alternatively, 3D reconstruction can be performed by tracing the soma with a series of contours, each at a different focal plane. In this case the volume of the soma is estimated by integrating over the series of individual contours. In general, we choose to use a single contour.

Figure 2a–c illustrates three different stages of the reconstruction workflow, with maximum intensity projections (MIPs) of the fluorescence image stack before (Fig. 2a) and after (Fig. 2b) deconvolution, and a projection of the final digital reconstruction (Fig. 2c). All projections have been overlaid on a single, representative image slice from the IR-LSGC channel (identical for panels a-c). The details of the dendritic arborization of the reconstructed neuron are more clearly displayed by the two-dimensional (2D) projection (shape plot) in Fig. 2d and the 3D visualization in Fig. 2e.

For general morphological analysis and quantification of dendritic branching of a complete reconstruction, there are many different options available in the program Neurolucida Explorer (64-bit, MBF Bioscience; https://www.mbfbioscience.com/neurolucida-explorer; https://www.mbfbioscience.com/help/neurolucida_explorer/Content/NeurolucidaExplorer.html). In addition, the freely available program L-measure (http://cng.gmu.edu:8080/Lm/; [49]) offers similar as well as additional analysis options compared to Neurolucida Explorer, but does not have visualization capabilities.

3.2 Special Considerations When Reconstructing Very Thin Processes

During manual reconstruction, it is essentially up to the operator to determine the thickness of the different fluorescent processes by visual inspection during the reconstruction, irrespective of whether this is done at the microscope or at the computer with an image stack. When morphological reconstruction is based on light microscopic imaging, it is a problem when the diameter of a neuronal process is below the resolution limit of light microscopy [14, 15]. For a self-luminous point object, as is the case when imaging sub-resolution structures with fluorescence light microscopy, the lateral (xy) Rayleigh two-point resolution (minimum resolved distance) is given by $0.61\lambda/NA$ (e.g., [26, 50]), where λ is the wavelength of the emitted light and NA is the numerical aperture of the microscope objective. With MPE microscopy, only the excitation wave-

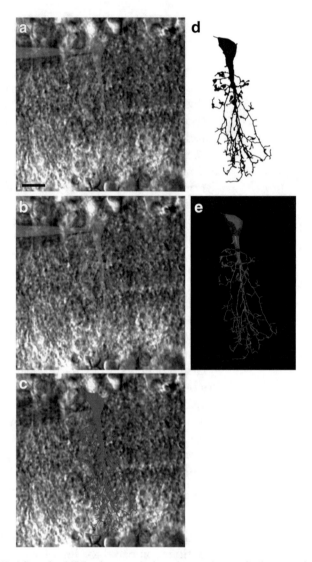

Fig. 2 Workflow for MPE microscopic imaging and quantitative morphological reconstruction of dye-filled AII amacrine cells. (**a**) Maximum intensity projection of raw image stack of AII amacrine cell filled with Alexa Fluor 594 during whole-cell recording (dye-filled pipette attached to the cell body) overlaid on image of retinal slice acquired with IR-laser scanning gradient contrast microscopy. (**b**) Same as in (**a**), but after deconvolution. (**c**) Shape plot generated from computerized morphological reconstruction of cell in (**a, b**). Brightness and contrast of background image of retina had to be re-adjusted for composite images in (**a–c**). (**d**) Shape plot of reconstructed cell showing details of dendritic arborization. (**e**) Three-dimensional view of morphological reconstruction. Bar, 10 μm (**a–d**). (Adapted from Zandt et al. 2017 [44])

length is important, and the resolution is improved by $\sqrt{2}$ (in the ideal, diffraction-limited case, assuming that the laser beam completely fills the back-focal plane of the objective) and the equation becomes $0.61\lambda / \left(NA\sqrt{2} \right)$ [51]. For example, in a situation with an excitation wavelength of 810 nm and an objective NA of 0.95, the resolution limit becomes approximately 0.37 μm in the ideal (diffraction-limited) case. If a neuronal process is thinner than this value, it can be detected if the fluorescence intensity is high enough to generate a signal above the noise level in the microscope hardware, but the diameter cannot be adequately resolved. To our knowledge, the only case where super-resolution (i.e., diffraction-unlimited) light microscopy has been used to image neuronal processes under conditions where the acquired images can be used for morphological reconstruction is the work of Sigal et al. [52]. Here, retinal neurons were imaged with stochastic optical reconstruction microscopy (STORM) applied to fluorescently labeled ultrathin sections, providing a lateral resolution of ~20 nm and an axial resolution of ~70 nm (limited by section thickness), which is sufficient for imaging some of the thinnest neuronal processes. However, sectioning of tissue for serial electron microscopy (EM) makes this approach very labor-intensive.

EM has a limit of resolution which is adequate for imaging the very thinnest neuronal processes, but has not routinely been applied for reconstructing complete arborizations of single neurons. The existence of ultrastructural data for a specific type of neuron, however, can often provide valuable information that can guide light microscopic reconstruction of processes with diameters below the limit of resolution of light microscopy [15]. An example is the detailed study of Tsukamoto & Omi of AII amacrine cells in mouse retina [53], which we used to guide our work with reconstruction of rat AII amacrine cells imaged with MPE microscopy [44]. We used the published 2D projections of complete EM reconstructions to make measurements from the thinnest processes illustrated. The estimated range of diameters were clearly below the expected limit of resolution for MPE microscopic imaging. We then subsequently corrected our morphological reconstructions digitally using a custom algorithm (for details, see [44]).

4 Semiautomatic and Automatic Morphological Reconstruction

There is a strong motivation to develop semiautomatic or even fully automatic methods for reconstructing the morphology of single neurons. The most important reason for this is undoubtedly the amount of time and work required for reconstructing even neurons with simple morphologies. During the course of manually reconstructing a large number of neurons from the mammalian retina, we became increasingly interested in supplementing the manual reconstructions with semiautomatic or fully automatic reconstructions. However, our success after trying out some of the available computer programs for automatic reconstruction was limited, both for freely available and commercial software. To compensate for this, we developed a method that combined elements of existing reconstruction procedures, in particular the algorithm implemented in the TREES toolbox [38]. To enhance the functionality, we extended this with an algorithm for connecting disconnected neuronal segments that typically result from a previous stage of image segmentation. This was implemented with the Fast Marching method for generating image-extracted paths [54].

In the following, we present a detailed account and workflow for semiautomatic reconstruction using our approach (Fig. 3), with the retinal AII amacrine cell as an example. This is currently the only type of neuron for which our method has been extensively tested. Although the method was developed for reconstructing neurons with densely branching dendritic trees, we believe that it is likely to work well for other types of neurons as well. The software was developed in MATLAB and is freely available, including the source code (see Sect. 8; additional details can be found in the original publication [41]). Specific parts of the code use functions in the TREES toolbox [37, 38] and the CellSegm toolbox [55], both of which are freely available.

4.1 Image Preprocessing

The preprocessing consists of two steps:

1. Image restoration by deconvolution. This is described in detail in Sect. 2.3 and is performed as for manual reconstruction. The Object Stabilizer module in Huygens is also used here to align images along the z-axis to compensate for drift and mechanical instabilities. Processed image stacks are saved in 16-bit TIFF file-format, utilizing the whole dynamic range, and converted to the NRRD format in Fiji (http://fiji.sc; [56]). Following the next step, all subsequent processing is performed with MATLAB, and the NRRD format makes file handling in MATLAB easier.

2. Manual removal of the pipette and extraneous fluorescence. Before automatic cell segmentation, it is necessary to remove the image of the recording pipette used to fill the cell with fluorescent dye (Fig. 4a). This step can also be used to remove any extraneous fluorescence that sometimes accumulates around the site of recording, caused by leakage from the pipette when the recording was established (Fig. 4b). Otherwise, as long as the intensity is higher than background, such false signals will be treated by the segmentation algorithm as being part of the cell. To remove both the pipette and extraneous fluorescence, our procedure depends on manual circumscription of the corresponding area in a frontal (xy) MIP of the image stack (using the *getline* function in MATLAB; Fig. 4c).

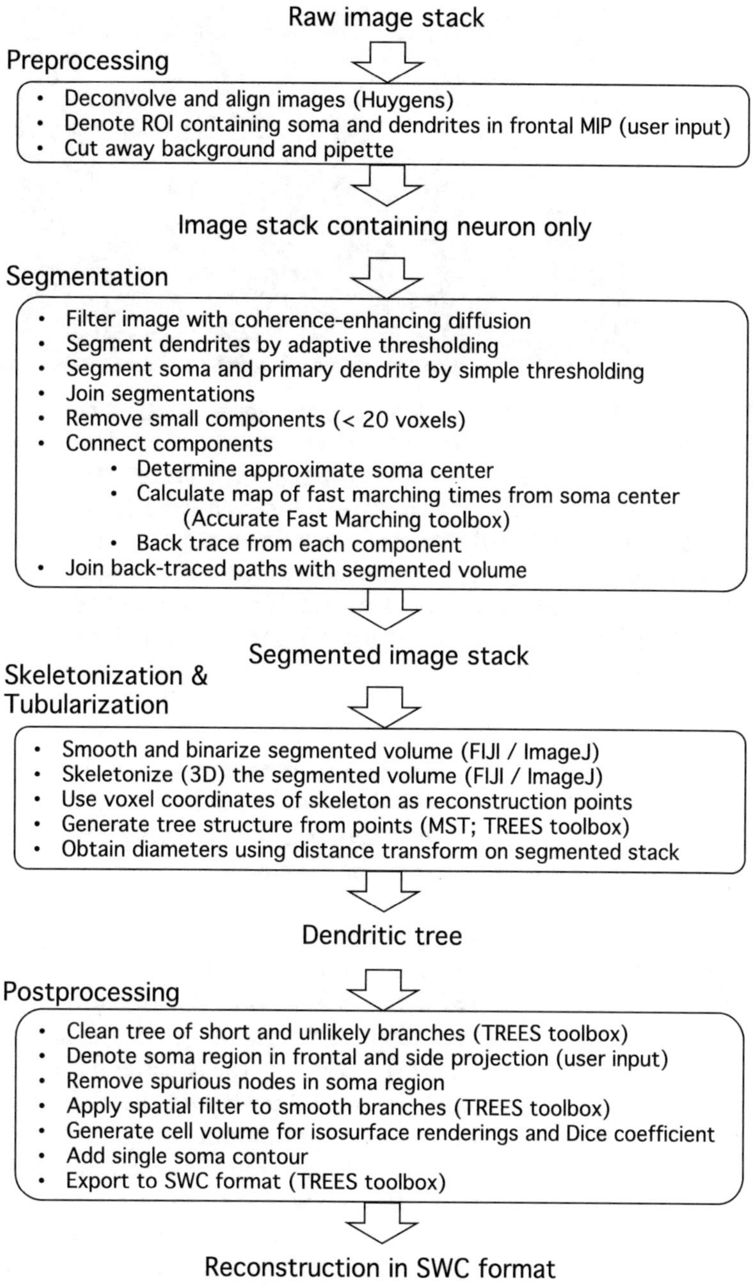

Raw image stack

Preprocessing

- Deconvolve and align images (Huygens)
- Denote ROI containing soma and dendrites in frontal MIP (user input)
- Cut away background and pipette

Image stack containing neuron only

Segmentation

- Filter image with coherence-enhancing diffusion
- Segment dendrites by adaptive thresholding
- Segment soma and primary dendrite by simple thresholding
- Join segmentations
- Remove small components (< 20 voxels)
- Connect components
  - Determine approximate soma center
  - Calculate map of fast marching times from soma center (Accurate Fast Marching toolbox)
  - Back trace from each component
- Join back-traced paths with segmented volume

Segmented image stack

Skeletonization &
Tubularization

- Smooth and binarize segmented volume (FIJI / ImageJ)
- Skeletonize (3D) the segmented volume (FIJI / ImageJ)
- Use voxel coordinates of skeleton as reconstruction points
- Generate tree structure from points (MST; TREES toolbox)
- Obtain diameters using distance transform on segmented stack

Dendritic tree

Postprocessing

- Clean tree of short and unlikely branches (TREES toolbox)
- Denote soma region in frontal and side projection (user input)
- Remove spurious nodes in soma region
- Apply spatial filter to smooth branches (TREES toolbox)
- Generate cell volume for isosurface renderings and Dice coefficient
- Add single soma contour
- Export to SWC format (TREES toolbox)

Reconstruction in SWC format

Fig. 3 Overview of the algorithm and workflow for automatic reconstruction of neuronal morphology. Each box summarizes a series of steps that together constitute a processing stage, the name of which is indicated at the upper left corner of each box. The user is prompted for input twice during the procedure (indicated by "user input" in the workflow). Abbreviations: MIP maximum intensity projection, 3D three dimensional, MST minimum spanning tree. (Reproduced from Zandt et al. 2017 [41] with permission from Elsevier)

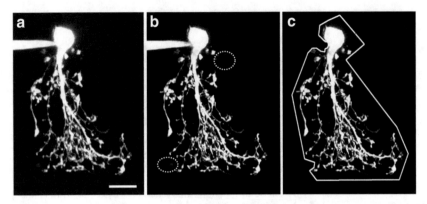

Fig. 4 Workflow for MPE microscopic imaging and preprocessing during morphological reconstruction of dye-filled AII amacrine cells. (**a**) Maximum intensity projection of raw image stack of cell filled with Alexa 594 during whole-cell recording (dye-filled pipette attached to the cell body). (**b**) Same as in (**a**), but after deconvolution and alignment. Areas with increased background fluorescence caused by leakage of fluorescent dye are circumscribed by dashed lines. (**c**) Same as in (**b**), but after removal of fluorescence corresponding to pipette and contaminating areas. Continuous line corresponds to manually delineated region containing the cell, outside of which fluorescence was removed. To enhance visibility, contrast in all panels was increased by 50%, leading to saturation of areas with higher intensity. Bar, 10 μm (**a–c**). (Reproduced from Zandt et al. 2017 [41] with permission from Elsevier)

In cases where the image of the pipette is projected on top of cellular processes in the MIP, this procedure will unfortunately not generate a satisfactory result and an alternative method must be developed.

4.2 Image Segmentation

First, before intensity segmentation with thresholding, the image stack is filtered (smoothed) with coherence-enhancing diffusion [57]. For segmenting the dendritic arborization, the procedure uses adaptive thresholding, with an adaptive filter size $r = [2.5, 2.5, 5]$ (μm). To reduce the likelihood that noise is segmented, we routinely increase the adaptive threshold by a constant value equal to 5% of the maximum image intensity. Sometimes adaptive thresholding generates indentations or holes in thicker structures with size comparable to the size of the adaptive filter. To compensate for this, we perform a second segmentation by applying simple global thresholding (by trial and error, the threshold is set to ~11% of the maximum image intensity), followed by morphological operations that identify the largest connected component. The goal of the simple second segmentation is to obtain the soma and, for the specific example of the AII amacrine cell, the neck of the thick primary (apical) dendrite. The results from the two segmentation operations (adaptive thresholding and global threshold) are joined and followed by removing small components consisting of <20 voxels (considered to be noise).

In the next step, the goal is to connect all disconnected components generated by the initial segmentation by computing the optimal paths by which they can be connected. To accomplish this, our procedure uses the Fast Marching method, generally considered to be a versatile path extraction technique [54]. The procedure uses a point in the soma as a starting point and computes the Fast Marching arrival times for all voxels in the image stack. For the so-called speed function, we use the original image stack (after preprocessing). Then, segments are connected by back tracing the Fast Marching arrival time map from each segment to the soma. To accomplish this, the procedure automatically selects a point approximately in the center of the soma (for details, see [41]). The final (binary) representations of the cells are created by joining the voxels traversed by the connection paths and the segments from the initial intensity segmentation.

4.3 Tree Representation by Skeletonization and Tubularization

In the next step, the goal is to transform the binary cell representation (generated by the automated segmentation) to a tree representation. Our procedure uses a tree representation that follows the definitions of the SWC format. This is a simple tabular text format that lists a set of (x, y, z) positions and associated radii [58]. First, all slices within the segmented volume of the image stack are smoothed (3×3 pixel mean filter) and binarized by the Smooth and Make Binary functions in Fiji. Then, the segmented volume is skeletonized, i.e., thinned by surface erosion [59], using the Skeletonize3D function in Fiji. Each of the functions in Fiji is called automatically from within MATLAB. The coordinates of the voxels in the resulting skeleton are then used to generate a minimum spanning tree (MST; [60]), using the TREES toolbox [37, 38]. Generating an MST simultaneously optimizes both the total branch length and the total path length to the soma for all points (for details, see [41]). The process requires input from the user of a value for the so-called balancing factor that controls the relative weight between the wiring cost and the path length cost. For the AII amacrine cells imaged in our experiments, a small balancing factor of 0.01 provides a good result and essentially corresponds to drawing the shortest possible connections between points.

In the last step of this section, the goal is to obtain the diameters of the neuronal branches. It is well known that for image stacks generated by light microscopy, the axial (z) resolution is inherently lower than the lateral (xy) resolution. In both the raw image stacks and those obtained by segmentation, this is reflected by the elliptical cross sections of the branches, i.e., they are much thicker along the z-direction. In our procedure, this imaging artifact is corrected by a process termed "tubularization" where the diameter associated with each reconstruction point is calculated as twice the distance from the point to the closest border of the segmented volume. This is estimated by fitting the largest possible sphere, centered at the reconstruction point, into the segmented volume. These results are obtained by calculating the distance transform of the segmented image [61, 62]. As for the manual reconstructions, our procedure enforces a minimum branch diameter (set by the user).

4.4 Post-processing and Correction of Reconstruction Errors

It is essentially unavoidable that an automatic reconstruction will contain errors. Some of these can be corrected in a series of semi-automatic post-processing steps, with some requiring more explicit user intervention.

The first post-processing step is to clean the reconstructed tree from short and most likely false branches introduced during skeletonization. This is done with the *clean_tree* routine in the TREES toolbox [37, 38]. For this step the user has to input a value for the cleaning factor. Based on empirical testing, we set this value to 0.2, but for other cells different values might generate better results. The skeletonization procedure also usually generates a number of false branches inside and close to the outside of the soma. We have not found a satisfactory way to remove these branches programmatically, and instead we have implemented a semiautomated procedure where the user is prompted to manually delineate a region around the soma where such "spurious" branches occur (using MATLAB's *getline* function). This is conveniently done by superimposing the generated skeleton on a MIP of the pre-processed image stack from two different perspectives (front and side views). All points with xy- and yz-coordinates within the delineated outlines are then removed.

The second post-processing step is to compensate for the introduction of discrete steps along the reconstructed branches. If left uncorrected, this will artificially increase the apparent branch length. It is a common problem that both manual and automatic reconstructions suffer from discrete steps along the z-dimension, but with the automatic reconstructions we have observed that a similar phenomenon can be observed along the x- and y-dimensions as well, most likely due to the large number of reconstruction points (per unit branch length). As a correction, our procedure spatially filters the reconstruction using the *smooth_tree* function in the TREES toolbox [38].

In the last step, the procedure generates a single contour to represent the soma by automatically tracing the outline of the soma in each slice of the image stack and selecting the contour close to the center, typically the contour with the largest area [41]. For calculating the surface area and volume of the soma, analysis programs such as L-measure and NEURON (https://neuron.yale.edu/neuron/) assume that the soma has a primary axis in the xy plane and is rotationally symmetric. In the final step, the complete reconstruction is exported and saved in the SWC format.

4.5 Proofreading and Editing for Semiautomatic Versus Automatic Reconstructions

Inevitably, a procedure for automatic morphological reconstruction will generate errors that a human observer immediately recognizes as such. It is perhaps a paradox that when procedures for automatic reconstruction are designed by giving the user extensive control of the different steps of the reconstruction by providing access to a number of different parameters that control the reconstruction process, the user might end up spending more time tuning and varying the different parameters and re-running the reconstruction, such that in the end, little time is saved relative to full manual reconstruction. Similarly, if automatic reconstruction has to be followed by difficult and time-consuming post-reconstruction manual editing, the time saved during automatic reconstruction is effectively lost and little real progress has been made. Naturally, this calls for powerful, real-time visualization tools that can be used for efficient evaluation and proofreading and also for correcting the reconstructions when needed (see [63, 64] for detailed discussions). Our procedure does not provide such functionality. Instead, we recommend visualization and proofreading using either commercial software like Neurolucida/Neurolucida 360 and Imaris or the freely available programs Vaa3D (http://www.alleninstitute.org/what-we-do/brain-science/research/products-tools/vaa3d/; [63, 65]), neuTube [40], and CVAPP (https://github.com/pgleeson/Cvapp-NeuroMorpho.org). Hopefully, even with an additional step of manual editing and post-reconstruction error correction, the total reconstruction time will be significantly shorter and less exhausting compared to a fully manual reconstruction.

5   Morphological Reconstruction for Compartmental Modeling

A major motivation for morphological reconstruction of any neuron with an arborizing dendritic tree is to perform computational modeling and simulations of signal propagation using realistic neuronal geometry (e.g., [36, 66]). Neuronal function can depend strongly on the dendritic tree morphology (e.g., [11]), and computer simulations based on accurate reconstruction of neuronal morphology and electrophysiological recording are considered necessary to understand the underlying mechanisms [67, 68]. Importantly, for a neuron with a complex dendritic arborization, the flow of electric current cannot be solved analytically and investigations must resort to compartmental representations where the neuronal structure is discretized. When the goal is to understand signal processing in a specific type of neuron and reproduce its electrical behavior in a series of conditions, the morphological reconstruction must be detailed and accurate. Ideally, such models should be developed from the same neurons from which the electrophysiological measurements are made [32, 33, 69]. Even though morphological reconstructions of the desired type of neuron might be available from a database, they will be of limited value for the development of compartmental models unless they are accompanied by high-quality electrophysiological data from the same cells. If the investigator assigns numerical values to the passive electrical parameters (specific membrane capacitance; C_m, specific membrane resistance; R_m, and cytoplasmic resistivity; R_i) of the model by guessing, there is no simple way of knowing how well the result will reproduce the behavior of the neuron under physiological conditions.

When electrophysiological recording is performed in parallel with MPE microscopy to develop compartmental models for investigating single neuron computation and signaling, potential artifacts associated with fixation and processing of the tissue are completely avoided. Tissue shrinkage and distortion during fixation and histological processing is often a major limitation during subsequent morphological reconstruction and can change both branch diameters, individual branch morphology, as well as the overall shape of the dendritic (and axonal) arborization [14, 30]. Live imaging in parallel with the electrophysiological recording also eliminates the need to embed the tissue which can lead to tissue compression and consequent distortions. Live imaging with MPE microscopy has the additional advantage that there is no need to remove the recording pipette before imaging (as would generally be the case for confocal microscopy). With longer-lasting electrophysiological recordings it is often difficult to remove the pipette from the cell without removing the whole cell body. This can be problematic for subsequent morphological reconstruction and compartmental modeling, especially for small neurons where the cell body constitutes a large part of the total surface area of the neuron.

6   Limitations and Comparison with Other Methods

Despite the amount of work required, light microscopic reconstruction is currently unrivaled when it comes to generating databases with quantitative information for a large population of cells of a given type. Such databases can be very useful for understanding the degree of variability of different morphological properties within a given type of neuron [12, 70]. Even a purely morphological database can be useful for judging the extent to which any individual neuron of the same kind is representative of the "typical" morphology.

Fully automatic reconstruction is not the only alternative to manual reconstruction. Semiautomatic procedures, e.g., methods that involve variants of "finger pointing" with user-guided input to determine the basic skeleton and branching pattern of a neuronal tree (dendritic, axonal), have been developed. This includes both commercial solutions (e.g., Neurolucida 360) and freely available, academic software such as neuTube [40] and hxskeletonize (developed for the Amira environment; [16, 71]).

It is currently not feasible to use EM for a larger population analysis with complete volumetric reconstruction of individual neurons. So far, complete EM reconstructions of single neurons have typically been limited to center-line skeletons, e.g., for large-scale reconstructions of amacrine and bipolar cells in mouse retina [72, 73].

7 Notes

1. The workflow for both manual and (semi)automatic reconstruction can also be applied for reconstruction of neuron morphology from confocal image stacks (with only a few minor modifications). When cells are injected with fluorescent dye using sharp (high-resistance) electrodes, either in live or fixed tissue, we generally recommend imaging with confocal microscopy as this allows the use of microscope objectives with higher NA and consequent higher resolution (both lateral and axial).

2. For structural imaging with subsequent morphological reconstruction and compartmental modeling, we have obtained good results with a pipette solution with either Alexa Fluor 488 hydrazide (50 or 100 μM) or Alexa Fluor 594 hydrazide (40 or 60 μM) as sodium salts (Thermo Fisher/Invitrogen). Additional details can be found in the chapter "Combining Multiphoton Excitation Microscopy with Fast Microiontophoresis to Investigate Neuronal Signaling" by Hartveit and Veruki (this volume).

8 Step-by-Step Procedure for Running the Automated Reconstruction in MATLAB

This section describes running the automated algorithm described above (for details, see [41]). All computer code required is freely available and can be downloaded from https://www.researchgate. net/profile/Bas-Jan_Zandt. Running the code requires installations of MATLAB (commercial; https://www.mathworks.com) and Fiji (freely available; https://imagej.net/Fiji). The description below assumes that the software runs under Mac OS X with Apple's version 6 (legacy) of Java.

8.1 Files Provided, Description of Folders and Subfolders

Code_reconstruction (folder with MATLAB files for the reconstruction algorithm, containing a number of subfolders).

Code (folder with files containing reconstruction code, including functions from the MATLAB toolboxes CellSegm [55] and Fast Marching [74]).

Miji (folder with files to run Fiji from MATLAB).

Xtra (folder with files from the TREES toolbox, files from the cbiNiftiRead toolbox to read/write NIFTI images and to scroll through image stacks).

Data (folder with an image stack for testing the algorithm, downsampled to $256 \times 256 \times 100$ voxels to speed up execution and avoid long processing times).

NLMorphologyConverter (folder with the program NLMorphologyConverter for converting between different file formats for representing the morphology of reconstructed neurons; freely available at http://www.neuronland.org/NL.html).

8.2 Outline of Workflow

The overall workflow consists of three parts:

1. Installing and configuring the necessary software.

2. Running the automated reconstruction on an image stack of an AII amacrine cell acquired with MPE microscopy and converting the resulting files such that they can be opened in Neurolucida and Neurolucida Explorer (for editing or analysis) or imported to NEURON (for compartmental modeling and simulations).

3. Optimizing reconstruction parameters and inspecting the results of the individual steps of the algorithm.

8.3 Installing and Configuring the Software

1. Running the program requires an installed version of MATLAB. The workflow described here has been tested on two different versions of MATLAB, version 2014b (running under Mac OS X 10.7.5) and version 2015b (running under Mac OS X 10.9.5). The Image Processing Toolbox and the Curve Fitting Toolbox are both required.

2. In MATLAB, go to Preferences - MATLAB - General - Java Heap Memory and set "Java Heap Size" to its maximum (4096 MB) to prevent memory issues.

3. Download and install Fiji, a distribution of ImageJ used for converting the format of the image stack and for the skeletonization step during reconstruction. Note: If too many plugins are installed for Fiji, this might cause MATLAB to crash. If this is a problem, it can be solved by moving a number of the plugins from "/Applications/Fiji.app/plugins" to a new folder in the same path, e.g., named "plugins2."

8.4 Running the Automated Reconstruction

It is assumed that the image data are available in the form of a single image stack in the TIFF format. For the example described here, we will use the file named m999999_9_002_ch0_DS.tif (available in the *Data* folder). Note that the algorithm assumes a specific folder structure and that the subfolders containing MATLAB code and image data reside in the same parent folder. The output data will be written to a new subfolder inside this parent folder (as detailed below).

1. Start by generating a folder for the image data and a scaling file:

   (a) Make a new folder inside the *Code_reconstruction* folder, and provide it with a name identical to that of the file containing the image stack, e.g., *m999999_9* for the example used here.

   (b) Copy the image stack file to the folder generated in step a.

   (c) For easier file handling, convert the image stack file to the NRRD format by opening it in Fiji and saving it as an *∗.nrrd* file (menu: File / Save As / Nrrd ...). Save the file used here as *m999999_9_002_ch0_DS.nrrd* in the folder generated in step a.

   (d) Create a *∗.mat* file in MATLAB with the appropriate voxel ($x \times y \times z$) dimensions (same as in the image stack), i.e., $250 \times 250 \times 800$ nm for the file used here. Open the Command Window in MATLAB and type on the command line:

   voxel_size = [250, 250, 800]*1e-6

   (for historical reasons, the dimension of voxel_size is in mm). In MATLAB's Workspace window, right-click on *voxel_size* and select "Save As..." to save as *voxel_size.mat* in the folder generated in step a above (*m999999_9*).

2. Generate an empty folder under *Code_reconstruction* and name it *results_m999999_9* for storing the results generated by the reconstruction algorithm. Now the reconstruction script can be run.

3. In MATLAB, open the file *Main.m* (*Code_reconstruction / Code / Main.m*).

4. Right-click the tab of *Main.m* and select Change Current Folder To / ... / Code

5. Set up the parameters for the reconstruction algorithm as follows:

   (a) Select the Editor tab in MATLAB's ribbon and edit *Main.m* in MATLAB's editor window. Set all parameters to the values recommended in Appendix A (tuning the parameters is described in Sect. 8.5).

   (b) In *Main.m*, set the FIJIpath and FIJIscriptpath variables to the file paths where the Fiji plugins and scripts, respectively, are stored (see Appendix A for an example).

   (c) Optionally save *Main.m* (this will be done automatically in the next step).

6. Run *Main.m* as follows: Select the Editor tab in MATLAB's ribbon and click *Run* (green arrow head). This also automatically saves the *Main.m* file. If Step 4 above was not executed,

MATLAB will open a dialog stating that the file is not in the current directory. Respond by selecting "Change Folder."

7. MATLAB displays a dialog with a request to select the image stack (∗.nrrd) file:

... / Code_reconstruction / m999999_9 / m999999_9_002_ch0_DS. nrrd

Note: this automatically selects and configures the output folders and files based on the name of the folder in which the file is contained (folder *m999999_9* for the example used here).

8. MATLAB generates a window displaying a MIP of the cell in the frontal (xy) image plane, requesting the user to draw a region of interest (ROI) containing the cell, but excluding the pipette and any spurious fluorescence. Use the computer mouse to draw a ROI loosely around the cell by clicking multiple times and try to keep spurious fluorescence "blobs" (if present) outside the ROI. Draw the ROI such that it separates the image of the recording pipette from the image of the cell (presumably the soma, as in the example used here; Fig. 4).

9. During execution, MATLAB will print information about the progress to the Command Window. Wait for the segmentation and skeletonization to finish. Note: the time required depends heavily on the size of the stack. For a stack with $1024 \times 1024 \times 100$ voxels, the process can take several hours and the time depends both on the number of voxels and the ratio between the parameters *voxel_size* and *adaptivefiltersize* (contained in the file *Main.m*). If the algorithm takes too long, it is a good idea to downsample the image stack (e.g., in Fiji or ImageJ) and/or change the size of *adaptivefiltersize* to a smaller value.

10. The preliminary skeleton that is generated and displayed in a separate window will typically contain several spurious branches inside the soma. MATLAB prompts the user to indicate the region in which these branches should be removed. The region has to be delineated both in front (xy) and side (yz) view. The reconstruction points that are inside the drawn ROIs will be removed. For the example cell used here, the selected ROIs should correspond to the soma in both projection planes such that only branches inside the soma will be removed.

11. After the ROIs with spurious branches have been indicated, program execution will continue and the algorithm will generate the tree, clean it, denote the soma, and export the results to SWC files. These steps take place automatically and do not require user intervention.

12. Potential warnings (printed to the Command Window) about trifurcations can be ignored.

13. When program execution has finished, MATLAB will have written four reconstruction files to the results folder (*results_m999999_9*) in addition to several *.mat files that contain the results of various intermediate steps:

m999999_9_auto.swc (contains the reconstructed branches)

m999999_9_auto_sm.swc (contains a smoothed version of the reconstructed branches)

m999999_9_auto_soma.swc (contains the reconstructed branches and a soma contour)

m999999_9_auto_sm_soma.swc (contains a smoothed version of the reconstructed branches and a soma contour)

14. The *.swc output files can be directly visualized in, e.g., Vaa3D. If desired, the user can convert the SWC format files to the Neurolucida ASC format with the NLMorphologyConverter program as follows:

    (a) Copy the three *.swc files in the *results_m999999_9* folder to the *NLMorphologyConverter* folder.

    (b) Open a Terminal window (Mac OS X program).

    (c) Change directory (with the *cd* command) to the *NLMorphologyConverter* folder.

    (d) Execute the following command in the Terminal window (the character ">"is the prompt and should not be typed):

    > ./NLMorphologyConverter m999999_9_auto.swc m999999_9_auto.ASC NeurolucidaASC

    (repeat this step for the two other SWC format files)

    The ASC format reconstruction files can be opened in Neurolucida, Neurolucida Explorer, and NEURON. Note that in Neurolucida, morphology reconstruction files contain a single or multiple contours to represent the soma and a single or multiple unconnected trees that represent the branching tree(s) emanating from the soma (irrespective of whether the files are saved in the DAT or ASC formats). In SWC files, branch roots (corresponding to origin nodes in Neurolucida) are denoted by setting their parent identifier to -1 (at the last entry of a given line). In principle, an SWC file can therefore contain multiple unconnected trees. However, several applications, including the TREES toolbox, that are used to generate and export the reconstruction do not allow this. Our workaround for this problem is twofold. First, for the intermediate reconstruction files that do not contain a soma contour (see step 13 above), the code connects all trees to a single point inside the soma (with the thickness of this point set to zero). Second, for the intermediate reconstruction files that contain a soma contour, the code connects all trees to a common point on the soma contour. As such, this might appear a bit odd in some of the figures generated to display the reconstruction. A more fundamental problem is that this property also prevents correctly importing the reconstruction to NEURON. Accordingly, we recommend using the ASC format reconstruction files generated with NLMorphologyConverter for this purpose.

8.5 Optimizing Reconstruction Parameters

When an image stack differs with respect to resolution, cell type, etc., from the one used in the procedure described above, the reconstruction parameters should be tuned to generate optimal results. Optimizing parameter values can be an art. The following section will be a guide to some of the tricks, with the goal of selecting optimal parameter values and settings in an informed way. The algorithm will be run step-by-step to optimize the corresponding parameters. For this, we will make use of the convenient option in MATLAB that permits the user to run individual sections of code without running the whole script. To enable this functionality, the relevant sections in the *Main.m* file are delimited by a double percentage sign (%%) and can be run by using the computer mouse to click anywhere within a section (this will highlight the code in the corresponding section) and clicking on the icon "Run and Advance" in the MATLAB ribbon displayed for the Editor tab (see https://www.mathworks.com/help/matlab/matlab_prog/run-sections-of-programs.html).

Detailed procedure:

1. Open *Main.m* in the MATLAB editor window (as described above).

2. Use the recommended parameter values, but for the purpose of the example workflow described here, edit *Main.m* to change two parameter values (from their default values) in the section "2. Parameters" (in this way, the script will run with parameters not optimized for the example image stack):

add2threshold = 0.2;

simpleThreshold = 0.2;

3. Select the section "1. Clear previous session" by positioning the text cursor within it (left mouse click). In the MATLAB ribbon (displayed after selecting the Editor tab), click "Run and Advance" to execute the section and advance to the next section. Run the sections up to and including "4. Segmentation (thresholding)."

4. Three graphs will appear with the results from the thresholding procedures. Inspect the graphs carefully. Compared to the result obtained with the default parameter values, it can be seen that many branches have been missed and the threshold values should be lowered.

5. Optimize the threshold parameters as follows:

(a) In MATLAB, open the file *optimize_threshold.m* (located in the *Code* folder) that will be used to optimize the threshold.

(b) Set two parameter values in the section "2. Calculate and plot result of thresholding" of this optimization script as follows:

add2threshold = 0.2;

simpleThreshold = 0.2;

(c) Note that the optimization script loads results from the adaptive threshold calculations. Therefore, for a new image stack, the script in *Main.m* needs to have been run, up to and including the section "4. Segmentation (thresholding)," at least once with any set of parameters before running the optimization script (see above). In *optimize_threshold.m*, run the section "1. Initiation" followed by the section "2. Calculate and plot result of thresholding."

(d) A graph displaying the results appears. The *add2threshold* parameter increases the adaptive threshold. Setting it to a high value, where 1 corresponds to the maximum image intensity, results in the weaker branches not being segmented, while setting it to a low value can result in segmentation of background / noise fluorescence.

(e) Note that the simple thresholding mainly contributes to the total result by closing the gaps that appear in the segmentation of the soma.

(f) Lower the value of *add2threshold* as follows:

add2threshold = 0.1;

and run the section "2. Calculate and plot result of thresholding" again. The result already looks better.

(g) Gradually reduce the value of *add2threshold* until reaching 0.005 (e.g., along the series 0.2, 0.1, 0.02, 0.005) and check the consequence for the segmentation for each value by re-executing the section code. Note that changing this parameter affects not only which branches are segmented, but also their thickness.

(h) Try to find a value for *add2threshold* such that the resulting segmentation represents a reasonable compromise between segmenting too much noise and missing too many branches.

(i) The image stack used in this example can be properly segmented using adaptive thresholding alone. However, we found that especially in stacks with larger cell bodies, adaptive thresholding can lead to gaps in the segmentation inside the soma and primary dendrite. In such cases, the value for the parameter *simpleThreshold* can be tuned (in a similar way as done for *add2threshold* above) to find a value for which the gaps are removed, without affecting the rest of the segmentation too much. For the tuning, gradually reduce the value of *simpleThreshold* from 1 to 0 (e.g., along the series 1, 0.7, 0.4, 0.2, 0.07, 0).

(j) The last segmentation parameter that can be tuned is *Nvoxels_segment_min*. It is used to remove small and isolated segmented clusters of voxels ("blobs"). Segmented clusters containing fewer voxels than the parameter value are removed. To clearly demonstrate the effect of this parameter (do not do this for a real optimization), lower the threshold value to deliberately segment blobs of low-intensity background fluorescence, e.g., by setting

add2threshold = 0.005;

Now gradually change the parameter *Nvoxels_segment_min* (located at the beginning of the section "2. Calculate and plot result of thresholding" in *optimize_threshold.m*) from 1 to 500 (e.g., along the series 1, 10, 50, 200, 500; the default value is 20). Observe that for the smallest values of *Nvoxels_segment_min* only the smallest loose segments will be removed, whereas the larger segments will be removed when larger values are selected. Again, a compromise has to be found between removing all unwanted background and not removing any smaller fragmented neuronal segments at all.

(k) Note the optimal values found for *add2threshold*, *simpleThreshold*, and *Nvoxels_segment_min*. Close the file *optimize_threshold.m* and its accompanying figure window. Enter the optimized values for the parameters in the file *Main.m*.

(l) In *Main.m*, verify that the parameter *redo_segmentation* is set to "true" and rerun *Main.m* from the top, section by section, up to and including the section "4. Segmentation (thresholding)." Verify that the resulting segmentation is reasonable.

6. Run the section "5. Segmentation (fast marching)." The result of this section depends on the parameter *NbrightPixelsSoma* which sets how many brightest pixels are initially used to determine the soma center (set to 100 by default; in section "2. Parameters"), but there is not much point in tuning this parameter, as it will not significantly influence the outcome. Inspect the results of the Fast Marching procedure after executing the code in the section "5. Segmentation (fast marching)." The various segments are displayed with different colors and are connected by the Fast Marching paths contoured with lines.

7. Run the section "6. Skeletonization." There is nothing to optimize here, as the skeletonization algorithm does not have any parameters. The result will be shown after executing the next section of code.

8. Run the section "7. Tree generation and conversion to SWC file." Here, the values for the tree parameters can be optimized:

    (a) Inspect the results from the skeletonization, displayed in the graph labeled "Skeletonization results" (generated after executing the code in the section "6. Skeletonization"), especially with respect to whether the mask for the soma was drawn correctly. The graph shows the skeleton with and without the voxels removed by the soma mask. Also inspect the graph labeled "Tree after cleaning and spurious branch removal" that can be rotated in 3D to see if any spurious branches remain. Note that for this example, relatively few branches have been generated inside the soma.

    (b) If required, redraw the mask demarcating the branches inside the soma as follows: Delete the file *somamask_manual.mat* (written to the folder *results_m999999_9*) and re-execute the section "7. Tree generation and conversion to SWC file." Note that the algorithm will always load the soma mask from this file if it exists. If an error is reported concerning the size of the mask, delete *somamask_manual.mat*.

    (c) In MATLAB's Command Window, type

clean_tree_scaling = 2;

and press Return and run the section "2. Parameters." Alternatively, change the value for *clean_tree_scaling* (the cleaning factor) directly in the file *Main.m.* and run the section "2. Parameters." Note that the majority of the small branches are removed, but so are many branches that seem perfectly valid. When *clean_tree_scaling* is set too low (e.g., at zero), small spurious tips that should be removed are kept. This can sometimes be hard to see in the graphs. In the MATLAB Command Window, type

sum(T_tree(tree))

to calculate and display the number of branch tips. Increase the value of the cleaning factor (e.g., along the series 0.01, 0.1, 0.5, 2) to reduce the number of tips. Decrease the value of the cleaning factor when the cleaning process removes (too many) valid branches.

    (d) The balancing factor (*MST.bf*) can be varied to change the pattern of connecting branches that are close to each other.

A small value for the balancing factor (e.g., 1e-3) favors short connections, whereas a larger value (e.g., 0.1) favors shorter routes to the soma. Try values ranging from 1e-4 to 1 and check the results when running the section "7. Tree generation and conversion to SWC file."

9. Run the section "8. Soma contour generation." Verify that the graph of the area of the individual soma contours, as a function of (axial) distance into the stack, looks approximately circular and that the peak is correctly detected (it is marked by a red diamond symbol in the displayed graph). Verify that the corresponding contour indeed denotes the soma in the figure where the contour is displayed on top of an intensity projection. Enlarge the value of the parameter *Dth* (set in section "2. Parameters"; the default value is 36°) to make the soma rounder or reduce it to include more of the extrusions into the soma contour.

10. Finally, run the section "9. Plotting" to render a series of cylinders representing the tree displayed on top of a MIP of the cell. The figure (labeled "3D Tree above MIP") can be rotated in 3D.

8.6 Additional Tips and Tricks for Running the Semiautomatic Reconstruction

1. When it is desirable to re-execute parts of the reconstruction algorithm with different parameter values, the redo switches at the top of the section "2. Parameters" (lines 24–27) in *Main.m* can be set to "false" (the default setting is "true") to load the results of previous processing written to files on the hard drive. For example, to generate the trees with a different cleaning factor, set the redo switches for segmentation (line 24), fast-marching (line 25), and skeletonization (line 26) to "false," and change the cleaning factor to the desired value and re-execute *Main.m*. This skips the long-lasting calculations corresponding to segmentation and skeletonization and can save much time during parameter tuning.

2. Note that if one section is re-executed, all the sections following it will have to be re-executed as well.

3. The files with the smoothed image (*im_cohendiff.mat*), adaptive threshold (*th.mat* and *imth.mat*), and manual masks (*neuronmask.mat, somamask3d.mat*) are always loaded from the hard drive if they exist. Delete these files from the results folder if their corresponding results need to be recalculated and/or redrawn.

Acknowledgments

This research was supported by The Research Council of Norway (NFR 182743, 189662, 214216 to EH; NFR 213776, 261914 to MLV).

Appendix A: Recommended Parameter Values for Automated Reconstruction

```
%% parameters


% % If the reconstruction is rerun, redo any previous steps?
redo_segmentation = true;
redo_fastmarching = true;
redo_skeletonization = true;
redo_treegeneration = true;
plot_results = true;


% % Segmentation
Nvoxels_segment_min = 20;  % Minimum nr of voxels in a segment (smaller
segments are discarded)
add2threshold = 0.05;      % Value added to adaptive threshold
adaptivefiltersize = [2.5 2.5 5];  % [um]
%adaptivefiltersize = [10 10 20];    % used for the Diadem Olfactory
Projection fibers
simpleThreshold = 0.11;  % Threshold for soma and apical dendrite (fraction
of maximum image intensity)
NbrightPixelsSoma = 100;


% coherence enhancing diffusion filtering parameters
cohenh.DT = 0.2;
cohenh.NITER = 15;
cohenh.KAPPA = 0.001;


% % Skeletonization
Nsample = 1;                  % every Nth point in tree is sampled
clean_tree_scaling = 0.2;     % larger values denote more aggressive cleaning
Nminpts_trees = 4;            % remove trees (branches extending from the
soma) with less then N points
MST.bf = 0.01;                % balancing factor for MST (should be small if
skeleton is accurate)
MST.maxdist = 10;             % [um], maximum distance between connected
points in MST


% % Soma contour generation
Dth = 36;            %degrees, a larger angle cuts off more branches and
makes the soma rounder
maxSomaSizeZ = 20; %[um], length of stack in z-direction to analyse: Set to
double the typical soma size.
SmoothLength = 8;  %[um], should be smaller than soma diameter: Set to
~half of typical soma diameter


% % FIJI path
FIJIpath = '/Applications/Fiji.app/plugins';
FIJIscriptpath = '/Applications/Fiji.app/scripts';
```

References

1. Waldeyer W (1891) Ueber einige neuere Forschungen im Gebiete der Anatomie des Centralnervensystems. Sonderabdruck aus der "Deutschen Medicinischen Wochenschrift", 1891, No. 44 u. ff. Georg Thieme, Leipzig

2. Shepherd GM (2016) Foundations of the neuron doctrine. 25th Anniversary Edition. Oxford University Press, New York

3. McKenna T, Davis J, Zornetzer SF (eds) (1992) Single neuron computation. Academic Press, Boston

4. Cuntz H, Remme MWH, Torben-Nielsen B (eds) (2014) The computing dendrite. From structure to function. Springer series in computational neuroscience, vol. 11. Series eds, Destexhe A, Brette R. Springer, New York

5. Shepherd GM, Grillner S (eds) (2018) Handbook of brain microcircuits, 2nd edn. Oxford University Press, New York

6. Golgi C (1873) Sulla struttura della sostanza grigia del cervello (Communicazione preventiva). Gazzetta Medica Italiana 33:244–246. Reprinted as: Sulla sostanza grigia del cervello, Opera Omnia, 1903, Vol. 1, Istologia Normale, pp. 91–98. Ulrico Hoepli, Milan

7. Cajal S Ramón y (1909) Histologie du Système Nerveux de l'Homme et des Vertébrés, vol. I. Maloine, Paris

8. Cajal S Ramón y (1911) Histologie du Système Nerveux de l'Homme et des Vertébrés, vol. II. Maloine, Paris

9. Segev I, Rinzel J, Shepherd GM (eds) (1995) The theoretical foundation of dendritic function. MIT Press, Cambridge

10. Rall W (2016) Modeling dendrites: a personal perspective. In: Stuart G, Spruston N, Häusser M (eds) Dendrites, 3rd edn. Oxford University Press, New York, pp 429–438

11. Mainen ZF, Sejnowski TJ (1996) Influence of dendritic structure on firing pattern in model neocortical neurons. Nature 382:363–366

12. Soltesz I (2006) Diversity in the neuronal machine. Order and variability in interneuronal microcircuits. Oxford University Press, New York

13. Meredith GE, Arbuthnott GW (eds) (1993) Morphological investigations of single neurons in vitro. IBRO handbook series: Methods in the neurosciences. General ed: Smith AD. Wiley, Chichester

14. Jaeger D (2001) Accurate reconstruction of neuronal morphology. In: De Schutter E (ed) Computational neuroscience: Realistic modeling for experimentalists. CRC Press, Boca Raton, pp 159–178

15. Jacobs G, Claiborne B, Harris K (2010) Reconstruction of neuronal morphology. In: De Schutter E (ed) Computational modeling methods for neuroscientists. MIT Press, Cambridge, pp 187–210

16. Evers JF, Duch C (2014) Quantitative geometric three-dimensional reconstruction of neuronal architecture and mapping of labeled proteins from confocal image stacks. In: Bakota L, Brandt R (eds) Laser scanning microscopy and quantitative image analysis of neuronal tissue, Neuromethods, vol 87. Springer, New York, pp 219–237

17. Cline HT (2016) Dendrite development. In: Stuart G, Spruston N, Häusser M (eds) Dendrites, 3rd edn. Oxford University Press, New York, pp 77–94

18. Parekh R, Ascoli GA (2013) Neuronal morphology goes digital: a research hub for cellular and system neuroscience. Neuron 77:1017–1038

19. Glaser JR, Glaser EM (1990) Neuron imaging with Neurolucida – a PC-based system for image combining microscopy. Comput Med Imaging Graph 14:307–317

20. Turner DA, Wheal HV, Stockley E, Cole H (1991) Three-dimensional reconstructions and analysis of the cable properties of neurons. In: Chad J, Wheal H (eds) Cellular neurobiology. A practical approach. IRL Press at Oxford University Press, Oxford, pp 225–246

21. Meijering E (2010) Neuron tracing in perspective. Cytometry A 77:693–704

22. Horikawa K, Armstrong WE (1988) A versatile means of intracellular labeling: injection of biocytin and its detection with avidin conjugates. J Neurosci Meth 25:1–11

23. Kita H, Armstrong W (1991) A biotin-containing compound N-(2-aminoethyl)biotinamide for intracellular labeling and neuronal tracing studies: comparison with biocytin. J Neurosci Meth 37:141–150

24. Dumitriu D, Rodriguez A, Morrison JH (2011) High-throughput, detailed, cell-specific neuroanatomy of dendritic spines using microinjection and confocal microscopy. Nat Prot 6:1391–1411

25. Blackman A, Grabuschnig S, Legenstein R, Sjöström PJ (2014) A comparison of manual reconstruction from biocytin histology or 2-photon imaging: morphometry and computer modeling. Front Neuroanat 8:65

26. Murphy DB, Davidson MW (2013) Fundamentals of light microscopy and electronic imaging, 2nd edn. Wiley-Blackwell, Hoboken

27. Denk W, Strickler JH, Webb WW (1990) Two-photon laser scanning fluorescence microscopy. Science 248:73–76

28. Denk W (2011) Introduction to multiphoton-excitation fluorescence microscopy. In: Yuste R (ed and series ed) Imaging. A laboratory manual. Cold Spring Harbor Laboratory Press, Cold Spring Harbor, pp 105–110

29. Tashiro A, Aaron G, Aronov D, Cossart R, Dumitriu D, Fenstermaker V, Goldberg J, Hamzei-Sichani F, Ikegaya Y, Konur S, MacLean J, Nemet B, Nikolenko V, Portera-Cailliau C, Yuste R (2006) Imaging brain slices. In: Pawley JB (ed) Handbook of biological confocal microscopy, 3rd edn. Springer, New York, pp 722–735

30. Groh A, Krieger P (2011) Structure-function analysis of genetically defined neuronal populations. In: Helmchen F, Konnerth A (eds) Yuste R (series ed) Imaging in neuroscience. A laboratory manual. Cold Spring Harbor Laboratory Press, Cold Spring Harbor, pp 377–386

31. Zandt B-J, Veruki ML, Hartveit E (2018) Electrotonic signal processing in AII amacrine cells: compartmental models and passive membrane properties for a gap junction-coupled retinal neuron. Brain Struct Funct 223:3383–3410

32. Major G (2001) Passive cable modeling – a practical introduction. In: De Schutter E (ed) Computational neuroscience. Realistic modeling for experimentalists. CRC Press, Boca Raton, pp 209–232

33. Holmes WR (2010) Passive cable modeling. In: De Schutter E (ed) Computational modeling methods for neuroscientists. MIT Press, Cambridge, pp 233–258

34. Donohue DE, Ascoli GA (2011) Automated reconstruction of neuronal morphology: an overview. Brain Res Rev 67:94–102

35. Acciai L, Soda P, Iannello G (2016) Automated neuron tracing methods: an updated account. Neuroinformatics 14:353–367

36. Losavio BE, Liang Y, Santamaría-Pang A, Kakadiaris IA, Colbert CM, Saggau P (2008) Live neuron morphology automatically reconstructed from multiphoton and confocal imaging data. J Neurophysiol 100:2422–2429

37. Cuntz H, Forstner F, Borst A, Häusser M (2010) One rule to grow them all: a general theory of neuronal branching and its practical application. PLoS Comput Biol 6:1–14

38. Cuntz H, Forstner F, Borst A, Häusser M (2011) The TREES toolbox – probing the basis of axonal and dendritic branching. Neuroinformatics 9:91–96

39. Myatt DR, Hadlington T, Ascoli GA, Nasuto SJ (2012) Neuromantic – from semi-manual to semi-automatic reconstruction of neuron morphology. Front Neuroinform 6:4

40. Feng L, Zhao T, Kim J (2014) neuTube 1.0: a new design for efficient neuron reconstruction software based on the SWC format. eNeuro. https://doi.org/10.1523/ENEURO.0049-14

41. Zandt B-J, Losnegård A, Hodneland E, Veruki ML, Lundervold A, Hartveit E (2017) Semi-automatic 3D morphological reconstruction of neurons with densely branching morphology: Application to retinal AII amacrine cells imaged with multi-photon excitation microscopy. J Neurosci Meth 279:101–118

42. Neher E (1992) Correction for liquid junction potentials in patch clamp experiments. In: Rudy B, Iverson LE (eds) Ion channels, Methods in enzymology, vol 207. Academic Press, San Diego, pp 123–131

43. Heintzmann R (2006) Band limit and appropriate sampling in microscopy. In: Celis JE (ed) Cell biology. A laboratory handbook, vol 3. Elsevier, Amsterdam, pp 29–36

44. Zandt B-J, Liu JH, Veruki ML, Hartveit E (2017) AII amacrine cells: quantitative reconstruction and morphometric analysis of electrophysiologically identified cells in live rat retinal slices imaged with multi-photon excitation microscopy. Brain Struct Funct 222:151–182

45. Pologruto TA, Sabatini BL, Svoboda K (2003) ScanImage: flexible software for operating laser scanning microscopes. Biomed Eng Online 2:13

46. Dorostkar MM, Dreosti E, Odermatt B, Lagnado L (2010) Computational processing of optical measurements of neuronal and synaptic activity in networks. J Neurosci Meth 188:141–150

47. Cannell MB, McMorland A, Soeller C (2006) Image enhancement by deconvolution. In: Pawley JB (ed) Handbook of biological confocal microscopy, 3rd edn. Springer, New York, pp 488–500

48. van der Voort HTM, Strasters KC (1995) Restoration of confocal images for quantitative image analysis. J Microsc 178:165–181

49. Scorcioni R, Polavaram S, Ascoli GA (2008) L-measure: a web-accessible tool for the analysis, comparison and search of digital reconstructions of neuronal morphologies. Nat Prot 3:866–876

50. Wouterlood FG, Beliën JAM (2014) Translation, touch, and overlap in multi-fluorescence confocal laser scanning microscopy to quantitate synaptic connectivity. In: Bakota L, Brandt R (eds) Laser scanning microscopy and quantitative image analysis of neuronal tissue, Neuromethods, vol 87. Springer, New York, pp 1–36

51. Cox G, Sheppard CJR (2004) Practical limits of resolution in confocal and non-linear microscopy. Microsc Res Tech 63:18–22

52. Sigal YM, Speer CM, Babcock HP, Zhuang X (2015) Mapping synaptic input fields of neurons with super-resolution imaging. Cell 163:493–505

53. Tsukamoto Y, Omi N (2013) Functional allocation of synaptic contacts in microcircuits from rods via rod bipolar to AII amacrine cells in the mouse retina. J Comp Neurol 521:3541–3555

54. Sethian JA (1996) A fast marching level set method for monotonically advancing fronts. Proc Natl Acad Sci U S A 93:1591–1595

55. Hodneland E, Kögel T, Frei DM, Gerdes HH, Lundervold A (2013) CellSegm – a MATLAB toolbox for high-throughput 3D cell segmentation. Source Code Biol Med 8:16

56. Schindelin J, Arganda-Carreras I, Frise E, Kaynig V, Longair M, Pietzsch T, Preibisch S, Rueden C, Saalfeld S, Schmid B, Tinevez JY, White DJ, Hartenstein V, Eliceiri K, Tomancak P, Cardona A (2012) Fiji: an open-source platform for biological-image analysis. Nat Methods 9:676–682

57. Weickert J (1997) A review of nonlinear diffusion filtering. In: ter Haar Romeny B, Florack L, Koenderink J, Viergever M (eds) Scale-space theory in computer vision, Lecture notes in computer science, vol 1252. Springer, Berlin, pp 3–28

58. Cannon RC, Turner DA, Pyapali GK, Wheal HV (1998) An on-line archive of reconstructed hippocampal neurons. J Neurosci Meth 84:49–54

59. Lee T-C, Kashyap RL, Chu C-N (1994) Building skeleton models via 3-D medial surface/axis thinning algorithms. CVGIP: Graph Models Image Process 56:462–478

60. Prim RC (1957) Shortest connection networks and some generalizations. Bell Syst Technol J 36:1389–1401

61. Paglieroni DW (1992) Distance transforms: properties and machine vision applications. CVGIP: Graph Models Image Process 54:56–74

62. Maurer CR, Qi R, Raghavan V (2003) A linear time algorithm for computing exact Euclidean distance transforms of binary images in arbitrary dimensions. IEEE Trans Pattern Analysis and Machine Intelligence 25:265–270

63. Peng H, Ruan Z, Long F, Simpson JH, Myers EW (2010) V3D enables real-time 3D visualization and quantitative analysis of large-scale biological image data sets. Nature Biotech 28:348–353

64. Peng H, Long F, Zhao T, Myers G (2011) Proof-editing is the bottleneck of 3D neuron reconstruction: the problem and solutions. Neuroinformatics 9:103–105

65. Peng H, Bria A, Zhou Z, Iannello G, Long F (2014) Extensible visualization and analysis for multidimensional images using Vaa3D. Nat Prot 9:193–208

66. Koch C, Segev I (2000) The role of single neurons in information processing. Nat Neurosci 3 Suppl:1171–1177

67. De Schutter E, Steuber V (2001) Modeling simple and complex active neurons. In: De Schutter E (ed) Computational neuroscience: realistic modeling for experimentalists. CRC Press, Boca Raton, pp 233–257

68. De Schutter E, van Geit W (2010) Modeling complex neurons. In: De Schutter E (ed) Computational modeling methods for neuroscientists. MIT Press, Cambridge, pp 259–283

69. Carnevale NT, Hines ML (2006) The NEURON Book. Cambridge University Press, Cambridge

70. Schneider CJ, Cuntz H, Soltesz I (2014) Linking macroscopic with microscopic neuroanatomy using synthetic neuronal populations. PLoS Comput Biol 10:e1003921

71. Evers JF, Schmitt S, Sibila M, Duch C (2005) Progress in functional neuroanatomy: precise automatic geometric reconstruction of neuronal morphology from confocal image stacks. J Neurophysiol 93:2331–2342

72. Helmstaedter M, Briggman KL, Denk W (2011) High-accuracy neurite reconstruction for high-throughput neuroanatomy. Nat Neurosci 14:1081–1088

73. Helmstaedter M, Briggman KL, Turaga SC, Jain V, Seung HS, Denk W (2013) Connectomic reconstruction of the inner plexiform layer in the mouse retina. Nature 500:168–174

74. Kroon DJ (2009) Accurate Fast Marching toolbox for MATLAB. www.mathworks.com/matlabcentral/fileexchange/24531-accurate-fast-marching

Chapter 9

Serial Multiphoton Tomography and Analysis of Volumetric Images of the Mouse Brain

Denise M. O. Ramirez, Apoorva D. Ajay, Mark P. Goldberg, and Julian P. Meeks

Abstract

Mapping the structural and synaptic organization of the central nervous system is fundamental to a principled understanding of neural circuit development and function and an important goal of modern neuroscience. A plethora of new imaging technologies and computational advances have made whole-brain mapping studies more widely accessible. One such volumetric imaging method is known as serial two-photon tomography (STPT). STPT is an automated block-face imaging method in which a brain or other whole-organ specimen is repetitively imaged using multiphoton illumination and physically sectioned using an integrated vibratome. The resultant tile images are stitched in two dimensions to form mosaic whole-section images, and the mosaic images need only be stacked in three dimensions to generate a whole-brain volumetric image. Automated image analysis pipelines may then be employed to mine quantitative information at the whole-brain scale across large cohorts of experimental animals. Here, we describe our methods optimized in the University of Texas Southwestern Whole Brain Microscopy Facility for STPT using the TissueCyte1000 platform and a custom pipeline for whole-brain image analysis including registration into the Allen Institute Common Coordinate Framework version 3.0 (CCF 3.0). Included is a description of the inclusion of supervised machine learning using a voxel-wise random forest model for classification of features of interest, including cell bodies and subcellular structures. The rapidly advancing pace of STPT and other complementary methods for whole-brain mapping and systematic analysis has the potential to generate transformative insights into brain circuitry in both health and disease.

Key words TissueCyte, Volumetric imaging, Automated image analysis, Block-face imaging, Serial two-photon tomography, Connectomics

1 Introduction

Three-dimensional (3D) reconstruction of intact organs can promote understanding of how they function in both health and disease, but the acquisition of volumetric whole-organ images has historically been difficult, if not impossible. In perhaps no other organ is this undertaking more appropriate than in the brain, since mapping specific connections between distant brain regions requires 3D reconstruction.

Espen Hartveit (ed.), *Multiphoton Microscopy*, Neuromethods, vol. 148,
https://doi.org/10.1007/978-1-4939-9702-2_9, © Springer Science+Business Media, LLC, part of Springer Nature 2019

"Connectomics," a popular term that describes research on brain connectivity, spans several scales [1, 2]. For example, some groups have focused on precisely mapping every synaptic connection within a small volume of neural tissue [3–5]. Others have focused on identifying functionally connected regions in the human brain based on temporal fluctuations in blood oxygen level-dependent signals in functional magnetic resonance imaging (BOLD fMRI) signals [6, 7]. Between these spatial extremes of connectomics is the so-called "meso" scale, which involves tracing the connections of populations of neurons across distant brain regions [1, 2, 4, 8, 9].

Advances in automated imaging and tissue sectioning technologies, including fast laser scanning two-photon microscopy [10], structured illumination microscopy (SIM; [11, 12]), high-precision piezoelectric positioners, and sectioning technology, have been combined into powerful instruments capable of performing serial tomographic imaging in intact tissue samples the size of whole rodent organs [13–17] ideal for mesoscale brain imaging. These technologies, combined with increasingly powerful computational resources and software [18–27], make it possible to generate whole-brain data at subcellular resolution with high throughput, enabling comprehensive assessment of potential new preventative or therapeutic agents for brain injury and disease [28].

Mapping the connections from one brain region to another is not a new achievement, as Golgi stain and tracer-based studies have, over many decades, built up the anatomical foundation of neuroscience research (reviewed in [29]). These techniques, though powerful, lack the precision available with modern genetic tools. For example, it is now commonplace for neuroscience laboratories to utilize genetic tools to label specific populations of neurons, for example, by expressing Cre recombinase under the control of a gene expressed by specific subsets of neurons [30, 31]. By introducing Cre-dependent fluorescent reporters into specific brain regions, one can now fluorescently label the entire cytoplasm (including dendrites, axons, and synaptic terminals) of a specific type of neuron in a specific region of the brain (e.g., using Cre-dependent adeno-associated viruses (AAV) (reviewed in [32]). These and other powerful new approaches have spurred a renaissance in brain anatomy research [14, 15, 33–37] that are beginning to produce new insights into brain function in both healthy and diseased states [8, 9, 13, 38–44].

It is now known that in brain disease and injury, damage may extend to areas distant from a localized lesion because long range axonal connections are also lost or impaired [45–47]. With traditional methods, most laboratories balance the ideal and the practical, often performing high-resolution imaging in just a few regions in several animals or performing whole-brain, high-resolution imaging in just a few individuals. It is becoming increasingly clear that fully assessing the therapeutic potential of new treatments in brain injury and disease will require assessment of brain structure at both the local and global levels; "the forest and the trees" [1]. The rise of accessible, high-throughput mesoscale volumetric brain imaging resources [8, 13, 25, 48–50] is enabling a new revolution in experimental design and analytical power related to brain research, with the goal of improving our understanding of brain function in health and disease.

1.1 Methods Used for Mesoscale Brain Imaging

1.1.1 Traditional (Slide Mounted) Imaging

Historically, neuroscience laboratories have used long-established slide-based methods for assessing brain morphology and pathology since the time of Ramon y Cajal. The tools for cutting and staining thin brain sections are common in most laboratories, and the increasing availability of automated whole slide imaging (WSI) instrumentation renders imaging large numbers of serial sections feasible. Although WSI imaging techniques are extremely useful for improving the throughput of slide-mounted brain samples, reconstruction of a whole-brain volume using any slide-based technique will necessarily suffer from artifacts of tissue processing (e.g., due to sections folding or tearing) and the difficulty of precisely aligning the section images together. Additionally, such methods for assessing large regions of healthy and diseased tissue are extremely labor intensive and expensive. As a result, broad and unbiased investigations into neuronal and circuit dysfunction over the entire brain across many experimental subjects have been rare. As we describe in detail below, these techniques, utilizing either optical clearing or physical sectioning methods, now allow one to interrogate the whole brain in 3D at single cell or even subcellular resolution without using traditional slide-based imaging methods.

1.1.2 Cleared Tissue Imaging

Optically clearing an organ of interest by making the refractive index uniform can allow volumetric imaging of the entire organ without the need to prepare thin tissue sections, due to the elimination of excitation light scattering present when imaging into opaque tissue. The first tissue clearing protocols involved the use of high refractive index organic solvents [51]. Such methods are quite efficient at clearing the tissue but are associated with significant tissue shrinkage due to dehydration and do not typically maintain endogenous fluorescent signals. In recent years, new tissue clearing [35, 36, 52–54] and light sheet-based imaging techniques [49, 55–57] have emerged which are greatly expanding the utility of high-resolution whole-organ imaging methods in terms of increased image resolution, decreased acquisition time, more streamlined protocols, and the ability to clear and image difficult organs such as the spleen, bone, or even the whole mouse body [34, 54]. Because volumetric imaging of the brain, allowing accurate reconstruction of long distance neuronal circuits, is arguably one of the most useful applications of tissue clearing, a large number of methods have been optimized for this tissue [34–36, 52, 54, 58]. Volumetric whole-organ imaging via tissue clearing has the notable benefit of allowing whole-mount immunostaining with traditional mammalian antibodies prior to imaging in some cases, as well as the ability to acquire images much faster via light sheet imaging than with confocal or multiphoton excitation. Nevertheless, tissue clearing methods can be challenging and/or expensive to implement in an individual laboratory, and tissues subjected to clearing often display morphological changes such as increases or decreases in organ size which can be non-isotropic in nature [34, 59].

1.1.3   Serial Two-Photon Tomography (STPT)

Physically sectioning and imaging throughout a tissue and reconstructing the 3D volume post hoc is the other available strategy for collecting whole-organ images. The TissueCyte 1000 serial two-photon system is designed for automated volumetric imaging of intact, uncleared organs [8, 13, 41, 60–65]. Indeed, the STPT method is limited only to tissues which are amenable to vibratome sectioning and structures that can be fluorescently labeled as described below. STPT overcomes many of the obstacles inherent in both traditional slide-based and cleared tissue imaging methods by imaging fluorescence just beneath the surface of a sample, then making a precise cut with a built-in vibrating microtome, and repeating this process until the entire sample has been imaged. The TissueCyte 1000 instruments are the first commercially available instruments to implement STPT, but additional tomographic imaging platforms such as the fluorescence micro-optical sectioning tomography (fMOST) system using widefield structured illumination [16, 48, 66, 67] are now available. Unlike volumetric imaging techniques which require optically clearing the sample, the STPT procedure utilizes standard paraformaldehyde-fixed brains, making this imaging modality readily accessible to laboratories accustomed to traditional sectioning methods. Fixed brain samples are embedded in an oxidized agarose gel and transferred to the fully motorized TissueCyte sample chamber. Block-face imaging eliminates issues associated with warping and section alignment, and greatly facilitates downstream processing of the images into a full 3D brain rendering. The microscope is completely automated after the sample has been installed. At typical resolutions, which have <1 μm precision in the x and y dimensions and 75 μm in the axial dimension, the entire mouse brain can be imaged in 10 hours, generating about 70 gigabytes (GB) of raw data. Samples can be imaged at up to 0.3 μm resolution in the x and y dimensions and axial resolutions of <5 μm with longer acquisition times and increased data storage needs (up to many TB per brain). The ability to increase resolution at the expense of acquisition time allows STPT data sets to be used to inform a variety of scientific questions. For example, brain-wide changes in cell density across regions of interest [60, 61, 65] can be studied at the standard resolution. At higher resolutions (voxel size approaching 1 μm³), more detailed information including axon trajectories and dendritic spine analysis could be obtained. In conjunction with continually maturing robust, automated computational tools, STPT-based methods can effectively generate massive amounts of high-resolution, whole-brain data to probe circuit changes in models of brain injury, neurodegenerative or psychiatric conditions. Here, we describe our methods for volumetric imaging of mouse brains using STPT and a custom-designed computational pipeline for image registration into the Allen Institute for Brain Science's Common Coordinate Framework, machine learning-based pixel classification and automated quantification of features of interest.

2 Materials

2.1 Agarose Embedding

1. Mouse brain or other tissue containing inherent fluorescent label of choice.

2. 4% Paraformaldehyde in Phosphate Buffered Saline (PBS) (Affymetrix, #199431LT).

3. 0.01% (w/v) NaN_3 (Sigma, #S2002) in PBS (Fisher Scientific, #BP39920); toxic, irritant, and environmental hazard. Use personal protective equipment when handling NaN_3 solid and solutions and check your local regulations for disposal of the solution.

4. Agarose (Type 1-A, Sigma, #A0169).

5. $NaIO_4$ (Sigma, #S1878, MW 213.89 g/mol); light-sensitive, hygroscopic, toxic; use in chemical fume hood.

6. Phosphate buffer (PB; 0.42 g/l monobasic sodium phosphate (NaH_2PO_4), 0.92 g/l dibasic sodium phosphate (Na_2HPO_4)). Must be filtered and degassed prior to use.

7. Borax ($Na_2B_4O_7 \cdot 10\ H_2O$; Sigma, Catalog # 221732, MW 381.37 g/mol); 0.05 M solution = 19 g/l.

8. Boric acid (H_3BO_3, Sigma, Catalog # B6768, MW 61.83 g/mol); 0.05 M solution = 3 g/l.

9. Sodium borohydride ($NaBH_4$; Sigma, Catalog # 452882, MW 37.83 g/mol); toxic, use in chemical hood.

10. Peel-a-way embedding molds, 22 × 22 × 20 mm (VWR, Catalog # 15160-215).

11. Frozen ice blocks (2), such as used for shipping.

12. 1 oz. (approximately 30 ml) wide-mouth glass jars (Uline #S-17073 M).

13. Vacuum filtration system (0.22 μm membrane, 1 l capacity bottle; Millipore Stericup-GP #SCGPU10RE).

14. Dessicating chamber (Fisher # 08-594-16C) and connection to house vacuum line.

2.2 Specimen Installation

1. Rectangular neodymium magnets (Indigo Instruments, #44226-2.5).

2. Glass microscope slides with frosted writing surface on one end (any type of charged or uncharged slide can be used).

3. Epoxy (5 minutes, Loctite #1365868 or similar); use in a well-ventilated area or in a chemical fume hood.

4. 1 ml pipette tips.

5. Superglue (water resistant, Loctite Ultra Liquid control #1647358).

6. PB, filtered and degassed.

7. Unbacked single edge razor blades (Razor Blade Company, #94-0180).

2.3 Image Acquisition

1. TissueCyte 1000 serial two-photon tomography system (TissueVision, Inc.; www.tissuevision.com) equipped with integrated precision vibratome, Nikon 16×, 0.8NA, 3 mm working distance water immersion lens, precision x-y-z stage to enable mosaic tiling and piezo-controlled objective height to enable optical sectioning. Three emission bands (<500 nm, 500–560 nm, and >560 nm) are detected by independent photomultiplier tubes (PMTs). Custom software (Orchestrator, TissueVision, Inc.) running on dedicated integrated workstations in a master-slave configuration controls the STPT system.

2. Spectra-Physics MaiTai Deep See tunable Ti:Sapphire near-infrared multi-photon excitation laser with dispersion compensation operating from 690 to 1040 nm. Custom software (MaiTai GUI version 2.00.23) controls the laser.

3. Network Attached Storage (NAS) and dedicated workstation running custom software (Autostitcher, v0.7.18; TissueVision, Inc.) enabling near real-time processing and stitching of the raw image tiles into full 2D mosaic TIF images.

2.4 Image Analysis

1. Dedicated workstation-class computer with >128 GB RAM or access to high-performance computing (HPC) cluster. Software environment supporting FIJI/ImageJ [68, 69], MATLAB (Mathworks), Python (3.4.X/Anaconda), and ilastik [70].

2. 10 GB/second networking between data stitching computers and analysis workstation (s).

3. "Average template" and "annotation" files defining the Allen Institute for Brain Science's Common Coordinate Framework, v3.0 [8].

4. GPU-enabled workstation for image processing/visualization/3D rendering (e.g., at least Windows 7 64-bit, Xeon E5-1660 v2 @3.7GHz processor, Nvidia Quadro K2000 (2 GB) graphics card, 96 GB RAM, 8 TB hard drive).

5. Suitable open source or commercial package to be used for 3D rendering/visualization of STPT whole-brain datasets (e.g., ClearVolume plugin for FIJI [71], Vaa3D [21], Imaris (Bitplane), arivis Vision4D (arivis AG), Amira (FEI)).

6. Hard drives for data backup and long-term storage (~0.5–1 TB per imaged sample).

3 Methods

3.1 Suitable Animal Models/Labeling Techniques

STPT is compatible with a wide variety of labeling strategies in which the fluorescent label of interest is either inherent to the tissue or can be introduced into the intact uncleared, fixed organ. Unlike many optical clearing methods, this method is typically not directly compatible with whole-mount immunostaining using traditional mouse or rabbit antibodies (but see [72]), though camelid single-domain antibodies (sdAbs, also known as nanobodies) [73–75] hold promise for potential use in whole-organ immunostaining and subsequent STPT due to their increased tissue permeability. Labeling methods that have been successfully used in our hands include transgenic expression of fluorescent proteins including GFP, YFP, tdTomato, and mCherry, injection of retrograde and anterograde viral tracing agents (i.e., AAV or pseudorabies virus; PRV) driving fluorescent protein expression, and injection of biotinylated dextran amine (BDA) tagged with Alexa Fluor dyes. In vivo labeling of amyloid beta plaques using methoxy-X04 injection has been recently used in conjunction with STPT imaging [41]. A variety of published methods for whole-mount fluorescent labeling of the vasculature which are retained after PFA fixation [39, 50, 76] are also compatible with this imaging modality. It is noteworthy to consider that the native autofluorescence of brains or other organs provides a great deal of structural detail in STPT images even in the absence of specific labels. We have successfully used such autofluorescence images for verification of electrode placement following long-term in vivo recording studies as well as in morphological studies of human brain specimens (data not shown). Additionally, the inherent autofluorescence signals are used for image registration as described in detail below. In terms of non-brain organs, the 3D organization of any feature of interest that can be suitably labeled, such as tumor growth and metastasis or peripheral nerve anatomical mapping, could be investigated with this imaging modality.

3.2 Agarose Embedding

Brains or other organs must be free of blood and very well fixed for use in STPT. While outside the scope of this chapter to describe in detail, a standard perfusion protocol is used. Briefly, under deep anesthesia, PBS is first perfused transcardially to remove blood, then 4% PFA in PBS is perfused and the tissues are dissected. Following tissue removal, samples should be further immersion fixed overnight at 4 °C. The fixed tissue is then transferred to PBS + 0.01% NaN$_3$ for storage at 4 °C. Brains or other organs are then embedded in an oxidized agarose matrix for support during vibratome sectioning.

Covalent cross-linking of the brain with the agarose matrix is important to keep the tissue firmly embedded during sectioning and to limit shadowing artifacts in the image by insufficiently cut meninges (see Sect. 4). Covalent linkage between the brain surface and agarose is activated by equilibrating the agarose-embedded brain in a sodium borohydride/borate solution [13, 77]. Oxidized 4.5% agarose solution in 10 mM NaIO$_4$ is prepared by mixing the following in a 250 ml beaker: 2.25 g Type 1 low-melting point agarose, 0.21 g NaIO$_4$, and 100 ml of PB (pH = 7.4). 1× PB should be filtered and degassed prior to use. A 10× concentrated stock solution can be prepared and diluted as needed. PB should be filtered with vacuum filtration (Millipore 0.22 μm vacuum filtration system) and degassed under vacuum after filtration for at least 2–3 hours prior to use in imaging. A standard dessication chamber and house vacuum line or a vacuum chamber can be used for degassing (See Sect. 4). Gently stir the agarose solution for 2–3 hours at room temperature (RT) in a chemical fume hood, protected from light. Do not exceed 3 hours, or the agarose will polymerize poorly and form a brittle block. Filter solution with vacuum suction and a 0.22 μm Millipore filter bottle. Wash out all remaining NaIO$_4$ with PB (3 washes ×50 ml/wash). Remove filter from suction, and then resuspend washed agarose by adding 50 ml of PB in a clean beaker. The agarose will remain as a suspension and will not dissolve. Any oxidized agarose not used immediately can be stored at 4 °C in a light-protected container for up to 2 weeks. Each brain or other tissue to be embedded will require ~10 ml of oxidized agarose solution, so the recipe amounts given above will be sufficient to embed five brains.

Use a microwave to bring agarose to a boil, then cool down to 60–65 °C while stirring on a hot plate. Embed each brain in a Peel-a-way mold as follows. Dry the brain with Kimwipes to remove excess PBS. While most of the meningeal layers over the mouse brain are not generally retained with the brain following extraction from the skull, with the exception of the pia mater, care should be taken to remove visible meningeal tissue from adult mouse brains if noted prior to embedding. If necessary, trim cervical spinal cord off to level of cerebellum—the brain will be embedded caudal side down, so it is best to have a flat surface at the back. Place a labeled mold on a flat frozen ice block, such as that used for shipping. Make sure the surface of the ice block is leveled horizontally (by eye is OK, or you can use a bubble level). If needed, you can level the block using paper towels underneath the lower side. Pick up the mold and pipette melted oxidized agarose into the pre-cooled mold using a plastic Pasteur pipette. Slightly overfill the mold all the way to the top so that agarose extends above the top edges of the mold. Place the filled mold back on the ice block. Quickly pick up the brain with forceps in a vertical orientation, and plunge it into the mold all the way to the bottom with cerebellum touching the bottom of the mold and olfactory bulbs facing up toward you. The agarose should quickly start to solidify and the brain should stay upright as long as the agarose is not too hot, but you may need to support it for a few seconds near the top of the block to make sure it keeps a vertical orientation. If the brain is not embedded vertically but leans to one side in the block, the resulting coronal sections will be oblique. Allow the mold to sit undisturbed on the ice block for a few minutes before transferring to a second ice block (or 4 °C refrigerator) to completely harden for 15–20 minutes. The agarose blocks should be allowed to harden in the dark. Once the block is completely solid (appearance will change from semi-transparent to more opaque), shave off excess agarose from the top of the block using a razor blade to create a flat surface. Slice the mold open at each corner to remove the agarose block. Place each block in individual wide-mouth glass jars.

Finally, the agarose must be crosslinked to the brain [77]. First, prepare borate buffer by adding 19 g borax and 3 g boric acid to 1 l of water. Stir until dissolved. Adjust pH to 9.0–9.5 with 1 N NaOH. This stock solution can be kept in a bottle at room temperature indefinitely. Next, prepare a working solution by adding sodium borohydride to borate buffer. Add 0.2 g $NaBH_4$ to 100 ml of borate buffer in a fume hood. Vigorous gas formation (CO_2) will be observed. Stir for 15–30 min at room temperature. Protect bottle from light. Leave the uncapped bottle in the hood overnight to allow the solution to degas. Do not cap the bottle—if capped it may explode due to CO_2 formation. Tighten the cap the next morning. Importantly, you should not use the solution the same day as it is made as gas pockets will form inside the agarose block and make it into a porous spongy structure instead of a solid block, which will severely reduce sectioning quality. The sodium borohydride solution can be stored for up to 1 week but using a fresher solution is better for covalent chemistry. Fill each jar containing the embedded brain blocks with borohydride-borate solution. Approximately 20 ml is needed for each block. Make sure the block is completely submerged in solution. Leave in solution overnight at 4 °C (no shaking; protected from light). Specimen crosslinking should be performed at least overnight but can be done over the weekend if desired (embedding on Friday and removal of borohydride-borate solution on Monday morning). The following morning, discard the used sodium borohydride-borate buffer according to your local regulations for hazardous waste disposal and replace with filtered, degassed PB. The blocks can now be stored at 4 °C until imaging, protected from light.

3.3 Specimen Installation

The agarose-embedded specimen block is attached to a magnetic slide, which in turn is used to attach the specimen to the magnetic plate at the bottom of the sample bath. The magnetic slides should be prepared ahead of time, and many slides can be made in a batch and stored for later use. In order to prepare magnetic slides, a long strip or sheet of magnetic metal can be used, or you can work on a metal table or filing cabinet. Dispense some of the two epoxy compounds from the syringe onto a disposable container or piece of paper, and mix using a plastic application tool (e.g., a 1 ml pipette tip). Take one glass microscope slide and place it on the benchtop or metal surface with the frosted writing surface facing down. Using the plastic tool, smear a small bit of epoxy approximately the size of the magnet close to the narrow edge of the slide on the side with the frosted writing surface. Place the magnet into the epoxy spot—it will be attracted to the metal behind the glass slide and thus stay in place. Continue placing slides along the strip with a single magnet at the top to make more slides. Allow the epoxy around the single magnet to solidify. Repeat the procedure, placing a second magnet beside the first, as close as possible without allowing the magnet to touch the first magnet (~1–1.5 cm distance). Two magnets will be glued using epoxy to each glass slide. The key is to have the metal surface hold the magnet in place, and keep it from attaching to the first magnet. Once you have a set of magnetized slides prepared, they can be used as templates to prepare more slides without having to work on a metal surface.

Next, the agarose block is attached to the magnetic slide, and the phosphate buffer bath is prepared. Remove the block from the jar containing PB. Dry the block with Kimwipes. Pay special attention to drying the sides and bottom of the block. Apply a drop of superglue to the frosted writing surface of the magnetic slide, which should be the end of the slide where the magnets are attached underneath, and place the block on the drop. Press down gently on the block to make sure there is good contact between the block and the slide. Using the frosted surface of the slide to attach the block allows for better adherence of the superglue to the agarose block. For brains, the block should be oriented so the cerebellum is facing upwards and the olfactory bulbs are on the bottom nearest the slide. Allow the superglue to dry for ~15 minutes at RT. Do not leave the block out of solution for more than an hour as the water in the agarose will evaporate and the block will begin shrinking.

The magnet slide with agarose block attached should now be installed on the central metal plate of the specimen bath (See Fig. 1). Care should be taken to align the sample as close to the middle of the bath as possible in both the x and y dimensions. A ruler can be used for this purpose. For brain imaging, the ventral surface of the brain should be oriented to the left side of the bath. The bath should be carefully filled with 1 l of filtered and degassed

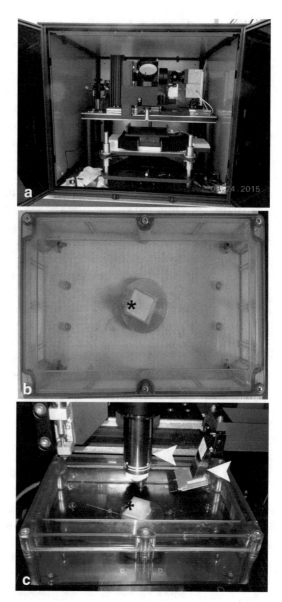

Fig. 1 Installation of agarose-embedded brain sample for serial two-photon imaging. (**a**) Front view inside the enclosure of the TissueCyte 1000 system. (**b**) Top-down view of sample bath showing properly installed embedded brain sample (asterisk) attached to a magnetized slide on the round metal plate in the center of the bath. Note the position of the brain: caudal side up, ventral side to the left and closely aligned to the midpoint of the bath in the x and y dimensions. (**c**) Placement of sample and bath on the TissueCyte 1000 stage in preparation for imaging. Note the position of the embedded brain (asterisk) directly underneath the objective (left white arrow), as well as the vibratome blade and blade holder (right white arrow)

PB and slid into place on the stage of the TissueCyte 1000. Make sure the laser shutter is closed, and the PMTs are off prior to opening the TissueCyte doors to place a new sample. Once the bath is anchored lightly in place with the red thumbscrew on the left side of the stage, an additional ~200 ml of PB is added to the bath. The razor blade is now installed in the blade holder and the blade holder attached to the vibratome arm using the provided hex screwdrivers. Razor blades obtained from Razor Blade Company as listed above do not require degreasing. A fresh razor blade is used for each specimen. Be sure to check the beveled edge of each blade prior to installation to make sure it is not nicked or damaged as this will disrupt the sections. The block is then centered directly beneath the objective using the x-y controls in the Orchestrator control software and advanced upward to the level of the plane of section by carefully moving the block up in the z dimension. As the block approaches the level of the vibratome blade, the objective will begin to dip into the bath solution. Check from the front of the bath whether any bubbles have become trapped underneath the objective. If so, you can use a transfer pipette to displace them. Also, be careful not to advance the block up too far in order to avoid physically contacting the objective. It is helpful to use a flashlight to see the block and bath better once in place on the TissueCyte stage.

3.4 Image Acquisition

The TissueCyte collects a series of image tiles which are stitched in 2D into whole-section images (see below). The sequence of events that occur during image acquisition—scanning, x-y stage movement, physical and optical sectioning, and data storage—are coordinated by the Orchestrator program. The mosaic size, lateral resolution, laser shutter voltage, pixel shift, and sectioning parameters including optical sectioning will all be adjusted appropriately in Orchestrator prior to an acquisition run for each individual sample. The number of mosaic tiles and the dimensions of the mosaic are dependent on the sample size as well as the specific objective installed. For mouse brain, we typically use a 9×13 mosaic. The lateral resolution can be specified in the Orchestrator program at three levels: 0.875 µm/px, 0.4375 µm/px, and 0.35 µm/px. The flexibility of the TissueCyte platform allows one to increase the resolution as needed at the expense of acquisition time and dataset size. We find that the lowest lateral resolution available in our system configuration, 0.875 µm/px, provides sufficient detail to appreciate individual neuronal processes (Fig. 2) and is naturally the most efficient in terms of acquisition time and data storage required.

To start an imaging run, sectioning is initiated at the desired thickness for trimming (usually 200–300 µm) until the sections begin to contain tissue. Trimming can be performed at a fast sectioning speed of 2 mm/second. After the tissue sections have

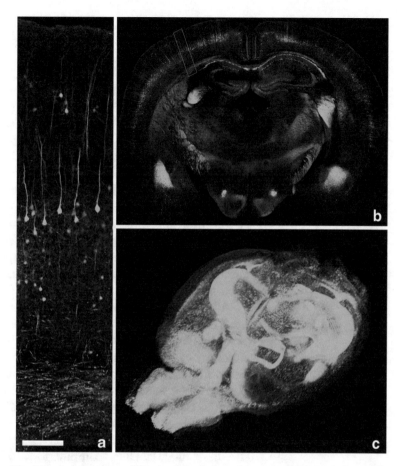

Fig. 2 Raw single channel fluorescence images from an adult YFP 2.2 transgenic mouse acquired using standard STPT procedures. (**a**) Representative field of view at native resolution (0.875 μm/px) of cortical layers 1–6. Robust YFP expression is visible in Layer 5 pyramidal neurons. Cell bodies and both dendritic and axonal projections can be resolved. Bar, 100 μm. (**b**) A single stitched mosaic two-dimensional coronal section image from the same YFP2.2 mouse brain. Boxed region corresponds to the area shown in (**a**). (**c**) Three-dimensional rendering of the whole brain

reached the desired plane, the section thickness is changed to the desired thickness for image acquisition (generally 75 μm) and three sections are collected to ensure the sections are produced at a consistent thickness. A fourth trim section is collected using the recommended brain sectioning speed of 0.5 mm/second. The last trim section should ideally contain brainstem and just start to show cerebellum (see Sect. 4). After ensuring the tissue is located directly under the objective (Fig. 1), the system doors should be closed and the room lights turned off once the trimming is complete. The PMTs are turned on and visible light (780 nm excitation) is used to determine the imaging plane by adjusting the position of the objective using the piezo controller and taking single tile snapshots. As you approach the surface of the block, the image will become dimmer and will display regular grooves from the vibratome.

The power of the excitation laser is modulated using an electro-optical device, referred to as an "excitation shutter" in the Orchestrator software. The excitation shutter voltage is set to allow for robust detection of the signal of interest without saturation and to detect sufficiently bright tissue autofluorescence to aid in visualization and analysis (see below). In practice, we image most samples at near maximal excitation shutter voltage and do not often observe oversaturation of fluorescently labeled structures. An excitation wavelength for imaging an individual fluorophore is chosen based on the appropriate two-photon cross section for that particular fluorophore. A range of excitation wavelengths spaced at short intervals can be tested during the image acquisition setup to empirically determine the most efficient excitation wavelength for a new fluorophore.

All three emission bands are automatically detected by dedicated PMTs for each imaging run. The PMT voltage settings are controlled by an independent custom GUI, and each channel's detector can be set to a different voltage if needed to accommodate differences in intensity between multiple signals of interest. Even in the absence of a specific signal of interest in one or more channels, the inherent autofluorescence of the tissue in each channel is useful for visualization and analysis (see below) by virtue of the significant amount of structural features visible in such images. The best dual labeling combinations of fluorophores for the TissueCyte using our specific emission filters (<500 nm KP, 500–560 nm BP, >560 nm LP) are GFP/tdTomato and Alexa Fluor 488/Alexa Fluor 594 due to the fact that they can be efficiently excited by the same wavelength but their emissions are spectrally separated. In practice, since the system is limited to one excitation laser, it is difficult to efficiently excite three such fluorophores simultaneously, but a triple labeling imaging experiment could be designed incorporating genetically expressed fluorescent proteins (e.g., Blue Fluorescent Protein, BFP; Evrogen) or dyes (e.g., Alexa Fluor 350, ThermoFisher Scientific) emitting in the blue range.

Proper sectioning parameters including speed, frequency, physical section thickness, and optical sectioning if desired must be specified for each image acquisition run. The TissueCyte utilizes a piezoelectric control for the objective lens that can be used to perform optical sectioning within the plane of an individual physical section if higher axial resolution is required (e.g., for tracing axons or localizing sparsely distributed cells) by virtue of its multiphoton excitation strategy allowing efficient penetration of photons hundreds of microns below the tissue surface. Typically, we collect images 50 μm below the tissue surface for single plane imaging using 75 μm physical sections and 25 μm, 50 μm, and 75 μm below the surface using 75 µm physical sections for increased axial resolution of 25 µm if needed. For mouse brain, a total of 190 coronal sections at 75 µm are imaged to encompass the whole-brain volume. The piezo controller must be set to the starting plane of imaging and the specifications for additional optical planes (distance between image planes and number of planes per physical section) is set in Orchestrator.

Once the imaging conditions have been set appropriately (see Sect. 4), the specimen is moved back and to the right relative to the center of the sample using the x-y controls such that the objective is located at the top left corner of the desired rectangular mosaic. It is convenient for mouse brain sectioned coronally to use the central canal and the border between the cerebellum and the brainstem to reference the sample position relative to the mosaic organization. If either of those anatomical landmarks is not present due to a poor dissection, or for imaging other types of tissue or alternative orientations of brain, the starting position of the mosaic must be determined visually. In this case, using slightly larger mosaic dimensions by adding additional rows or columns to the acquisition is advisable to compensate for less accuracy in referencing the sample position. The Orchestrator settings must all be thoroughly checked and saved before beginning an acquisition run, because these settings cannot be modified once the run is started. However, the absolute position of the objective can be adjusted using the piezo controller and both the excitation wavelength of the laser and the PMT voltages can be changed during an acquisition run if needed because they are controlled by separate GUIs.

Once the system begins to collect the 3D mosaic, the process is entirely automated, and no further user intervention is needed until the acquisition is completed. A mouse brain imaged at our standard resolution of $0.875 \times 0.875 \times 75$ µm takes around 10 hours. The room lights must be left off during acquisition because the enclosure is not completely sealed and turning on the room lights while the PMTs are on may damage them. Therefore it is best for the TissueCyte 1000 system to be installed in a dedicated room (see Sect. 4).

Upon completion of imaging, the PMTs must first be turned off, and the laser shuttered before opening the enclosure doors to remove the sample bath and vibratome blade. The stage should be lowered using Orchestrator controls to its maximal extent (3.7 cm) before removing the bath and blade. The blade holder is removed from the vibratome and the blade is removed and discarded safely in a sharps container. The bath is slid out and the magnet slide is discarded in a broken glass receptacle. The PB containing tissue sections can be strained using a sieve to recover sections for later use if desired. However, transferring the bath contents to a large tray in order to recover the intact sections of interest is recommended for best section quality to use in subsequent immunostaining experiments. Store sections in PB + 0.01% (w/v) NaN$_3$ at 4 °C if desired. The bath should be rinsed with Milli-Q water and stored upside down to dry in preparation for the next imaging run.

The TissueCyte 1000 system and the MaiTai DeepSee lasers require periodic maintenance tasks as described below. Our systems are typically run 2 to 4 days weekly. Full time imaging of 7 days per week may not be advisable for best laser performance in the long term, and the TissueCyte 1000 and lasers will likely require maintenance more frequently.

3.4.1 Align Lasers/Check Laser Power

The laser alignment should be checked regularly depending on system use and at any time the image quality is poor or images are dimmer than normal. The laser alignment can be checked visually outside the enclosure by tuning the laser to a visible wavelength such as 780 nm and using conveniently installed pinholes. The mirror angles can then be adjusted manually. Inside the enclosure, the laser alignment can be easily checked using the laser alignment module from TissueVision and appropriate mirrors are adjusted manually. A live camera feed and laser target are used to monitor laser mirror adjustments in real time. A laser power meter such as Newport 843-R should also be available to verify laser power at various points along the laser path if needed.

3.4.2 Clean Vibratome Bearings

It is necessary to periodically remove the vibratome bearings and re-grease them with the provided grease to maintain section quality. Depending on system usage, monthly or bimonthly greasing of the bearings should be sufficient. If section quality still remains poor after greasing the bearings, they should be replaced.

3.4.3 Replace Laser Chiller Algae Filter and Coolant

The MaiTai laser utilizes a chiller system to maintain the proper laser temperature (21 °C). Twice yearly removal and replacement of the coolant (distilled water, not deionized water) and the algae filter is required.

3.4.4 Clean Blade Holders

The vibratome blade holders can build up oxidation or mineral products from repeated immersion in PB. The blade holders can be cleaned by soaking in dilute phosphoric acid in a bath sonicator for several minutes followed by manual removal of the loosened buildup with a brush or razor blade. Two or more sonicating baths may be necessary to remove tough buildup. Only the part of the holder that contacts the blade (the silver blade clamps) should be immersed in solution, not the black part which is attached to the vibratome arm. Once the buildup has been removed the blade holders should be rinsed with Milli-Q water and allowed to dry.

3.5 Image Registration and Analysis Pipeline

We have developed a pipeline to analyze and extract quantitative results from raw whole-brain STPT image data, which has been custom-designed to emphasize speed and reproducibility. An overview of the image acquisition and analysis pipeline is depicted in schematic format in Fig. 3, and representative image data at each stage of the pipeline are shown in Fig. 4. The collected image tiles are saved directly on network attached storage as 16-bit TIF images. Each section consists of a 9×13 mosaic of image tiles of 832×832 pixels each at the standard resolution ($0.875\ \mu m/px$). The RGB image tiles collected for each physical and optical section are saved sequentially in folders for every physical section (generally 190 sections for a mouse brain sectioned coronally). The metadata related to image acquisition are written to individual text "mosaic" header files for each individual section as well as a combined "mosaic" text file containing metadata for the overall dataset.

3.5.1 2D Mosaic Stitching

The task of stitching image tiles into full 2D mosaic section images is performed by the Autostitcher software that is installed on dedicated workstations directly connected to network attached storage (see Fig. 2 and Sect. 4). Autostitcher runs as a daemon on the host computer, identifying the presence of a "trigger" file that initiates stitching. The software internally defines appropriate data structures based on the "mosaic" header file to arrange and retrieve tiles to be stitched into mosaic images. The user must specify input arguments to the Autostitcher program that identify the locations of the input directory containing the raw tiles and output directories to store the stitched mosaic images. The user must also specify stitching parameters, including the percentage of overlap between tiles (associated with choices in the Orchestrator image acquisition program; typically 5% overlap is used), the number of pixels to crop from each edge (typically 15px), and the number of tiles to

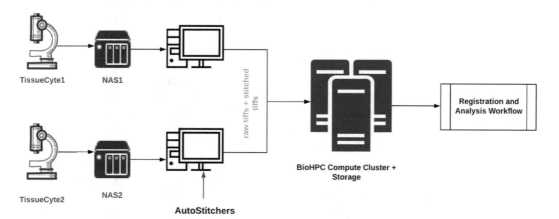

Fig. 3 Schematic overview of workflow for TissueCyte data acquisition and analysis. The raw image tiles are acquired on the TissueCyte1000 system, stored on NAS systems, and stitched into a mosaic two-dimensional tissue section image via the Autostitcher program running on dedicated workstations. The raw stitched two-dimensional section images are output to the compute cluster for processing, registration, and analysis

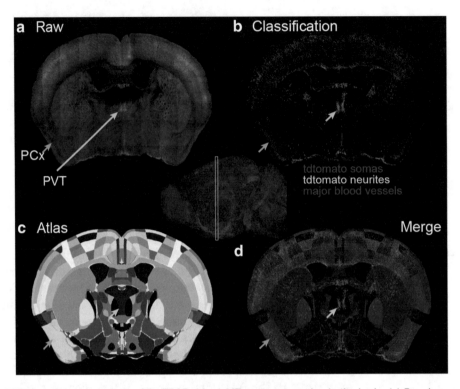

Fig. 4 STPT imaging and analysis of FosTRAP-driven tdTomato expression in the brain. (**a**) Raw image of red, green, and blue fluorescence of 1 of the 190 frames of a whole-brain STPT image registered to the Allen Institute Common Coordinate Framework version 3.0 (CCF 3.0). Orange and cyan arrows mark the piriform cortex (PCx) and paraventricular nucleus of the thalamus (PVT), respectively. (**b**) Supervised machine learning output (voxel-wise random forest model) identifying tdTomato-positive cell bodies (red), tdTomato-positive neurites (yellow), and major blood vessels (blue). Arrows mark same structures as in (**a**). (**c**) Annotated regions according to the CCF 3.0, with arrows marking structures as in (**a**). (**d**) Merged image overlaying the CCF 3.0 annotation on top of the machine learning output, with arrows marking structures as in (**a**). Center inset shows a three-dimensional rendering of the entire dataset, viewed from the left surface. The section highlighted in (**a–d**) is outlined by the white box. (Sample credit: Wen Mai Wong, Jie Cao, and Julian Meeks)

analyze to perform flat-field correction (typically at least 20% of the total tiles) [78]. These latter steps are necessary to compensate for illumination artifacts at the lateral edges of the scanned images, and slight differences in illumination intensity near the edges of the field of view. The corrected image tiles are then uniformly blended using feathering and alpha blending of adjacent tiles by Autostitcher [13, 78]. To assist in rapid evaluations, Autostitcher automatically outputs downsampled thumbnails of each mosaic section. Problematic datasets can be manually stitched using a manual version of the program (Manual Stitch) that maintains all of the features of the Autostitcher. For typical imaging runs, the Autostitcher produces complete stitching within several hours of the completion of the imaging (i.e., near-real time).

3.5.2 Mosaic Image Pre-processing

Mosaic sections produced by Autostitcher should be stored directly on a networked (i.e., mounted) storage system, including cloud storage via 10 GB/second networking (e.g., at UT Southwestern, the "BioHPC" multi-petabyte networked cloud storage system). Custom software can be written in MATLAB or Python to process the raw stitched sections for input into the registration pipeline described below. Critical steps include choosing the minimum resolution needed to resolve features of interest (allowing image downsampling, which dramatically reduces computational time) and compensation of the intensities of the red, green, and blue channels for the particular sample of interest. The latter is especially important when preparing for machine learning of image batches, because across-sample mismatches in the relative intensities of these color channels can preclude efficient model training (see below). Three-dimensional image stacks can be created in ImageJ, MATLAB, or Python for convenience at this stage. Image stacks can be further processed via Gaussian or median filtering to reduce the impact of noise. A convenient operation at this step is to downsample image stacks in the x- and y-directions to match the Allen Institute CCF 3.0, which has 10 μm^3 voxels in its highest-resolution version.

3.5.3 Image Registration

A significant benefit of using the TissueCyte 1000 platform over alternative volumetric imaging strategies (e.g., traditional slide-based serial sectioning and imaging, or optical clearing of brains followed by lightsheet imaging) is the relative ease of image registration. This is because the Allen Institute CCF 3.0 "average template" was created using the autofluorescence contrast in the "red" color channel of ~1600 mouse brains imaged using the TissueCyte 1000 [8]. The basic components of any registration framework consist of two input images (in this case, the CCF 3.0 average template and an experimental whole-brain image set; see Fig. 4), a transform function, an optimization metric, an interpolator, and an optimization function. Each software registration framework has each of these functionalities, which can be flexibly altered by users to achieve best results. In our current pipeline, we use the NiftyReg registration library that was developed within the Translational Imaging Group at the University College of London [79]. The specific commands used in our pipeline are *reg_aladin*, *reg_f3d*, and *reg_resample*. *Reg_aladin* is based on a block matching algorithm for global registration. This tool is capable of doing a combined rigid and affine transformation to optimize 12 degrees-of-freedom translation, rotation, scaling, and shearing. *Reg_f3d* uses cubic B-splines to deform the source image to optimize an objective function based on a similarity measure and a cost function (Normalized Mutual Information and bending energy, respectively, in this case). *Reg_resample* is used to resample the floating (experimental input image) with reference to the fixed image (atlas image) using the transformation file generated by *reg_aladin* and *reg_f3d*. Logistical steps in the image registration process include conversion of input images to a NiftyReg-compatible file format, execution of NiftyReg commands (*reg_aladin*, *reg_f3d*, and *reg_resample*), and conversion of output from Nifty to user-accessible image formats (e.g., TIF). These processes are highly conserved and are well-suited to software automation (e.g., via a nextflow workflow).

3.5.4 Machine Learning-Based Pixel Classification

Manual quantification of different signals of interest in large image datasets is a laborious process and can sometimes lead to counting errors, but automated quantification via machine learning-based pixel classification methods allows for faster and less biased image analysis. The main goal of a machine learning model is to learn to properly classify signals of interest from a subset of images in a representative dataset by assigning labels interactively. The same model can then be applied to all the images comprising the entire dataset (e.g., volumetric whole-brain data from large cohorts of animals) to obtain the probability maps of various objects of interest (such as fluorescently labeled cells). To effectively analyze our STPT data, we use an open-source toolkit called ilastik (Interactive Learning and Segmentation toolkit) [70] that uses voxel-wise random forest model-based machine learning techniques to classify pixels and objects by learning from "label" annotations, sparsely placed by the user, that represent the features present in the images that should be quantified [80, 81]. In our image analysis pipeline, we use the "Pixel Classification" function in ilastik to produce a voxel-wise map of predicted features (e.g., fluorescent cells or other features of interest; Fig. 4). To facilitate user interactivity (i.e., speed of model generation/training), images can be downsampled (for our workflow, images are downsampled to ~1.5 um/pixel lateral resolution). Users can then choose representative 2D sections from each sample in an experimental cohort as the input selection for model training. Ilastik allows several spatial features to be evaluated in model training (i.e., Gaussian filters, spatial derivatives of Gaussian filtered images, etc.). Once appropriate features have been identified, the user chooses appropriate labels for the desired components of the image worthy of quantification (e.g., black background, autofluorescence, fluorescent cell bodies, etc.). The user can then interactively train the random forest model by "painting" over image features that are presented in the ilastik graphical user interface. During model training, it is important to identify and label image features which are both desirable for quantification, such as fluorescent cell bodies, and those features which should ideally be excluded from quantification, such as tissue autofluorescence and high contrast noise artifacts. Typically, after an initial set of label definitions are established, the user inspects the sample images, correcting mis-assigned pixels until an acceptable model has been trained. In this process, we have found that model training and performance are optimized when label assignments are used sparingly (i.e., keeping the number of labeled pixels to a minimum, focusing on difficult conditions rather than easy ones, etc.). Once complete, ilastik can be used to export model probabilities for each label, transforming raw red, green, blue (RGB) values into a probability that each pixel belongs to a particular previously identified label (e.g., "black background," "autofluorescence," and "fluorescent cell bodies," etc.). These exported probabilities form a "probability map" that can be used for future quantification and 3D visualization. The major benefits of the machine learning step are enhancement of the signal-to-noise ratio and elimination of common artifacts associated with STPT imaging on the TissueCyte 1000 such as microbubbles or autofluorescence due to paraformaldehyde fixation and the presence of residual red blood cells present in vessels and capillaries. Importantly, using this strategy, all of the whole-brain samples in a cohort can be evaluated by the same random forest model, increasing the level of objectivity in macroscopic analyses. This process can be automated using shell scripts to run ilastik in "headless" mode on a workstation or HPC cluster. Finally, each machine learning probability map can be aligned to the CCF 3.0 reference atlas using the exact same registration parameters identified previously for the raw fluorescence images.

3.5.5 Meta-analysis

Any set of image data registered to the Allen Institute CCF 3.0 can be automatically quantified using MATLAB, Python, or an image analysis package of the user's choice. Any such script or function should take as input the Allen CCF 3.0 annotation file, a list of regions of interest, and a 3D image file to be quantified. This can include raw red, green, and/or blue channel STPT images or derivative (e.g., probability maps established via ilastik machine learning described above). An important practical consideration is that, at this stage, each 3D image to be quantified should match the image dimensions of the Allen CCF 3.0 annotation file, which may require resizing via FIJI/ImageJ or other image management software. Because the Allen Institute CCF 3.0 annotation file assigns a unique integer to each annotated brain region, any analysis script or function can simply generate a binary mask matching every unique integer in the 3D annotation file, then multiplicatively apply the mask to the image data to be quantified. In doing so, a matrix can be produced that associates fluorescence (or a derivative measure of the fluorescence such as the machine learning-derived probability map) with every annotated brain region in the CCF 3.0. Because every sample in a cohort can be analyzed using the same process, a matrix of $(n\_samples) \times (n\_regions) \times (n\_channels$ or $n\_labels)$ can be readily produced and utilized for brain network analysis (Fig. 5).

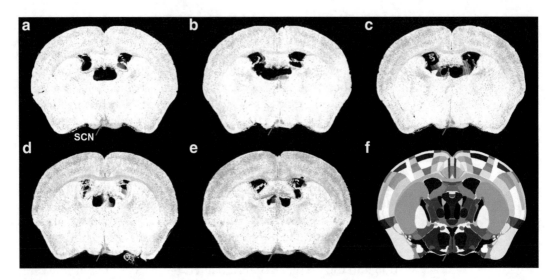

Fig. 5 Registered STPT images from five FosTRAP mice treated similarly to the animal shown in Fig. 4. (**a–e**) Single frames from registered whole-brain images showing supervised machine learning output. Predicted tdTomato-positive cell bodies, tdTomato-positive neurites, and major blood vessels are depicted in red, green, and blue, respectively. White background indicates the brain boundaries. Blue arrows point to a consistent region of high tdTomato expression, the suprachiasmatic nucleus of the hypothalamus (SCN), a light-sensitive brain region associated with circadian rhythms. (**f**) Annotated regions (CCF 3.0) of the same positions shown in (**a–e**). (Sample credit: Wen Mai Wong, Jie Cao, and Julian Meeks)

3.5.6 Volumetric Visualization

Due to the inherently pre-aligned nature of the images collected on the TissueCyte1000, unregistered or registered output can be used for volumetric visualization. There are many software tools that are available for visualization, including open-source solutions such as the ClearVolume plugin for ImageJ [71] or Vaa3D [21] for volumetric rendering, of the output data. Commercial solutions may also be utilized if available, such as Imaris (Bitplane), Arivis Vision4D (arivis AG), or Amira (FEI). Importantly, with our typical imaging strategy which results in one image plane collected every 75 μm, or even using optical sectioning within each physical section to increase axial resolution to 25 μm or greater, the assembled 3D volume will necessarily display gaps between each section unless interpolation between each plane is employed during visualization.

3.6 Example Data: Identification of Behaviorally Relevant Neuronal Populations with Fos-TRAP Labeling

Transgenic tools that transiently or permanently label neuronal populations that express immediate-early genes (IEGs; e.g., *Arc* or *Fos*; reviewed in [82, 83]) are inherently suited to STPT imaging. One of these, called "targeted recombination in active populations (TRAP), allows one to pair a single injection of the estrogen receptor agonist 4-hydroxytamoxifen (4-OHT) with a behavioral paradigm of interest, resulting in permanent fluorescent labeling of IEG-expressing neuronal populations [84]. TRAP mice, which express the tamoxifen-sensitive CreER^{T2} protein under control of the endogenous *Arc* or *Fos* promoter, can be mated to cre-dependent transgenic reporter mice [30], resulting in body-wide permanent expression. Such datasets are ideally suited to STPT imaging and whole-brain image analysis. Indeed, several studies have already utilized similar strategies to assess brain-wide circuit changes during various behaviors [85], retrieval of fear memories [61], sex-specific social behaviors [60], and in mouse models of depression [65].

An example dataset arising from the use of the FosTRAP animals was generated by mating mice heterozygous for the FosTRAP allele (Fos-CreER^{T2}) [84] with mice homozygous for the cre-dependent "Ai9" allele (Rosa26-lox-STOP-lox-tdtomato) [30] (Figs. 4 and 5). Adult male or female mice were single-housed for 5–7 days prior to performing behavioral "TRAPing" experiments. On the day of the experiment, mice were injected with 50 mg/kg 4-OHT intraperitoneally under light anesthesia 15 minutes prior to a 1-hour bout of chemosensory exposure. The mice were then returned to their home cage for 4–5 hours in a dark, quiet room prior to being returned to their normal housing location. The reason for this 4–5 hour period of rest is that the 4-OHT dose remains active in the brain for approximately 6 hours post-injection [84], so it is important to limit exogenous stimulation during this time period. Seven days following the TRAPing procedure (to allow sufficient time for tdTomato protein to accumulate in TRAPed cells), animals were deeply anesthetized and perfused with 4% paraformaldehyde according to standard procedures. Whole-brain STPT imaging was performed at 0.875×0.875 µm x-y resolution, with three optical planes spaced 25 µm apart (25, 50, 75 µm from the cut surface) per physical section (spaced 75 µm apart). The resulting images were subjected to the registration and analysis pipeline as described above. The raw data were downsampled by performing a within-physical-section maximum intensity projection (MIP), then downsampling the MIP image to a voxel size of 1.5 µm $\times 1.5$ µm in the x-y dimension. This resulted in a set of 16-bit three-color compressed TIF images totaling $5801 \times 4023 \times 190$ voxels (approximately 2 GB). A supervised random forest machine learning model was trained on selected sections from these data using ilastik. Because tdTomato fluorescence was so strong, neurites and cell bodies of TRAPed neurons were readily distinguished from brain autofluorescence throughout all active brain regions, so specific machine learning "labels" were trained to identify tdTomato-positive cell bodies and tdTomato-positive neurites including axons and dendrites independently (Fig. 5).

Summary
Whole-brain imaging via STPT is quickly advancing our knowledge of the brain in enabling researchers to quickly and quantitatively produce assessments of brain anatomy in healthy and diseased

states. With the ever-expanding set of fluorescent tools available to neuroscientists studying brain structure and function, many researchers' capacity to generate rich fluorescent samples has outstripped their capacity for thorough analysis via traditional imaging methods. STPT automates nearly all aspects of fluorescence imaging and analysis, removing major impediments for such laboratories in their quests for new knowledge about their area of interest. In the coming decades, STPT and other whole-brain methods seem likely to produce a paradigm shift in the way that neuroscience laboratories explore questions that require analysis of both "the forest and the trees."

4 Notes and Troubleshooting

4.1 System Installation

The TissueCyte 1000 system requires a dedicated room which can be completely darkened. The system must be installed on a high-quality vibration isolation system (e.g., Newport S-2000 High Performance Laminar Flow Isolators and RS 2000 table top with tuned damping) with minimum table dimensions of 6 ft. long by 4 ft. wide by 8 in. thick (1.83 m by 1.22 m by 20.32 cm). It is strongly recommended to locate the system in an area free of excessive vibrations from any source including nearby equipment or the building itself. Suitable precautions should be taken in earthquake prone areas to minimize earthquake related damage to the system and ensure personnel safety, such as installation of earthquake restraint devices (Newport ERS 2010) and vibration snubbers (Newport VS-1-I2000) as part of the anti-vibration system. An industrial type powered rack system (Newport ATS-4 or similar) is not included with the system but is also needed to house the system control components.

4.2 PB Preparation

The PB solution must be filtered first, then degassed. If it is filtered after degassing, the filtering process will re-introduce gas bubbles into the solution. The PB solution should be as free of particulate matter and bubbles as possible to avoid the introduction of high contrast noise artifacts in the images. For this reason, it is important to pour the PB into the bath slowly so as not to introduce bubbles while filling, and PB used for filling the bath must be at room temperature to avoid the formation of bubbles upon warming.

4.3 Mosaic Stitching

An earlier less automated pipeline may also be used to stitch the tile images into the 2D mosaic section images in the absence of the NAS/Autostitcher system. Custom MATLAB and FIJI scripts written by TissueVision, Inc. to produce the stitched images may be run on a local workstation, but will require all of the image tiles to be collected prior to beginning the stitching and also require more user intervention during the stitching process.

4.4 Imaging Non-brain Organs

Our standard workflow is optimized for mouse brains, but essentially any organ or tissue that is amenable to vibratome sectioning methods can be imaged using STPT. Other mouse organs we have imaged include the spinal cord, liver, kidney, lung, intestine, and spleen. We have also imaged brain tissue from songbirds, rats, and humans. In the case of spinal cords, we have noted that it is essential to remove the meninges as completely as possible prior to embedding and STPT imaging. Very soft tissues such as early postnatal brain, tissues with fibrous components (e.g., liver, intestine, kidney), and/or a heterogeneous composition (e.g., tumor-bearing tissues) are often more difficult to section and thus image using this method, but modifications to the embedding protocol (i.e., using acrylamide instead of agarose) and/or sectioning parameters including reduced speed, increased frequency, and increased thickness can improve the section quality and optimize conditions for non-brain organs. It is also important to limit the size of the tissue specimen to fit comfortably within the standard 22 × 22 × 20 mm (7/8 × 7/8 × 25/32″) embedding mold. If the tissue is too large for the block, there may not be enough agarose to support the tissue during sectioning. In this case, a larger mold can be used or the tissue size can be reduced. The TissueCyte has a limit on the total mosaic size (imageable area) of the block surface, approximately 27 mm by 27 mm. Another consideration when imaging non-brain organs is that while machine learning-based pixel classification methods as described above may still be employed for such images, automated quantification strategies will be limited in the absence of a corresponding annotated tissue atlas.

4.5 Common Imaging Problems

4.5.1 Poor Section Quality

In our experience, poor section quality is most commonly due to poor perfusion or bad fixation of the tissue. Brain lesions such as stroke or traumatic brain injury can also result in suboptimal section quality, as expected. Adult mouse brain tissue is generally sectioned very well using this method. Other softer or partially fibrous organs have proven more challenging to section and image on the TissueCyte system, such as intestine and liver or early postnatal brain (see above). The sections produced during vibratome sectioning using the TissueCyte should be of visibly high quality when they are released from the block (flat, unwrinkled, uniform thickness, and no disruption of the tissue such as shredding or tearing). If the section quality is poor, the resulting STPT images will show irregular missing/blank areas from each image plane.

4.5.2 Dim Images

A number of variables can influence the brightness of the images. If the laser appears to be functioning properly, and laser alignment is correct, another explanation for a dim image could be that the surface of the block may be set incorrectly on the piezo objective control (either too high or too low above the focal plane will produce dim images). With repeated use, the absolute focal plane can shift over time and, to a smaller extent, between individual samples in the short term. While minor adjustments can be made using the piezo controller for the objective height (total range of 400 μm), larger adjustments require manually advancing or retracting the objective in its attachment to the microscope itself. Alternatively, the fluorescent signal of interest may simply be inherently dim. In this case, a higher excitation shutter voltage and/or higher PMT voltage may be used during acquisition. It is also possible for air bubbles sitting at the objective lens, or a section that was incompletely severed from the rest of the block to impede detection and produce a very dim image. Bubbles at the objective should be displaced with a transfer pipette. If a section is incompletely severed from the block and does not properly detach after sectioning and sink to the bottom of the bath, the sample length (total travel distance of the vibratome blade) parameter may need to be adjusted in Orchestrator.

4.5.3 Wrong Mosaic Values

If the stitched 2D section image displays the tissue at the far side of the image and/or part of your tissue is cut off from the image, the mosaic dimensions or the starting position of the mosaic need to be adjusted. For mouse brain sectioned coronally, easily recognizable anatomical landmarks such as the central canal of the medulla and the border between the cerebellum and brainstem can be used to reference the mosaic starting position. For other section orientations of brain or other organs, it is often necessary to estimate the approximate center of the tissue once installed in the TissueCyte system to reference where to begin collecting the mosaic image tiles. In this case, it is best to set up a larger mosaic to be confident that all of the tissue will be present in the image.

Acknowledgements

The University of Texas Southwestern Whole Brain Microscopy Facility is funded by the Texas Institute of Brain Injury and Repair (TIBIR) and receives additional support from the University of Texas Southwestern Department of Neurology and Neurotherapeutics and the University of Texas Southwestern Center for Alzheimer's and Neurodegenerative Disease (CAND). We would like to thank Drs. Amy Bernard, Anh Ho, Lydia Ng, and Hongkui Zeng of the Allen Institute for Brain Science for helpful discussions and training in TissueCyte operations and data analysis. We also thank Drs. Timothy Ragan, Phil Knodle, and Adam Bleckert of TissueVision, Inc. for assistance in optimizing Autostitcher performance.

References

1. Zeng H (2018) Mesoscale connectomics. Curr Opin Neurobiol 50:154–162

2. Cazemier JL, Clasca F, Tiesinga PH (2016) Connectomic analysis of brain networks: novel techniques and future directions. Front Neuroanat 10:110

3. Helmstaedter M, Briggman KL, Turaga SC, Jain V, Seung HS, Denk W (2013) Connectomic reconstruction of the inner plexiform layer in the mouse retina. Nature 500:168–174

4. Kasthuri N, Hayworth KJ, Berger DR, Schalek RL, Conchello JA, Knowles-Barley S, Lee D, Vazquez-Reina A, Kaynig V, Jones TR, Roberts M, Morgan JL, Tapia JC, Seung HS, Roncal WG, Vogelstein JT, Burns R, Sussman DL, Priebe CE, Pfister H, Lichtman JW (2015) Saturated reconstruction of a volume of neocortex. Cell 162:648–661

5. White JG, Southgate E, Thomson JN, Brenner S (1986) The structure of the nervous system of the nematode Caenorhabditis elegans. Philos Trans R Soc Lond B Biol Sci 314:1–340

6. Fox MD, Snyder AZ, Vincent JL, Corbetta M, Van Essen DC, Raichle ME (2005) The human brain is intrinsically organized into dynamic, anticorrelated functional networks. Proc Natl Acad Sci U S A 102:9673–9678

7. Buckner RL, Andrews-Hanna JR, Schacter DL (2008) The brain's default network: anatomy, function, and relevance to disease. Ann N Y Acad Sci 1124:1–38

8. Oh SW, Harris JA, Ng L, Winslow B, Cain N, Mihalas S, Wang Q, Lau C, Kuan L, Henry AM, Mortrud MT, Ouellette B, Nguyen TN, Sorensen SA, Slaughterbeck CR, Wakeman W, Li Y, Feng D, Ho A, Nicholas E, Hirokawa KE, Bohn P, Joines KM, Peng H, Hawrylycz MJ, Phillips JW, Hohmann JG, Wohnoutka P, Gerfen CR, Koch C, Bernard A, Dang C, Jones AR, Zeng H (2014) A mesoscale connectome of the mouse brain. Nature 508:207–214

9. Jiang X, Shen S, Cadwell CR, Berens P, Sinz F, Ecker AS, Patel S, Tolias AS (2015) Principles of connectivity among morphologically defined cell types in adult neocortex. Science 350:aac9462

10. Denk W, Strickler JH, Webb WW (1990) Two-photon laser scanning fluorescence microscopy. Science 248:73–76

11. Neil MA, Juskaitis R, Wilson T (1997) Method of obtaining optical sectioning by using structured light in a conventional microscope. Opt Lett 22:1905–1907

12. Mertz J (2011) Optical sectioning microscopy with planar or structured illumination. Nat Methods 8:811–819

13. Ragan T, Kadiri LR, Venkataraju KU, Bahlmann K, Sutin J, Taranda J, Arganda-Carreras I, Kim Y, Seung HS, Osten P (2012) Serial two-photon tomography for automated ex vivo mouse brain imaging. Nat Methods 9:255–258

14. Economo MN, Clack NG, Lavis LD, Gerfen CR, Svoboda K, Myers EW, Chandrashekar J (2016) A platform for brain-wide imaging and reconstruction of individual neurons. Elife 5:e10566

15. Seiriki K, Kasai A, Hashimoto T, Schulze W, Niu M, Yamaguchi S, Nakazawa T, Inoue KI, Uezono S, Takada M, Naka Y, Igarashi H, Tanuma M, Waschek JA, Ago Y, Tanaka KF, Hayata-Takano A, Nagayasu K, Shintani N, Hashimoto R, Kunii Y, Hino M, Matsumoto J, Yabe H, Nagai T, Fujita K, Matsuda T, Takuma K, Baba A, Hashimoto H (2017) High-speed and scalable whole-brain imaging in rodents and primates. Neuron 94:1085–1100

16. Gong H, Zeng S, Yan C, Lv X, Yang Z, Xu T, Feng Z, Ding W, Qi X, Li A, Wu J, Luo Q (2013) Continuously tracing brain-wide long-distance axonal projections in mice at a one-micron voxel resolution. Neuroimage 74:87–98

17. Gong H, Xu D, Yuan J, Li X, Guo C, Peng J, Li Y, Schwarz LA, Li A, Hu B, Xiong B, Sun Q, Zhang Y, Liu J, Zhong Q, Xu T, Zeng S, Luo Q (2016) High-throughput dual-colour precision imaging for brain-wide connectome with cytoarchitectonic landmarks at the cellular level. Nat Commun 7:12142

18. Dorn JF, Danuser G, Yang G (2008) Computational processing and analysis of dynamic fluorescence image data. Methods Cell Biol 85:497–538

19. Sabouri-Ghomi M, Wu Y, Hahn K, Danuser G (2008) Visualizing and quantifying adhesive signals. Curr Opin Cell Biol 20:541–550

20. Kvilekval K, Fedorov D, Obara B, Singh A, Manjunath BS (2010) Bisque: a platform for bioimage analysis and management. Bioinformatics 26:544–552

21. Peng H, Ruan Z, Long F, Simpson JH, Myers EW (2010) V3D enables real-time 3D visualization and quantitative analysis of large-scale biological image data sets. Nat Biotechnol 28:348–353

22. Tsai CL, Lister JP, Bjornsson CS, Smith K, Shain W, Barnes CA, Roysam B (2011) Robust, globally consistent and fully automatic multi-image registration and montage synthesis for 3-D multi-channel images. J Microsc 243:154–171

23. Eliceiri KW, Berthold MR, Goldberg IG, Ibanez L, Manjunath BS, Martone ME, Murphy RF, Peng H, Plant AL, Roysam B, Stuurman N, Swedlow JR, Tomancak P, Carpenter AE (2012) Biological imaging software tools. Nat Methods 9:697–710

24. Burel JM, Besson S, Blackburn C, Carroll M, Ferguson RK, Flynn H, Gillen K, Leigh R, Li S, Lindner D, Linkert M, Moore WJ, Ramalingam B, Rozbicki E, Tarkowska A, Walczysko P, Allan C, Moore J, Swedlow JR (2015) Publishing and sharing multidimensional image data with OMERO. Mamm Genome 26:441–447

25. Freeman J (2015) Open source tools for large-scale neuroscience. Curr Opin Neurobiol 32:156–163

26. Niedworok CJ, Brown AP, Jorge Cardoso M, Osten P, Ourselin S, Modat M, Margrie TW (2016) aMAP is a validated pipeline for registration and segmentation of high-resolution mouse brain data. Nat Commun 7:11879

27. Berger DR, Seung HS, Lichtman JW (2018) VAST (Volume Annotation and Segmentation Tool): efficient manual and semi-automatic labeling of large 3D image stacks. Front Neural Circuits 12:88

28. Bohland JW, Wu C, Barbas H, Bokil H, Bota M, Breiter HC, Cline HT, Doyle JC, Freed PJ, Greenspan RJ, Haber SN, Hawrylycz M, Herrera DG, Hilgetag CC, Huang ZJ, Jones A, Jones EG, Karten HJ, Kleinfeld D, Kotter R, Lester HA, Lin JM, Mensh BD, Mikula S, Panksepp J, Price JL, Safdieh J, Saper CB, Schiff ND, Schmahmann JD, Stillman BW, Svoboda K, Swanson LW, Toga AW, Van Essen DC, Watson JD, Mitra PP (2009) A proposal for a coordinated effort for the determination of brainwide neuroanatomical connectivity in model organisms at a mesoscopic scale. PLoS Comput Biol 5:e1000334

29. Cowan WM (1998) The emergence of modern neuroanatomy and developmental neurobiology. Neuron 20:413–426

30. Madisen L, Zwingman TA, Sunkin SM, Oh SW, Zariwala HA, Gu H, Ng LL, Palmiter RD, Hawrylycz MJ, Jones AR, Lein ES, Zeng H (2010) A robust and high-throughput Cre reporting and characterization system for the whole mouse brain. Nat Neurosci 13:133–140

31. Taniguchi H, He M, Wu P, Kim S, Paik R, Sugino K, Kvitsiani D, Fu Y, Lu J, Lin Y, Miyoshi G, Shima Y, Fishell G, Nelson SB, Huang ZJ (2011) A resource of Cre driver lines for genetic targeting of GABAergic neurons in cerebral cortex. Neuron 71:995–1013

32. Nassi JJ, Cepko CL, Born RT, Beier KT (2015) Neuroanatomy goes viral! Front Neuroanat 9:80

33. Jeong M, Kim Y, Kim J, Ferrante DD, Mitra PP, Osten P, Kim D (2016) Comparative three-dimensional connectome map of motor cortical projections in the mouse brain. Sci Rep 6:20072

34. Pan C, Cai R, Quacquarelli FP, Ghasemigharagoz A, Lourbopoulos A, Matryba P, Plesnila N, Dichgans M, Hellal F, Erturk A (2016) Shrinkage-mediated imaging of entire organs and organisms using uDISCO. Nat Methods 13:859–867

35. Mano T, Albanese A, Dodt HU, Erturk A, Gradinaru V, Treweek JB, Miyawaki A, Chung K, Ueda HR (2018) Whole-brain analysis of cells and circuits by tissue clearing and light-sheet microscopy. J Neurosci 38:9330–9337

36. Murakami TC, Mano T, Saikawa S, Horiguchi SA, Shigeta D, Baba K, Sekiya H, Shimizu Y, Tanaka KF, Kiyonari H, Iino M, Mochizuki H, Tainaka K, Ueda HR (2018) A three-dimensional single-cell-resolution whole-brain atlas using CUBIC-X expansion microscopy and tissue clearing. Nat Neurosci 21:625–637

37. Swanson LW, Lichtman JW (2016) From Cajal to connectome and beyond. Annu Rev Neurosci 39:197–216

38. Liu JY, Ellis M, Brooke-Ball H, De Tisi J, Eriksson SH, Brandner S, Sisodiya SM, Thom M (2014) High-throughput, automated quantification of white matter neurons in mild malformation of cortical development in epilepsy. Acta Neuropathol Commun 2:72

39. Lugo-Hernandez E, Squire A, Hagemann N, Brenzel A, Sardari M, Schlechter J, Sanchez-Mendoza EH, Gunzer M, Faissner A, Hermann DM (2017) 3D visualization and quantification of microvessels in the whole ischemic mouse brain using solvent-based clearing and light sheet microscopy. J Cereb Blood Flow Metab 37:3355–3367

40. Luo Y, Wang A, Liu M, Lei T, Zhang X, Gao Z, Jiang H, Gong H, Yuan J (2017) Label-free brainwide visualization of senile plaque using cryo-micro-optical sectioning tomography. Opt Lett 42:4247–4250

41. Whitesell JD, Buckley AR, Knox JE, et al. (2019) Whole brain imaging reveals distinct spatial patterns of amyloid beta deposition in three mouse models of Alzheimer's disease. J Comp Neurol 527:2122–2145.

42. Grandjean J, Zerbi V, Balsters JH, Wenderoth N, Rudin M (2017) Structural basis of large-scale functional connectivity in the mouse. J Neurosci 37:8092–8101

43. Bienkowski MS, Bowman I, Song MY, Gou L, Ard T, Cotter K, Zhu M, Benavidez NL, Yamashita S, Abu-Jaber J, Azam S, Lo D, Foster NN, Hintiryan H, Dong HW (2018)

Integration of gene expression and brain-wide connectivity reveals the multiscale organization of mouse hippocampal networks. Nat Neurosci 21:1628–1643

44. Weber MT, Arena JD, Xiao R, Wolf JA, Johnson VE (2018) Clarity reveals a more protracted temporal course of axon swelling and disconnection than previously described following traumatic brain injury. Brain Pathol 29:437–450. https://doi.org/10.1111/bpa.12677

45. Carmichael ST, Kathirvelu B, Schweppe CA, Nie EH (2017) Molecular, cellular and functional events in axonal sprouting after stroke. Exp Neurol 287:384–394

46. Harris NG, Verley DR, Gutman BA, Thompson PM, Yeh HJ, Brown JA (2016) Disconnection and hyper-connectivity underlie reorganization after TBI: a rodent functional connectomic analysis. Exp Neurol 277:124–138

47. Fornito A, Zalesky A, Breakspear M (2015) The connectomics of brain disorders. Nat Rev Neurosci 16:159–172

48. Zheng T, Yang Z, Li A, Lv X, Zhou Z, Wang X, Qi X, Li S, Luo Q, Gong H, Zeng S (2013) Visualization of brain circuits using two-photon fluorescence micro-optical sectioning tomography. Opt Express 21:9839–9850

49. Chen BC, Legant WR, Wang K, Shao L, Milkie DE, Davidson MW, Janetopoulos C, Wu XS, Hammer JA 3rd, Liu Z, English BP, Mimori-Kiyosue Y, Romero DP, Ritter AT, Lippincott-Schwartz J, Fritz-Laylin L, Mullins RD, Mitchell DM, Bembenek JN, Reymann AC, Bohme R, Grill SW, Wang JT, Seydoux G, Tulu US, Kiehart DP, Betzig E (2014) Lattice light-sheet microscopy: imaging molecules to embryos at high spatiotemporal resolution. Science 346:1257998

50. Watson AM, Rose AH, Gibson GA, Gardner CL, Sun C, Reed DS, Lam LKM, St Croix CM, Strick PL, Klimstra WB, Watkins SC (2017) Ribbon scanning confocal for high-speed high-resolution volume imaging of brain. PLoS One 12:e0180486

51. Spalteholz W (1914) Über das Durchsichtigmachen von menschlichen und tierischen Präparaten (About the transparency-generation of human and animal preparations). S. Hirzel, Leipzig

52. Richardson DS, Lichtman JW (2015) Clarifying tissue clearing. Cell 162:246–257

53. Susaki EA, Ueda HR (2016) Whole-body and whole-organ clearing and imaging techniques with single-cell resolution: toward organism-level systems biology in mammals. Cell Chem Biol 23:137–157

54. Jing D, Zhang S, Luo W, Gao X, Men Y, Ma C, Liu X, Yi Y, Bugde A, Zhou BO, Zhao Z, Yuan Q, Feng JQ, Gao L, Ge WP, Zhao H (2018) Tissue clearing of both hard and soft tissue organs with the PEGASOS method. Cell Res 28:803–818

55. Huisken J, Swoger J, Del Bene F, Wittbrodt J, Stelzer EH (2004) Optical sectioning deep inside live embryos by selective plane illumination microscopy. Science 305:1007–1009

56. Girkin JM, Carvalho M (2018) The light-sheet microscopy revolution. J Opt 20:053002

57. Dean KM, Fiolka R (2017) Lossless three-dimensional parallelization in digitally scanned light-sheet fluorescence microscopy. Sci Rep 7:9332

58. Epp JR, Niibori Y, Liz Hsiang HL, Mercaldo V, Deisseroth K, Josselyn SA, Frankland PW (2015) Optimization of CLARITY for clearing whole-brain and other intact organs. eNeuro 2(3). pii: ENEURO.0022-0015.2015

59. Hama H, Kurokawa H, Kawano H, Ando R, Shimogori T, Noda H, Fukami K, Sakaue-Sawano A, Miyawaki A (2011) Scale: a chemical approach for fluorescence imaging and reconstruction of transparent mouse brain. Nat Neurosci 14:1481–1488

60. Kim Y, Venkataraju KU, Pradhan K, Mende C, Taranda J, Turaga SC, Arganda-Carreras I, Ng L, Hawrylycz MJ, Rockland KS, Seung HS, Osten P (2015) Mapping social behavior-induced brain activation at cellular resolution in the mouse. Cell Rep 10:292–305

61. Vousden DA, Epp J, Okuno H, Nieman BJ, Van Eede M, Dazai J, Ragan T, Bito H, Frankland PW, Lerch JP, Henkelman RM (2015) Whole-brain mapping of behaviourally induced neural activation in mice. Brain Struct Funct 220:2043–2057

62. Zapiec B, Mombaerts P (2015) Multiplex assessment of the positions of odorant receptor-specific glomeruli in the mouse olfactory bulb by serial two-photon tomography. Proc Natl Acad Sci U S A 112:E5873–E5882

63. Quina LA, Tempest L, Ng L, Harris JA, Ferguson S, Jhou TC, Turner EE (2015) Efferent pathways of the mouse lateral habenula. J Comp Neurol 523:32–60

64. Amato SP, Pan F, Schwartz J, Ragan TM (2016) Whole brain imaging with serial two-photon tomography. Front Neuroanat 10:31

65. Kim Y, Perova Z, Mirrione MM, Pradhan K, Henn FA, Shea S, Osten P, Li B (2016) Whole-brain mapping of neuronal activity in the learned helplessness model of depression. Front Neural Circuits 10:3

66. Li A, Gong H, Zhang B, Wang Q, Yan C, Wu J, Liu Q, Zeng S, Luo Q (2010) Micro-optical

sectioning tomography to obtain a high-resolution atlas of the mouse brain. Science 330:1404–1408

67. Gang Y, Liu X, Wang X, Zhang Q, Zhou H, Chen R, Liu L, Jia Y, Yin F, Rao G, Chen J, Zeng S (2017) Plastic embedding immunolabeled large-volume samples for three-dimensional high-resolution imaging. Biomed Opt Express 8:3583–3596

68. Schindelin J, Arganda-Carreras I, Frise E, Kaynig V, Longair M, Pietzsch T, Preibisch S, Rueden C, Saalfeld S, Schmid B, Tinevez JY, White DJ, Hartenstein V, Eliceiri K, Tomancak P, Cardona A (2012) Fiji: an open-source platform for biological-image analysis. Nat Methods 9:676–682

69. Schneider CA, Rasband WS, Eliceiri KW (2012) NIH image to ImageJ: 25 years of image analysis. Nat Methods 9:671–675

70. Somner C, Straehle C, Kothe U, Hamprecht FA (2011) Ilastik: interactive learning and segmentation toolkit. Proceedings of the Eighth IEEE International Symposium on Biomedical Imaging (ISBI). pp 230–233

71. Royer LA, Weigert M, Gunther U, Maghelli N, Jug F, Sbalzarini IF, Myers EW (2015) ClearVolume: open-source live 3D visualization for light-sheet microscopy. Nat Methods 12:480–481

72. Gleave JA, Lerch JP, Henkelman RM, Nieman BJ (2013) A method for 3D immunostaining and optical imaging of the mouse brain demonstrated in neural progenitor cells. PLoS One 8:e72039

73. Hamers-Casterman C, Atarhouch T, Muyldermans S, Robinson G, Hamers C, Songa EB, Bendahman N, Hamers R (1993) Naturally occurring antibodies devoid of light chains. Nature 363:446–448

74. Perruchini C, Pecorari F, Bourgeois JP, Duyckaerts C, Rougeon F, Lafaye P (2009) Llama VHH antibody fragments against GFAP: better diffusion in fixed tissues than classical monoclonal antibodies. Acta Neuropathol 118:685–695

75. Li T, Vandesquille M, Koukouli F, Dudeffant C, Youssef I, Lenormand P, Ganneau C, Maskos U, Czech C, Grueninger F, Duyckaerts C, Dhenain M, Bay S, Delatour B, Lafaye P (2016) Camelid single-domain antibodies: a versatile tool for in vivo imaging of extracellular and intracellular brain targets. J Control Release 243:1–10

76. Tsai PS, Kaufhold JP, Blinder P, Friedman B, Drew PJ, Karten HJ, Lyden PD, Kleinfeld D (2009) Correlations of neuronal and microvascular densities in murine cortex revealed by direct counting and colocalization of nuclei and vessels. J Neurosci 29:14553–14570

77. Sallee CJ, Russell DF (1993) Embedding of neural tissue in agarose or glyoxyl agarose for vibratome sectioning. Biotech Histochem 68:360–368

78. Preibisch S, Saalfeld S, Tomancak P (2009) Globally optimal stitching of tiled 3D microscopic image acquisitions. Bioinformatics 25:1463–1465

79. Modat M, Cash DM, Daga P, Winston GP, Duncan JS, Ourselin S (2014) Global image registration using a symmetric block-matching approach. J Med Imaging (Bellingham) 1:024003

80. Breiman L (2001) Random forests. Mach Learn 45:5–32

81. Haubold C, Schiegg M, Kreshuk A, Berg S, Koethe U, Hamprecht FA (2016) Segmenting and tracking multiple dividing targets using ilastik. Adv Anat Embryol Cell Biol 219:199–229

82. Barth AL (2007) Visualizing circuits and systems using transgenic reporters of neural activity. Curr Opin Neurobiol 17:567–571

83. Reijmers L, Mayford M (2009) Genetic control of active neural circuits. Front Mol Neurosci 2:27

84. Guenthner CJ, Miyamichi K, Yang HH, Heller HC, Luo L (2013) Permanent genetic access to transiently active neurons via TRAP: targeted recombination in active populations. Neuron 78:773–784

85. Renier N, Adams EL, Kirst C, Wu Z, Azevedo R, Kohl J, Autry AE, Kadiri L, Umadevi Venkataraju K, Zhou Y, Wang VX, Tang CY, Olsen O, Dulac C, Osten P, Tessier-Lavigne M (2016) Mapping of brain activity by automated volume analysis of immediate early genes. Cell 165:1789–1802

Chapter 10

Studying a Light Sensor with Light: Multiphoton Imaging in the Retina

Thomas Euler, Katrin Franke, and Tom Baden

Abstract

Two-photon imaging of light stimulus-evoked neuronal activity has been used to study all neuron classes in the vertebrate retina, from the photoreceptors to the retinal ganglion cells. Clearly, the ability to study retinal circuits down to the level of single synapses or zoomed out at the level of complete populations of neurons has been a major asset in our understanding of this beautiful circuit. In this chapter, we discuss the possibilities and pitfalls of using an all-optical approach in this highly light-sensitive part of the brain.

Key words Vertebrate retina, Mouse, Zebrafish, Two-photon microscopy, Biosensor, Activity probes, Visual stimulus-evoked activity, Laser-evoked retinal activity

1 Introduction

Unlike most other neuronal tissues, the retina directly responds to light. Accordingly, using any form of light microscopy to study its function inevitably leads to superimposition of neuronal activity driven by the "intended" visual stimulus with activity driven by the optical imaging system itself. Since single-photon (1P) excitation of fluorescence, as used in traditional wide-field charge-coupled device (CCD) camera systems or for confocal microscopy, usually uses excitation light within the visual spectrum, this conflict is near insurmountable. Typically, the excitation light of these imaging systems would saturate or even permanently blind the photoreceptors of the retina, thus leaving little room for modulation of their activity using additional sources of light. Two-photon (2P) microscopy [1] dramatically ameliorates – though never eliminates – this problem by using an infrared laser outside the visual spectrum for fluorescence excitation. This way, direct activation of retinal photoreceptor cells by the imaging system becomes much less of a problem [2–4]. As a result, 2P microscopy as a tool for studying neuronal function revolutionised retinal research over the past decades. At last, researchers

Espen Hartveit (ed.), *Multiphoton Microscopy*, Neuromethods, vol. 148,
https://doi.org/10.1007/978-1-4939-9702-2_10, © Springer Science+Business Media, LLC, part of Springer Nature 2019

could draw on the power of optical imaging to complement their physiological toolkit. Here, following a brief recap of retinal structure and function, we will highlight the advantages and remaining pitfalls of using 2P imaging to study this highly light-sensitive tissue with optical recording techniques, and how the light-induced side effects of the actual measurement can be estimated and overcome.

1.1 Probing Retinal Function with Optical Methods

The retina comprises five principal classes of neurons that are organised into three nuclear and two synaptic layers (Fig. 1a). In the outer plexiform layer (OPL), photoreceptors connect to the dendrites of horizontal cells and bipolar cells. The latter in turn carry the signal to the inner plexiform layer (IPL) where they contact amacrine cells as well as the dendrites of retinal ganglion cells (RGCs), the eye's projection neurons to the brain. Most computational "power" of the retinal network derives from its two synaptic layers as well as from final signal integration by the RGCs. Accordingly, most studies of retinal function have focused on these sites. As for neuronal networks in other parts of the brain, the ability of 2P microscopy to resolve neuronal activity down to the level of single synapses here really comes to shine.

With a few notable exceptions, most studies looking at vertebrate retinal function using 2P microscopy have focussed on the retinae of mice, rabbits and larval zebrafish. Initially dominated by the use of synthetic fluorescent probes introduced by single-cell injections (Fig. 1b), nowadays, genetically encoded indicators of neuronal activity ("biosensors") are used in genetically accessible model systems such as mouse and zebrafish (Fig. 1c–f). Such biosensors include, for instance, the GCaMP family of calcium (Ca^{2+}) probes [5, 6] or, more recently, iGluSnFR, a sensor for extracellular glutamate [7–9]. For example, Dreosti et al. [10] expressed SyGCaMP2 under the RibeyeA promoter to drive expression of this Ca^{2+} biosensor in bipolar cells of the larval zebrafish retina. This was followed by a series of studies using this approach and its subsequent iterations using other forms of GCaMP or SypHy to study zebrafish inner retinal processing at great depth [11–17]. In parallel, Ca^{2+} imaging using biosensors also found its way into the mouse retina (Fig. 1c, e) [5, 18]. This enabled investigating more complex synaptic interactions with multiple partners [19, 20], such as those at the core of the intensely studied "direction-selective" (DS) circuit in the retina [21]. Even intracellular chloride (Cl^-) levels – indicative of inhibitory synaptic inputs mediated by γ-aminobutyric acid (GABA) or glycine receptors – can be monitored in this way. For example, Duebel et al. [22] used Cl^- imaging to reveal in certain bipolar cell types a standing Cl^- gradient, which enables these cells to process GABAergic inputs differentially at the dendrites and the axon terminals. More recently, the ascent of glutamate imaging further expanded the available toolset for studying retinal function (Fig. 1d, f; [7, 23]). All major excitatory neurons of the retina use glutamate as their neurotransmitter. Accordingly, the ability to directly monitor their synaptic output using glutamate

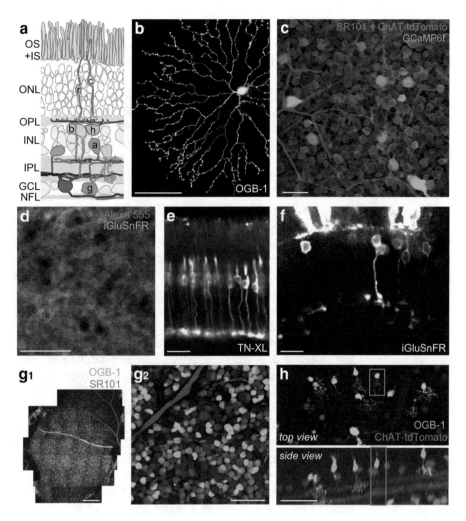

Fig. 1 Retina and retinal labelling. (**a**) Schematic cross section of a mammalian retina (c, cone photoreceptor; r, rod photoreceptor; h, horizontal cell; b, bipolar cell; a, amacrine cell; g, retinal ganglion cell; OS+IS, outer and inner segments; ONL, outer nuclear layer; OPL, outer plexiform layer; INL, inner nuclear layer; IPL, inner plexiform layer; GCL, ganglion cell layer; NFL, nerve fibre layer). The following panels illustrate methods for labelling retinal neurons (for details, see text). (**b**) Top view of rabbit starburst amacrine cell (SAC) injected with the synthetic Ca^{2+} indicator Oregon Green 488 BAPTA-1 (OGB-1). (**c**) Whole-mount retina of transgenic mouse expressing tdTomato (red) in starburst amacrine cells (SACs) and ubiquitously expressing the genetically encoded Ca^{2+} biosensor GCaMP6f (green) introduced by viral infection. The tissue was counterstained with sulforhodamine (SR101, red), here visualising the blood vessels (cf. Fig. 3). (**d**) Scan of the axon terminal system of a bipolar cell filled with Alexa 555 (red) via a sharp electrode in the flat-mounted mouse retina, with the iGluSnFR staining overlaid in green. (**e**) Retinal slice from HR2.1:TN-XL mouse line with cones expressing the Ca^{2+} biosensor TN-XL. (**f**) iGluSnFR sparsely expressed under the RibeyeA promoter labels both photoreceptors and bipolar cells in larval zebrafish in vivo (7 days post-fertilisation; courtesy of T. Yoshimatsu, Univ. Sussex, Brighton, UK). (**g₁,₂**) Mouse retinal whole-mount electroporated with OGB-1 (green) and counterstained with SR101 (red), here visualising the vessels. (**h**) Retina from a same mouse line as in (**c**), with SACs expressing tdTomato (red) and bipolar cells sparsely loaded with OGB-1 (green) via dye incubation (courtesy of J. Diamond, NINDS, Bethesda, MA, USA; for protocol, see [34]). Scale bars: (**b**), (**c**), (**g₂**), (**h**), 50 μm; (**d**), (**e**), 10 μm; (**g₁**), 300 μm; (**f**), 20 μm. Panel (**d**) adapted from Franke et al. 2017 [24]. Panel (**e**) adapted from Baden et al. 2013 [59] with permission from Elsevier. Panel (**g**) adapted from Baden et al. 2016 [30]

imaging – rather than "just" the presynaptic Ca^{2+} signals that drive release – turned out to be yet another game changer. For example, Franke et al. [24] surveyed light-driven glutamate release across the entire mouse IPL to provide a functional fingerprint of each bipolar cell type and in turn used this to study global rules of inhibition in the inner retina [24]. Meanwhile, others used this approach to study receptor types underlying transient and sustained Off bipolar cell responses [25] and the role of excitation in direction selectivity [26].

Imaging of genetically encoded and targeted probes aside, retinal research also made major headway using single-cell or population labelling of retinal neurons using synthetic fluorescent Ca^{2+} indicators (Fig. 1b, g, h). For example, in an early study, Euler et al. [27] imaged starburst amacrine cells in the rabbit retina following single-cell dye loading (Fig. 1b) to uncover a major mechanism in these cells that supports the extraction of motion information in the aforementioned DS circuit. Multiple studies followed suit using similar techniques, for example, most recently demonstrating species differences in these cells' dendritic wiring with the presynaptic partners [28]. Finally, given that the retina is a thin sheet of tissue, also bulk loading of synthetic indicators, in particular via electroporation (Fig. 1g, h), can be used to address questions that require complete labelling of neuronal populations, such as the synaptic terminals of photoreceptors in the OPL [29] or the cell bodies in the ganglion cell layer (GCL) [30–32]. Bulk loading with synthetic indicators may also be tweaked to result in sparse labelling useful for imaging individual neurons [33, 34], avoiding the need for technically challenging single-cell injections.

Taken together, 2P imaging has to date been used to study all major sites in the inner and outer retina, including cone pedicles [4, 29, 30, 35], horizontal cell dendrites [36], bipolar cell axon terminals [20, 26, 34, 37], amacrine cell circuits [17, 19, 27, 28, 38–41] and ganglion cells [20, 30, 32, 34, 42, 43]. Clearly, the ability to study retinal circuits down to the level of single synapses, or zoomed out at the level of complete populations of neurons, has been a major asset in our understanding of this beautiful circuit. However, as we shall see, we must continue to be vigilant about the remaining pitfalls of using an optical approach.

1.2 Direct and Indirect Excitation of the Retina in 2P Microscopy

Optical access to the retina from the photoreceptor side is limited, because here several layers of photoreceptor cell bodies form a light-scattering lenslet array (cf. Fig. 1a; [44]). Instead, the isolated and "flat-mounted" retina is typically accessed from the retinal ganglion cell side. However, as the retinal tissue itself is only around 150–200 μm thick and transparent, the main challenge is not depth penetration – as it is in other parts of the brain – but rather light-sensitivity. Perhaps contrary to intuition, the infrared (IR) laser used in 2P microscopy is not at all invisible to the retina [2, 3].

The typical wavelength of the 2P excitation laser ranges between 850 and 1050 nm, far beyond the peak of the spectral sensitivity curve of even the long-wavelength (L)-sensitive photopigments (L-opsins; peak sensitivity around 600 nm, depending on the species [45]). However, all opsins have a long, ultralow-sensitivity tail that reaches far into the IR range [46]. Combined with the high laser power required for generating fluorescence signals, this direct (1P) excitation of opsins can be substantial (Fig. 2a). In addition, the 2P effect [47] can also excite opsins [2] (Fig. 2b) – just like the fluorescent probes. Due to the quadratic laser intensity dependence of 2P excitation (see Sect. 3.4) and the fact that scanning usually occurs in a plane several tens of micrometres away from the photoreceptors, 2P excitation of opsins appears highly unlikely. However, because of the extreme opsin concentration in the photoreceptor outer segments and the high gain of the phototransduction cascade, even sparse 2P opsin activation can contribute strongly to the laser-evoked response of the retina. Finally, the fluorescence that is meant to be generated within the tissue adds to this laser response: when scattered towards the photoreceptors, it hits the opsins right in the bulk of their sensitivity curve (Fig. 2c). This effect can cause rhythmic activity as the laser focus repetitively scans across a cell's receptive field. This effect is referred to as "indirect excitation" (of the photoreceptors) and further contributes to setting up a background activation of the retinal network.

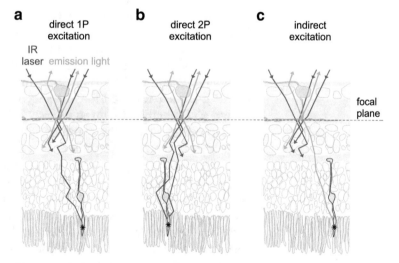

Fig. 2 Direct and indirect excitation of photoreceptors. Retinal cross sections with schematic trajectories of exemplary single photons of infrared (IR) excitation laser (red) and fluorescence emitted from activity probe (green). Photoreceptors can get activated directly by the IR laser via single-photon (**a**) and two-photon events (**b**), as well as indirectly by emitted fluorescence in the visible range (**c**). Arrows indicate photon propagation direction, and stars symbolise photoisomerisation events in the highlighted photoreceptor. Dotted line, focal plane of the scan

Together, direct 1P and 2P excitation as well as indirect excitation of photoreceptors generates a "low-photopic" level of background activation in the retina [3]. This makes it difficult, if not impossible, to study retinal processing outside the photopic regime using 2P microscopy. How to estimate and minimise these effects is discussed in Sect. 3.4.

1.3 Visual Stimulation

Besides the imaging system interfering with the photoreceptors and thereby triggering the retinal network ("laser effect", see Sect. 1.2), the possibility of "opposite" crosstalk presents a further complication: any light introduced in addition to the excitation laser – for example, to provide a controlled visual stimulus to the retina – is potentially "seen" by the microscope's light detectors (some variant of photomultiplier tube (PMT)). In other words, the visual stimulus can easily "swamp" the detection part of the 2P-system, thus making it impossible to image anything at all (or even permanently damage the PMTs). Two main approaches can be used to ameliorate this problem: (*i*) segregation of the spectrum into specific bands for fluorescence emission (from the probe) and light stimulation, for instance, using customised dichroic mirrors, or (*ii*) rapid temporal separation of image acquisition and visual stimulation. Ideally, both are used in conjunction [3].

Spectral separation is usually easy to achieve as visual stimulation does not need to cover the retina's full spectral sensitivity range. Depending on the species, two to four narrow wavelength bands are typically sufficient to differentially drive the spectrally different opsins (photoreceptors) and thus address basic questions in colour vision. In this way, conflicts between visual stimulation and imaging can be avoided, so long as two spectral windows for detection of the common fluorescent probes, which usually exhibit their fluorescent peaks around 530 nm ("green") and 630 nm ("red"), are left reserved for imaging. Example arrangements for mouse and zebrafish are given in Sect. 3.3.

However, due to the extreme light sensitivity of the microscope's PMTs and because filters never completely block light outside their pass-band(s), spectral segregation alone is usually not sufficient to satisfactorily ameliorate stimulation artifacts in the scans. Even if these are barely detectable, such stimulation artifacts hamper the reliable detection of the actual light stimulus-evoked activity. This problem can be addressed by segregating periods of image acquisition from those of light stimulation, for instance, by stimulating solely during the scan retrace (see Sect. 3.3). In addition, "gated" PMTs that, for instance, contain a shutter that can be driven with microsecond precision (see, e.g., the H11526 series by Hamamatsu) may become an option to prevent stimulation light from entering the actual detectors.

1.4   Recording Configurations

Though principally transparent itself, the retina embedded in the pigmented eyeball remains difficult to optically access in vivo. As a result, most work on retinal physiology has been done in so-called retinal explants where the retina is removed in its entirety from the eyeball (see Sect. 3.1). In its "flattened" form, this in vitro configuration provides for straightforward optical access of any retinal neuron and synaptic layer, and it simplifies spatially structured visual stimulation. Since the retinal network is largely feedforward with most mammalian species receiving little or no centrifugal projections from the brain [48], this "whole-mount" configuration is considered to provide a reasonable approximation for most retinal functions.

An alternative in vitro configuration is the retinal slice – here, the retina is, for instance, mounted on a filter paper and sliced vertically into 200–300-μm-thick sections using a mechanical chopper [49]. Such slices provide direct electrophysiological access to all classes of retinal neurons and enable powerful combinations of single-cell 2P imaging and electrical recording (e.g., [38]). However, with techniques for targeting and visualising neurons deep in the tissue, whole-mounts are quickly catching up in this respect. In terms of imaging, slices have the additional advantage that layers close to the light-sensitive photoreceptor outer segments, such as the OPL, become more accessible for optical recordings [4, 50, 51]: with the axis of the excitation laser then running perpendicular to the retinal layering, direct 1P and 2P opsin excitation (see Sects. 1.2 and 3.4) is substantially reduced. The main downside of the slice is that crucial lateral connections in the retinal network are severed by the sectioning. Therefore, from a functional point of view, whole-mounts are preferable as here the long-range inhibitory interactions (i.e., polyaxonal amacrine cells [52] and horizontal cells [53]) remain intact.

In principle, in vivo imaging of the mammalian retina is also possible. This can be achieved using adaptive optics which permit optically corrected access of the imaging system through the cornea and lens and has been demonstrated, for example, in mice and macaques [54, 55], in the latter, though, not yet for 2P imaging of neuronal activity. However, this in vivo approach is technically challenging and therefore not yet widely used. In contrast, small, semi-transparent model species, such as the larval zebrafish, offer a simpler alternative for many research questions. Here, genetic (e.g., "crystal fish", [56]) or chemical removal of some of the eye's pigment [10] allows direct optical access to most of the retinal network in vivo. One remaining downside is that any removal of the eye's screening pigment around the eyeball also interferes with the spatial profile of light reaching different parts of the eye. Of course, this is also an issue in any form of retinal explant.

2 Materials

2.1 Two-Photon Microscope

In principle, any type of upright 2P microscope that allows visualising the selected fluorescent probe(s) at sufficient spatial and temporal resolution can be used. In our case, a through-the-objective (TTO) solution for the visual stimulator is employed [3]. For that, the scope must have the option for coupling in the stimulus into the main optical path. Visual stimulation from below the recording chamber – through-the-condenser (TTC) – is somewhat easier to set up but requires access to the sub-stage space and a condenser (or objective) lens of sufficient optical quality for projecting the stimulus onto the retina in the recording chamber (for details, see Sect. 2.2). Moreover, if the PMTs are positioned above the objective lens, as is often the case, sub-stage visual stimulation may run the risk of photon-swamping the detectors with the stimulus (see above).

In our labs, we use MOM-type 2P microscopes, originally designed by W. Denk (MPI for Neurobiology, Martinsried, Germany) and now built and further developed by Sutter Instrument (Novato, CA, USA). We chose the MOM – the "movable objective microscope" – because it can be easily customised, e.g., by adding different types of visual stimulators (cf. Sect. 2.2). For a detailed description of the MOM design, see elsewhere [3]. In brief, our systems are equipped with a mode-locked Ti:Sapphire laser (MaiTai-HP DeepSee, Newport Spectra-Physics, Darmstadt, Germany) tuneable to wavelengths roughly between 800 and 1030 nm, a water immersion objective (e.g., W Plan-Apochromat 20×/1.0 DIC M27 VIS-IR, Zeiss, Oberkochen, Germany) and two fluorescence detection channels. The latter are optimised for detection of green and red fluorescent probes by using appropriate bandpass filters positioned in front of the PMTs (cf. Fig. 5a; for example, HQ 510/84 and HQ 630/60 dichroic filters, respectively; AHF, Tübingen, Germany). For image acquisition, we use custom software ("ScanM" by M. Müller, Max Planck Gesellschaft, Martinsried and T.E.) running under IGOR Pro 6.3 for Windows (Wavemetrics, Lake Oswego, OR, USA) on a PC equipped with fast multichannel I/O cards (PCI-6110 or PCI-6115 and PCIe-6363, National Instruments, München, Germany), which control scanners and image acquisition and generate blanking and trigger signals (cf. Sect. 3.3). However, other software packages (e.g., ScanImage, Vidrio Technologies, Ashburn, VA, USA) can be used on the same hardware as well. In addition, ScanImage is quite flexible and can be adapted to a range of imaging hardware and I/O cards.

2.2 Visual Stimulation

The advantages of TTO stimulation (cf. Sect. 2.1) are (*i*) that the high quality of the objective lens used for imaging is also exploited for stimulus projection; (*ii*) that, assuming proper alignment, the stimulation centre is always registered to the centre of the imaging field (the stimulus is "carried around" with the imaging field when moving across the retinal surface); and (*iii*) that only a relatively small fraction of the stimulation light (the portion that is reflected from the tissue) re-enters the objective lens and needs to be prevented from reaching the PMTs, which in the case of the MOM are positioned above the objective. The main disadvantage of the TTO configuration is that the objective lens must provide sufficient spatial resolution (i.e., a high NA) while having a large field-of-view (FOV) and, thus, large stimulation area. Here, a 20× objective like the selected one (cf. Sect. 2.1) represents a good compromise for many research questions in the retina.

The TTO solution presented in this chapter uses a digital mirror device (DMD)-based video projector that is coupled into the optical path of the microscope after the scan mirrors. Commercially available "consumer" devices can be employed; however, depending on the species studied and, thus, the wavelengths needed for effective opsin stimulation, the internal LEDs may need to be replaced by appropriate LED/filter combinations (for mouse and zebrafish solutions, see Sect. 3.3). For minimising crosstalk between light stimulator and PMTs (cf. Sect. 1.3), a modification of the LED control may be necessary to blank the stimulator light during the scanning (see Sect. 3.3). While such consumer devices are very affordable, they undergo fast revision cycles, such that only a year later it may be difficult to find the same model for which the LED modification was developed. Therefore, employing more expensive but well-documented and controllable evaluation platforms, with longer revision cycles, such as the LightCrafter™ (e.g., DLP LightCrafter 4500, Texas Instruments, Dallas, TX, USA), may be considered as an alternative. These devices are also available with customer-specified LEDs and appropriately optimised internal optics (e.g., EKB Technologies Ltd., Bat-Yam, Israel) and even as a version with a port for a light-guide, which facilitates adding other light sources. The LightCrafter can be driven like a computer monitor via an HDMI link from any visual stimulation software (e.g., Psychtoolbox, http://psychtoolbox.org [57]; or our own software, QDSpy, https://github.com/eulerlab/QDSpy; see Sect. 3.3).

For full-field stimuli, a simple, LED-based light stimulator in TTC configuration may be sufficient. Such a contraption can be easily built from a set of filters and LEDs driven by off-the-shelf electronics, like an Arduino Uno microcontroller board (https://www.arduino.cc/). It enables easy control of the LEDs' intensities via pulse-width modulation (PWM), precise timing of the stimulus time course and simple synchronisation between image acquisition and stimulus presentation. For a possible design, see elsewhere [58, 59].

3 Methods

3.1 Tissue Preparation

The tissue preparation procedures for retinal imaging experiments are very similar to those used for standard physiological recordings in the retina, which is why we will not go into detail here. A step-by-step description of mouse retina dissection, including retinal slicing and whole-mount preparation, as well as a list of tools and materials, is available elsewhere (e.g., [50, 60]). In the following, we will discuss a few guidelines regarding tissue preparation that are important for successfully imaging light-driven activity in the mouse retina. While most of these "rules" apply to any retinal experiment using light stimulation, some are critical for 2P imaging of the retina.

To maximise the amount of available light-sensitive visual pigment in the in vitro retina, animals should be dark-adapted for ≥1 hour prior to dissection. In our experience, dark-adaptation also facilitates detaching the neural retina from pigment epithelium (PE), which is advantageous when isolating the retina from the eyecup. While helpful in the mouse, this step is essential in many non-mammalian preparations such as birds and fish where the outer segments of photoreceptors markedly move in and out of the pigment epithelium depending on the adaptation state [61]. To minimise visual pigment bleaching, we carry out all steps of the dissection procedure using illumination light at the long-wavelength edge of the species' visible spectrum: in mice, with their visual spectrum shifted to shorter wavelengths [62], very dim red light (e.g., ~650 nm LEDs with ~100 μW power), still well visible to humans, can be used. Minimising bleaching is less critical for preparations with the PE attached (e.g., eyecup), as in vertebrates, the chromophore is mainly regenerated in the PE [63] and, therefore, the continuous supply with restored chromophore is ensured also in vitro. While the dissection of the retina with attached PE is possible and routinely done in some species, including, e.g., Macaque monkeys [64], it is more challenging in rodents (for procedure, see [65]) and therefore rarely used.

Depending on the research question, registering the orientation of the retina throughout the preparation procedure and the experiment may be critical. This is because in most, if not all species, the distribution of neurons across the retina is anatomically and/or functionally inhomogeneous (e.g., [66, 67]). For example, in mice, short- and medium-wavelength sensitive photoreceptor types are unequally distributed across the retina [59, 68, 69]. In addition, recent studies demonstrate that different types of mouse RGCs show distinct distributions across the retina to differentially sample visual space [70–73]. One possibility to keep track of the retina's orientation during the dissection is to use an orienting mark (e.g., dorsal) on the eye while still in the animal and make a little cut in the eyecup (with the retina still in) at the corresponding location. Alternatively, the stereotypic pattern of retinal blood vessels may be used (for mice, see Fig. 2 in [74]). As a result, this allows monitoring the retinal position and orientation of the recording site(s), which can be crucial for data interpretation.

As discussed in Sect. 1.4, retinal whole-mounts or slices can be chosen. While both preparations allow keeping track of the recording position (for preparing slices, see e.g., [36]), whole-mounts make it particularly easy to map positions in the explanted retina onto spherical coordinates, thereby reconstructing the recording site(s) in the intact, three-dimensional retina (e.g., using the "Retistruct" algorithm, [75]). This presents the researcher with the opportunity to relate retinal coordinates, for example, to motion axes [76] or visual coordinates, thus providing insights into the functional role and, potentially, behavioural relevance of the retinal cells and circuits studied.

When mounting the retina in the recording chamber, the stimulus direction needs to be considered. The TTO configuration – with the stimulus projected from above and the GCL up – allows "more intact" preparations, such as an eyecup or a PE-attached whole-mount. Here, the retina can be mounted on a non-transparent filter paper (e.g., nitrocellulose membrane, 0.8 μm pore size, Millipore, Ireland) or on IR-transmissive ceramic discs ("Anodisc", #13, 0.2 μm pore size, GE Healthcare, USA). For the TTO configuration, however, the stimulated area is limited by the field of view of the objective (cf. Sect. 2.2). In contrast, when the stimulus is projected from below through a condenser, the retina needs to be detached from PE and sclera. In addition, the tissue needs to be mounted on filter paper (see above) in which a small window has been cut (to allow the stimulus from below reaching the photoreceptors). Alternatively, a "harp" (e.g., a "U"-shaped Pt-wire with nylon strings, self-made or from companies such as Warner Instruments, Hamden, CT, USA) can be used to keep the retina flat and steady in the recording chamber. Another option is using the transparent ceramic discs described above – however, one needs to consider that these discs scatter short-wavelength light, resulting in a blurred stimulus. Specifically, we found that 385 nm stimuli smaller than 50 μm were blurred, whereas 576 nm stimuli of the same size were accurately projected on the retina.

If compatible with the other fluorescent (activity) probes used, 0.1 μM of the red dye sulforhodamine-101 (SR101, #S359, Invitrogen) can be added to the perfusion. SR101 diffuses in the extracellular space and thereby visualises the retinal structure under the 2P microscope (Fig. 3a). It also enters damaged cells and brightly labels these [3, 77], thereby allowing to monitor tissue health during recordings. In addition, SR101 strongly stains retinal blood vessels (Fig. 3a), which can be used as a reliable stratification marker in the IPL (Fig. 3b, c; [24]) and as landmarks in the

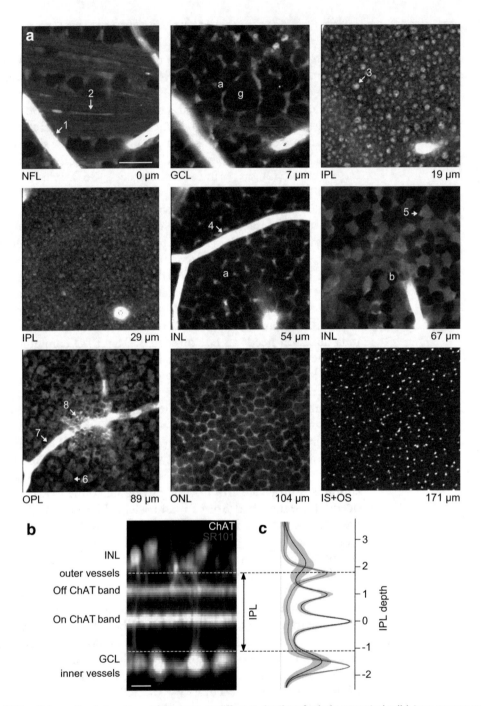

Fig. 3 Visualising retinal structure. (**a**) Images at different depths of whole-mounted wild-type mouse retina recorded using a 2P microscope. The living, light-sensitive tissue was stained with SR101 to visualise retinal structure and blood vessels. Layers, depths and examples for retinal cells are indicated (*a*, amacrine cell; *g*, retinal ganglion cell; *b*, bipolar cell; arrows: 1,4,7, blood vessels; *2*, ganglion cell axon bundles; *3*, rod bipolar cell terminal; *5*, rod bipolar cell soma; *6*, cone pedicle; *8*, rod spherule; NFL, nerve fibre layer; GCL, ganglion cell layer; IPL inner plexiform layer; INL, inner nuclear layer; OPL, outer plexiform layer; ONL, outer nuclear layer; OS+IS, outer and inner segments). (**b**) Vertical projection of stack from a ChAT:Cre × Ai9[tdTomato] mouse counterstained with SR101 illustrating the position of the choline acetyltransferase (ChAT) immunopositive bands (white) relative to blood vessels (red). Dashed lines mark IPL borders. (**c**) Average depth profiles of blood vessels (red) and ChAT bands (black; ± 1 s.d. shading). Scale bars: (**a**), 20 μm; (**b**), 10 μm. Panels (**b**) and (**c**) modified from Franke et al. 2017 [24]

whole-mount, for example, to match recorded with subsequently immunolabelled cells [30]. Note that synaptically very active cells slowly take up SR101 by endocytosis (Fig. 3a); this has, for instance, been used to identify cone axon terminals in slices [36]. The level of SR101 staining usually allows distinguishing between this uptake-dependent labelling and labelling of damaged cells: the latter are stained quickly and become much brighter [77].

Finally, two more general points are in our experience critical for ensuring high-quality recordings in the in vitro retina. First, recordings should be carried out at physiological temperatures (~37 °C for mouse or typically ~28 °C for zebrafish – although as ectotherms, the retinae of fish are more temperature resistant) to maximise photoreceptor light responses. Second, the perfusion should run with at least 3–4 ml per minute to keep the tissue healthy and responsive to light. In our experience, both artificial cerebral spinal fluid (ACSF) solution (for recipe, see, e.g., [30]) and Ames' medium (e.g., A1420, Sigma-Aldrich/Merck, Darmstadt, Germany) work well for the isolated mouse retina.

3.2 Labelling the Retina with Activity Probes

In recent years, the diversity of available indicators to measure different aspects of neuronal activity has increased tremendously (e.g., [78]). This presents the opportunity to not only record neuronal activity per se but a specific biological process like neurotransmitter release (cf. Sect. 1.1). Here, we will briefly describe a selection of labelling techniques that were successfully employed to introduce activity probes into the retina and that enable recordings of light-driven responses across all retinal cell classes.

Genetically encoded fluorescent probes (biosensors), such as the GCaMP family [5, 6] or iGluSnFR [7], can be introduced into the tissue by a viral approach or stably expressed in transgenic animal lines. The great advantage of biosensors over synthetic dyes is the possibility to address neuronal subsets selectively, including single, genetically defined cell types. For example, we use mouse lines expressing the Ca^{2+} sensor TN-XL and GCaMP3 in cone photoreceptors and horizontal cells [4, 36], respectively, to image light-evoked Ca^{2+} changes in cone axon terminals (Fig. 4a) and horizontal cell dendrites (Fig. 4b) in the outer retina. While transgenic mouse lines have the advantage of stable expression suitable for longitudinal studies (e.g., [79]), their generation is very time-consuming. Here, viral delivery of the biosensors, for instance, by transfection with adeno-associated virus (AAV), constitutes a more flexible alternative [80]. Viral vectors can be delivered to the retina via two routes [81]: they can be injected into the vitreous body of the eye ("intravitreal"), which predominantly targets inner retinal neurons, or into the subretinal space ("epiretinal"), targeting outer retinal neurons including photoreceptors.

We successfully injected AAVs purchased from Penn Vector Core (Philadelphia, PA, USA) encoding, e.g., the Ca^{2+} sensor GCaMP6f ([5], e.g., AAV9.hSyn.GCaMP6f) or the glutamate sen-

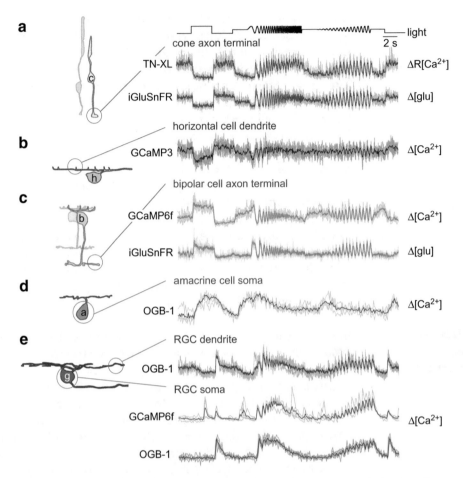

Fig. 4 Exemplary global (full-field) chirp stimulus-evoked light responses recorded with different activity probes in retinal neurons in mouse (3–5 individual trials overlaid with mean trace in darker colours, respectively). (**a**) Ca^{2+} (top) and glutamate signal (bottom) measured in/at a cone photoreceptor axon terminal using the ratiometric Ca^{2+} biosensor TN-XL ([4, 94]; HR2.1:TNXL mouse line) and the glutamate sensor iGluSnFR ([7]; ubiquitous AAV-mediated expression), respectively. (**b**) Ca^{2+} signal in a horizontal cell dendrite recorded with the Ca^{2+} biosensor GCaMP3 ([95]; transgenic Cx57$^{+/cre}$ × Ai38 mouse line). (**c**) Ca^{2+} (top) and glutamate (bottom) responses in/at axon terminals of different On cone bipolar cell types measured with the Ca^{2+} biosensor GCaMP6f ([6]; floxed AAV construct in PcP2^{Cre} mouse line) and iGluSnFR (ubiquitous AAV-mediated expression), respectively. (**d**) Somatic Ca^{2+} response of a starburst amacrine cell measured using OGB-1 introduced by bulk electroporation [31]. (**e**) Dendritic Ca^{2+} signals of an RGC injected with OGB-1 using a sharp electrode (1st row) and somatic Ca^{2+} responses recorded in RGCs using GCaMP6f (ubiquitous AAV-mediated expression, 2nd row), and OGB-1 (bulk electroporation; 3rd row). Panels (**a**; top), (**b**) and (**e**; top) courtesy of S. Pop, Y. Ran and C. Chapot, respectively (all: Univ. Tübingen, Germany)

sor iGluSnFR ([7], e.g., AAV2.hSyn.iGluSnFR) intravitreally using micromanipulators (M3301, WPI, Friedberg, Germany) and a Hamilton injection system (Hamilton Messtechnik GmbH, Höchst, Germany); for a description of the procedure, see, e.g., [24]. Red fluorescent versions of Ca^{2+} and glutamate biosensors, such as RGECO [82] and R-iGluSnFR1 [8], respectively, are also available. Depending on the combination of viral vector and mouse line used, the expression can be targeted to a subpopulation of cells or (more or less) ubiquitously expressed across cell types (see also below). For example, by injecting the "floxed" GCaMP6f virus into the Pcp2-cre line (JAX 006207, The Jackson Laboratory, Bar Harbor, USA), the Ca^{2+} sensor can be selectively expressed in a subset of bipolar cells, allowing, e.g., recording light-driven Ca^{2+} changes in individual bipolar cell axon terminal systems (Fig. 4c). In contrast, ubiquitous expression of a Ca^{2+} indicator such as GCaMP6f allows recording somatic light responses across a population of RGC types in the scan field (Fig. 4e). Similarly, the glutamatergic output of all bipolar cell types and photoreceptors can be systematically recorded by expressing the iGluSnFR ubiquitously across the retina (Fig. 4a, c). In general, the retinal transduction efficiency strongly depends on the serotype of AAV used: while some serotypes like AAV2 or AAV2.7m8 efficiently pass the inner limiting membrane [83], other serotypes, such as AAV9 or AAV1, require puncturing the inner limiting membrane with the injection needle to access the retina.

While synthetic dyes lack the cell-type selectivity and stable expression of biosensors (see above), their easy delivery to the tissue (cf. Sect. 1.1) as well as the variety of available dyes (reviewed in [84]) makes them a good alternative for imaging neuronal activity. Membrane-impermeable forms of synthetic dyes can be introduced into single cells or population of cells by electroporation (e.g., [31]). For example, injecting individual RGCs with the Ca^{2+} sensor Oregon Green 488 BAPTA-1 (OGB-1) labels single dendritic trees, which allows to record light-driven responses in different dendritic segments of the same cell (Fig. 4e). In contrast, bulk-electroporation of the retina using established Ca^{2+} dyes like OGB-1 and Rhod-2, or more recent ones, such as Cal-520 [85], allows imaging somatic responses of complete populations of GCL cells [30], including RGCs (Fig. 4e) and displaced amacrine cells (Fig. 4d).

3.3 Light Stimulation

As a visual stimulator for mice, we employ a customised DMD projector (DLP LightCrafter 4500; see Sect. 2.2) (Fig. 5a), equipped with a UV (λ_{peak} = 390 nm) and a green (λ_{peak} = ~520 nm) LED, whose emission spectra are shaped by a dual-band dichroic filter (DM2, F59–003, AHF, Tübingen, Germany). The projector is connected via HDMI to a Microsoft Windows PC running the visual stimulation software QDSpy (https://github.com/euler-lab/QDSpy) written in Python. It allows presentation of arbitrary visual stimuli and movies controlled by user-written Python scripts and supports generation of trigger signals (for off-line analysis of synchronisation of stimulus presentation and acquired imaging data) via a dedicated I/O card (e.g., PCI-DIO24, Measurement Computing, Bietigheim-Bissingen, Germany) or any Arduino microcontroller. The visual stimulus is coupled into the main opti-

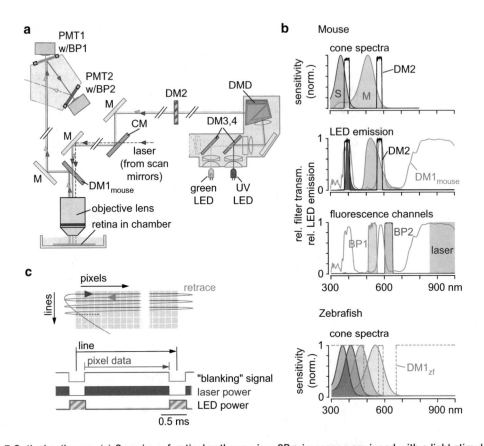

Fig. 5 Optical pathways. (**a**) Overview of optical pathways in a 2P microscope equipped with a light stimulator designed for chromatic stimulation of mouse retina. For simplicity, most lenses and some silver mirrors (M) were omitted (for a complete description, see [3]). Excitation/detection pathway: the laser (dashed red line) passes a cold mirror (CM), dichroic mirror 1 (DM1, here the mouse version, DM1$_{mouse}$) and the objective lens, before exciting the probe in the recording chamber. Fluorescence emitted from the probe (red/green lines) is reflected by DM1 into the detection head containing two GaAsP photomultiplier tubes (PMT), each equipped with a band-pass (BP) filter. Stimulus pathway: lights from a green and a UV LED are combined (DM3, 4), illuminating a digital mirror device (DMD). The grey area represents the "light engine" of a digital light processing (DLP) device (e.g., DLP LightCrafter 4500, Texas Instruments, Dallas, TX, USA), adapted for green/UV projection (by EKB Technologies Ltd., Israel). The resulting light stimulus is filtered by a dual-band filter (DM2) and coupled via CM into the laser path. (**b**) Spectral separation of stimulation and excitation/fluorescence detection: *First row*: sensitivity curves of the mouse's short- (S) and medium- (M) wavelength-sensitive opsins with stimulus filter DM2. *Second row*: stimulus LED emission spectra with DM1$_{mouse}$ and DM2. *Third row*: fluorescence detection bands and range of excitation laser. *Bottom row*: possible filter design for zebrafish (DM1$_{zf}$). (**c**) Temporal separation of stimulation and detection, illustrated for simple frame scans (top; grey squares represent pixels): each scan line is divided into two sections; one for acquiring pixel data (blue) and one for allowing the scan mirrors to move to the start of the next line ("retrace", orange). Bottom: signal generated by the scan software ("blanking", top) to synchronise laser and stimulus LEDs; during retrace, laser power is blanked (centre; e.g., by a Pockels cell), while LEDs are turned on (bottom). Schematic in (**a**) adapted from Euler et al. 2008 [3] with permission from Springer Nature

cal path of the 2P microscope and projected into the tissue via the objective lens (TTO arrangement). To this end, a cold mirror (CM in Fig. 5a) is placed just in front of the scan mirrors, where it transmits the excitation laser and reflects the light stimulus towards the scan lens (for details on the optical arrangement, see [3]).

The combination of the band-pass filter (DM2), fluorescence emission filters (BP1, BP2) and the custom-designed dichroic mirror (DM1, a mouse-version $DM1_{mouse}$ is available as z400/580/890Tpc from AHF on request) directly above the objective lens takes care of *spectral separation* between (*i*) stimulus light and (*ii*) IR excitation laser, both of which are transmitted by DM1 in different bands, and (*iii*) the probe's fluorescence, which is reflected by DM1 towards the PMTs (Fig. 5b). Mice feature two spectral types of cone photoreceptor [69], expressing a short wavelength-sensitive (S-, "UV") and a medium wavelength-sensitive (M-, "green") opsin with peak sensitivities at 360 and 510 nm [62], respectively. Mouse rod sensitivity peaks very close to that of its M-cones. The "mouse" version $DM1_{mouse}$ features two narrow transmission bands for stimulation centred at ~400 and 580 nm (Fig. 5b, top). The transmission bands are not aligned with the peak sensitivities of the mouse opsins to allow for two broad detection bands (see broad "troughs" in $DM1_{mouse}$ transmission at approx. 450–550 and 600–700 nm; Fig. 5b, bottom) for standard "green" and "red" fluorescent probes. While not ideal because higher LED intensities are needed to drive the cones, this "compromise" has been successfully used for chromatic stimulation of the mouse retina (e.g., [59, 86]). To increase stimulation efficiency of the mouse's S-cone, using shorter wavelength LEDs (e.g., 360 nm) is tempting; however, we recommend checking first if the chosen projector sufficiently modulates/transmits such short wavelengths – in the DMD projectors we tested (e.g., DLP LightCrafter 4500; K11 by Acer Computer GmbH, Ahrensburg, Germany), transmission sharply drops at ~380 nm. In any case, careful calibration of the effective photoisomerisation rates induced by the stimulator channels is mandatory, in particular when studying chromatic processing (for details and procedures, see, e.g., [59, 86, 87]).

It is possible to adapt the filter/LED combinations for TTO stimulation also to the specific spectral requirements of other species. A possible filter design for zebrafish ($DM1_{zf}$; Fig. 5b, bottom) could use the transmission band at ~590 nm of $DM1_{mouse}$ for stimulation of the fish red opsin, whereas the ~400 nm transmission of $DM1_{mouse}$ would need to be broadened to allow stimulating the fish UV, blue and green opsins. As long as temporally modulated visual stimuli suffice, such a tetrachromatic stimulator can be easily designed (cf. Sect. 2.2). For tetrachromatic, spatio-temporally structured stimuli, two DMD projectors equipped with a complementary set of LEDs can be combined. The selection of dichroic filter combinations for restricting the sometimes quite broad LED emission spectra to narrow bands – to accommodate for the necessary fluorescence detection channels – can become here more difficult, as it may necessitate (costly) custom filter designs.

Unfortunately, spectral separation only reduces, but does not eliminate stimulus light-evoked artifacts in the PMTs (cf. Sect. 2.2), mainly because filters are imperfect and PMTs are highly light-sensitive. Therefore, we also use *temporal separation* of fluorescent signal detection and visual stimulation (Fig. 5c) by enabling the LEDs only during the retrace period of the microscope's scan mirrors. This "blanking" signal is generated by the imaging software (in parallel to the signals that drive the scanners) and can also be used to decrease the laser power during retrace (e.g., with a Pockels cell), helping to minimise the exposure of the retina to the excitation laser. To avoid that the resulting "flicker" of the LEDs modulate the retina's activity, we use a total scan line duration of 2 ms (e.g., 1.6 ms for signal acquisition, 0.4 ms for the retrace/visual stimulation); the resulting blanking signal frequency of 500 Hz is about one order of magnitude faster than the flicker fusion frequency of most mammals [88]. The LEDs can be switched either electronically (e.g., a "breaker" circuit) or mechanically (e.g., using a "chopper" wheel that is synchronised with the blanking signal; see MC2000B, Thorlabs GmbH, Dachau, Germany). If the LEDs' current requirements are high, building a breaker circuit with the required temporal precision becomes increasingly difficult; in this case, an optical chopper solution may be considered. The latter, however, tends to be less flexible with respect to the timing and a potential source of vibrations which can be ameliorated by using rubber feet.

A focus adjustment lens in the stimulus pathway (not shown in Fig. 5a; for details, see [3]) introduces a vertical offset between focal plane (e.g., in the IPL) and stimulation plane in the photoreceptor outer segment layer (cf. Fig. 6a). In principle, it is possible to couple this adjustment lens to the movement of the objective (e.g., by using an electrically tuneable lens) to allow for changing the focal plane while keeping the stimulation plane fixed. In practice, we found that small changes of the focal plane – for example, by ~50 μm to scan different levels of the IPL – have a negligible effect on stimulus quality.

3.4 Estimating and Ameliorating Laser Effects on Retinal Activity

As discussed in Sect. 1.2, the laser used to excite the fluorescent probes can evoke substantial light responses in the retina's photoreceptors via three mechanisms: direct 1P and 2P excitation, as well as indirect excitation by the emitted fluorescence generated in the tissue. Light responses due to direct 1P and 2P excitation were first described when imaging whole-mounted salamander retina [2] and later quantified together with indirect excitation in more detail in the rabbit retinal whole-mount [3]. In the following, we

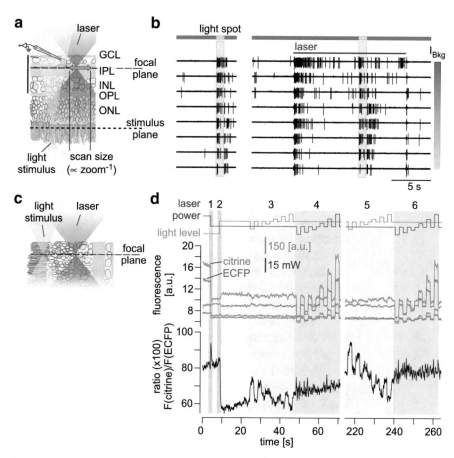

Fig. 6 Two-photon effects. (**a**) Schematic cross section of a whole-mount retina, illustrating scan size and focal planes of light stimulus (yellow) and scanning excitation laser (red). (**b**) Electrophysiological cell-attached recording of a rabbit ON-OFF ganglion cell, with traces showing spiking response to a 200 μm light spot alone (left) and during laser scanning (right; λ = 927 nm) for increasing background intensities (I_{Bkg}). Note the cell's responses to both laser scanning start and end. (**c**) Schematic cross section of a retinal slice. (**d**) Ca²⁺ signals recorded in axon terminals of mouse cone photoreceptors (slice preparation) exclusively expressing the ratio-metric Ca²⁺ biosensor TN-XL [4, 94] (all traces represent averages of n = 8 cells). Top: traces of light stimulus (yellow) and scanning laser power (red; λ = 860 nm). Centre: fluorescence signals of the two TN-XL fluoro-phores, citrine (green) and ECFP (cyan); lower trace pair represents background fluorescence measured out-side the photoreceptors. Bottom: ratiometric signal (black), as reported by TN-XL, representing the Ca²⁺ level in the terminals. Reducing the laser power (1) elicited a transient in the ratio signal. Turning on the light stimu-lator (2) evoked a strong decrease in Ca²⁺ (=light response). Cones responded strongly to bright and dark flashes (3) but barely to "flashes" in scanning laser power. Laser "flashes" are strongly reflected in citrine and ECFP fluorescence but hardly in the ratio signal. Second trial (5, 6) after cones had adapted to combined back-ground caused by laser scanning and stimulator (GCL ganglion cell layer, IPL inner plexiform layer, INL inner nuclear layer, OPL outer plexiform layer, ONL outer nuclear layer, F fluorescence, ECFP enhanced cyan fluores-cent protein). Panel (**b**) adapted from Euler et al. 2008 [3] with permission from Springer Nature; for details, see there. Panel (**d**) adapted from Baden et al. 2013 [59] with permission from Elsevier; for additional details, see there

discuss the extent of photoreceptor stimulation by the different mechanisms and give pointers for minimising these effects.

Generally, *direct 1P excitation* of the photopigment by the laser cannot be avoided, as the opsins' spectral sensitivity curves represent probability distributions, which means the chance of excitation is non-zero even for wavelengths in the IR range. However, this probability rapidly falls with increasing laser wavelengths [89]. For example, at 930 nm, mouse rhodopsin is in the order of 10^{-11} times less sensitive than at 497 nm, the peak of its sensitivity curve. This amounts to a rate of around 20 photoisomerisations per second per rod (R^*s^{-1} rod^{-1}) for a 5 mW laser beam scanning a whole-mounted retina with the focal plane in the IPL (Fig. 6a, b; for the detailed calculation, see [3]). Of course, when scanning closer to the photoreceptor layer, the laser-illuminated area in the photoreceptor layer decreases quadratically with decreasing distance, resulting in a higher photon density and, therefore, in an increased 1P-dependent photoisomerisation rate in photoreceptors within that area.

Two-photon excitation requires a high photon density, which in practice is generated by pulsing an IR laser that is "compressing" the laser's power into very short light pulses. As 2P excitation probability falls with the fourth power with increasing distance from the focal plane, fluorophores are only excited in a tiny volume (<1 fl) within the focal plane (2P-based optical sectioning; see [1, 90]). For the same reason, it might seem that *direct 2P excitation* of the photoreceptor layer is unlikely. However, due to the extreme photopigment concentration in the photoreceptor outer segment (e.g., mouse rods contain in the order of $7 \cdot 10^7$ rhodopsin molecules [91]), combined with the high gain of the phototransduction cascade [92], even a few 2P excitation events can trigger sizable light responses, as have been observed in isolated rods [93]. A back-of-the-envelope calculation for mouse rods suggests 2P excitation-related photoisomerisation rates of $2 \cdot 10^4$ R^*s^{-1}rod^{-1} at 5 mW laser power [3], which is 1000 times higher compared to the estimate for 1P excitation. Experimentally, the relative effect sizes of 1P vs. 2P photoreceptor excitation can be evaluated, for instance, by electrically recording spiking activity in RGCs in the absence of any fluorescent dye while scanning the tissue with the excitation laser in mode-lock (pulsing) or continuous mode (see Fig. 7 in [3]). Without the laser being pulsed, 2P absorption events becomes so unlikely that any remaining laser effect on the RGC activity should result from the photoreceptors "seeing" the 930 nm laser (1P excitation).

In addition to the dependence on absolute laser power, area and depth of the scanned field (or, more precisely, volume) within the retinal whole-mount are important factors that determine the laser-evoked activity observed under a specific experimental condition. Generally, the smaller the distance between the scanning plane and the layer of the photoreceptor outer segments, the stronger the laser-evoked 1P and, in particular, 2P effects. In practice, light stimulus-evoked activity in the OPL fades within a couple of seconds when using mouse retinal whole-mounts, but activity imaging in any other layer proximal to the OPL (that is, closer to or in the GCL) works quite well. The lack of discernible, stable light responses in the OPL may be due to excessive photopigment bleaching by the laser and/or that stimulus-evoked activity simply gets drowned in the laser-evoked activity. Finally, the combination of scan field depth and size determines the extent of the laser-illuminated area at the level of the photoreceptor outer segments; if this area is in the same size range as the receptive fields (RFs) of the recorded neurons, comparatively small changes in scan depth and/or area may have also rather unexpected consequences for the laser-evoked activity: for example, when moving the focal plane more proximally, the laser-illuminated area at the outer segment level grows and the electrically recorded laser-evoked response in, say, an RGC first increases until the laser-illuminated area covers the cell's RF centre and then decreases when the laser-illuminated area enters the cell's antagonistic RF surround. Eventually, the response might even change polarity if the laser stimulation of the RF surround overwhelms the RF centre response (e.g., see Fig. 7 in [3]).

Any fluorescence emitted within the tissue can lead to *indirect excitation* of the photoreceptors. This effect, however, is quite variable but tends to be in the range of that for direct 1P excitation. The amount of activity caused by indirect excitation is difficult to predict, because it depends on many factors related to the experimental situation, including dye concentration, the scanned volume from which the fluorescence is emitted and how much of the fluorescence is scattered towards the photoreceptor layer. Also, the homogeneity of the labelling within the scanned tissue volume matters; for instance, the presence of a highly fluorescent isolated structure (e.g., a dye-filled soma) in that volume can lead to rhythmic laser-evoked activity in the retinal network caused by the periodic modulation of emitted fluorescence, in particular when using low scan rates. While most of these factors cannot be easily optimised, it may be possible to reduce the fluorescence generated to the minimum required for a sufficient signal-to-noise ratio (S/N) in the signal of interest by carefully adjusting the laser power and restricting the scanned volume to the necessary minimum. The success of these efforts, of course, also depends on the sensitivity and fidelity of the fluorescence detection system (for an example of how to estimate the effect of indirect excitation as a function of S/N, see [3]).

Because slicing unavoidably damages the retinal network, whole-mounts are preferable over retinal slices – at least for circuit-oriented research questions (cf. Sects. 1.4 and 3.1). However, if optical recordings in the OPL are needed, vertical slices represent a viable alternative (e.g., [4, 51, 59]). By restricting the scan field to the OPL, the perpendicular orientation of the laser beam to the slice surface eliminates in-path laser illumination of the photoreceptor layer (Fig. 6c). Since only scattered laser light can reach the photopigments, direct excitation-related effects are strongly reduced compared to the whole-mount situation, enabling direct recordings of light stimulus-driven photoreceptor activity (Figs. 4a and 6d). Under the assumption that the focal volume roughly behaves like a point source (with fluorescence radiating equally likely in all directions), indirect laser effects are expected to be similar in the slice and the whole-mount.

Independent of the recording configuration of choice, we highly recommend checking carefully for laser-evoked contributions to the signal of interest. Here, it is also important to acknowledge that not all neuron types may have the same sensitivity to laser effects. Wherever feasible, the combined effect of direct and indirect excitation may best be first evaluated using electrical recordings (e.g., with and without fluorescent labelling) (cf. Fig. 6b). This also allows to check if the cells' activity returns to the pre-laser baseline after the initial laser-induced activity transient. Here, a crucial measure for reducing the impact of laser effects on the recorded activity is to provide sufficiently strong, steady background illumination via the light stimulator. By comparing the activity before and during laser scanning while titrating with different background levels, it is possible to reduce the effects to a short laser onset-evoked transient that settles to the pre-laser baseline level within seconds while preserving the characteristics of the light responses without the laser (e.g., third row from the bottom in Fig. 6b). In practice, this also means that stimulus presentations should always be preceded by an adequate period of time of laser scanning and background illumination (cf. Fig. 6d).

In conclusion, several measures can be taken to reduce direct 1P and 2P, as well as indirect laser-evoked effects:

(i) Limit the laser power to levels adequate to achieve sufficient S/N in the signal of interest.

(ii) Present a steady background illumination using the stimulator to reduce the overall light sensitivity of the retina.

(iii) After laser onset, give the retinal network sufficient time to adapt to the combined laser/stimulator background.

(iv) Restrict scan volume size as much as possible.

(v) If feasible, avoid highly fluorescent structures and/or inhomogeneities in fluorescent labelling within the scan volume.

By considering these measures, laser-evoked responses can often be reduced to a transient that no longer interferes with the actual light stimulus-evoked activity. Nevertheless, due to the unavoidable presence of background illumination and, in particular, the large contribution of 2P photopigment excitation to the laser effects, imaging of light-evoked retinal activity remains restricted to the mesopic/photopic range.

References

1. Denk W, Strickler JH, Webb WW et al (1990) Two-photon laser scanning fluorescence microscopy. Science 248(4951):73–76

2. Denk W, Detwiler PB (1999) Optical recording of light-evoked calcium signals in the functionally intact retina. Proc Natl Acad Sci U S A 96:7035–7040

3. Euler T, Hausselt SE, Margolis DJ et al (2009) Eyecup scope–optical recordings of light stimulus-evoked fluorescence signals in the retina. Pflugers Arch Eur J Physiol 457:1393–1414

4. Wei T, Schubert T, Paquet-Durand F et al (2012) Light-driven calcium signals in mouse cone photoreceptors. J Neurosci 32:6981–6994

5. Akerboom J, Chen TW, Wardill TJ et al (2012) Optimization of a GCaMP calcium indicator for neural activity imaging. J Neurosci 32:13819–13840

6. Chen T-W, Wardill TJ, Sun Y et al (2013) Ultrasensitive fluorescent proteins for imaging neuronal activity. Nature 499:295–300

7. Marvin JS, Borghuis BG, Tian L et al (2013) An optimized fluorescent probe for visualizing glutamate neurotransmission. Nat Methods 10:162–170

8. Wu J, Abdelfattah AS, Zhou H et al (2018) Genetically encoded glutamate indicators with altered color and topology. ACS Chem Biol. https://doi.org/10.1021/acschembio.7b01085

9. Marvin JS, Scholl B, Wilson DE et al (2017) Stability, affinity and chromatic variants of the glutamate sensor iGluSnFR. bioRxiv. https://doi.org/10.1101/235176

10. Dreosti E, Odermatt B, Dorostkar MM et al (2009) A genetically encoded reporter of synaptic activity in vivo. Nat Methods 6:883–889

11. Baden T, Esposti F, Nikolaev A et al (2011) Spikes in retinal bipolar cells phase-lock to visual stimuli with millisecond precision. Curr Biol 21:1859–1869

12. Esposti F, Johnston J, Rosa JM et al (2013) Olfactory stimulation selectively modulates the OFF pathway in the retina of zebrafish. Neuron 79:97–110

13. Nikolaev A, Leung K-M, Odermatt B et al (2013) Synaptic mechanisms of adaptation and sensitization in the retina. Nat Neurosci 16:934–941

14. Odermatt B, Nikolaev A, Lagnado L (2012) Encoding of luminance and contrast by linear and nonlinear synapses in the retina. Neuron 73:758–773

15. Rosa JM, Ruehle S, Ding H et al (2016) Crossover inhibition generates sustained visual responses in the inner retina. Neuron 90(2):308–319

16. Zimmermann MJ, Nevala NE, Yoshimatsu T et al (2017) Zebrafish differentially process colour across visual space to match natural scenes. bioRxiv. https://doi.org/10.1101/230144

17. Antinucci P, Suleyman O, Monfries C et al (2016) Neural mechanisms generating orientation selectivity in the retina. Curr Biol 26:1802–1815

18. Hasan MT, Friedrich RW, Euler T et al (2004) Functional fluorescent Ca^{2+} indicator proteins in transgenic mice under TET control. PLoS Biol 2:e163

19. Chen Q, Pei Z, Koren D et al (2016) Stimulus-dependent recruitment of lateral inhibition underlies retinal direction selectivity. Elife 5:1–19

20. Yonehara K, Farrow K, Ghanem A et al (2013) The first stage of cardinal direction selectivity is localized to the dendrites of retinal ganglion cells. Neuron 79:1078–1085

21. Vaney DI, Sivyer B, Taylor WR (2012) Direction selectivity in the retina: symmetry and asymmetry in structure and function. Nat Rev Neurosci 13:194–208

22. Duebel J, Haverkamp S, Schleich W et al (2006) Two-photon imaging reveals somatodendritic chloride gradient in retinal ON-type bipolar cells expressing the biosensor clomeleon. Neuron 49:81–94

23. Borghuis BG, Marvin JS, Looger LL et al (2013) Two-photon imaging of nonlinear glutamate release dynamics at bipolar cell synapses in the mouse retina. J Neurosci 33:10972–10985

24. Franke K, Berens P, Schubert T et al (2017) Inhibition decorrelates visual feature representations in the inner retina. Nature 542:439–444

25. Borghuis BG, Looger LL, Tomita S et al (2014) Kainate receptors mediate signalling in both transient and sustained OFF bipolar cell pathways in mouse retina. J Neurosci 34:6128–6139

26. Park SJ, Kim IJ, Looger LL et al (2014) Excitatory synaptic inputs to mouse on-off direction-selective retinal ganglion cells lack direction tuning. J Neurosci 34:3976–3981

27. Euler T, Detwiler PB, Denk W (2002) Directionally selective calcium signals in dendrites of starburst amacrine cells. Nature 418:845–852

248   Thomas Euler et al.

28. Ding H, Smith RG, Poleg-polsky A et al (2016) Species-specific wiring for direction selectivity in the mammalian retina. Nature 535:1–17

29. Jackman SL, Babai N, Chambers JJ et al (2011) A positive feedback synapse from retinal horizontal cells to cone photoreceptors. PLoS Biol 9:e1001057

30. Baden T, Berens P, Franke K et al (2016) The functional diversity of retinal ganglion cells in the mouse. Nature 529:345–350

31. Briggman KL, Euler T (2011) Bulk electroporation and population calcium imaging in the adult mammalian retina. J Neurophysiol 105:2601–2609

32. Briggman KL, Helmstaedter M, Denk W (2011) Wiring specificity in the direction-selectivity circuit of the retina. Nature 471:183–188

33. Baden T, Berens P, Bethge M et al (2013) Spikes in mammalian bipolar cells support temporal layering of the inner retina. Curr Biol 23:48–52

34. Poleg-Polsky A, Diamond JS (2016) Retinal circuitry balances contrast tuning of excitation and inhibition to enable reliable computation of direction selectivity. J Neurosci 36:5861–5876

35. Kemmler R, Schultz K, Dedek K et al (2014) Differential regulation of cone calcium signals by different horizontal cell feedback mechanisms in the mouse retina. J Neurosci 34:11826–11843

36. Chapot CA, Behrens C, Rogerson LE et al (2017) Local signals in mouse horizontal cell dendrites. Curr Biol 27:3603–3615.e5

37. Baden T, Nikolaev A, Esposti F et al (2014) A Synaptic mechanism for temporal filtering of visual signals. PLoS Biol 12:e1001972

38. Grimes WN, Zhang J, Graydon CW et al (2010) Retinal parallel processors: more than 100 independent microcircuits operate within a single interneuron. Neuron 65:873–885

39. Hausselt SE, Euler T, Detwiler PB et al (2007) A dendrite-autonomous mechanism for direction selectivity in retinal starburst amacrine cells. PLoS Biol 5:e185

40. Hsiang J-C, Johnson K, Madisen L et al (2017) Local processing of visual information in neurites of VGluT3-expressing amacrine cells. Elife e31307

41. Lee S, Zhang Y, Chen M et al (2016) Segregated glycine-glutamate co-transmission from vGluT3 amacrine cells to contrast-suppressed and contrast-enhanced retinal circuits. Neuron 90:27–34

42. Auferkorte ON, Baden T, Kaushalya SK et al (2012) GABA(A) receptors containing the alpha2 subunit are critical for direction-selective inhibition in the retina. PLoS One 7:e35109

43. Oesch N, Euler T, Taylor WR (2005) Direction-selective dendritic action potentials in rabbit retina. Neuron 47:739–750

44. Solovei I, Kreysing M, Lanctôt C et al (2009) Nuclear architecture of rod photoreceptor cells adapts to vision in mammalian evolution. Cell 137:356–368

45. Jacobs GH (2012) The evolution of vertebrate color vision. In: López-Larrea J (ed) Sensing in nature. Advances in experimental medicine and biology, vol 739. Springer, New York, pp 156–172

46. Lewis PR (1955) A theoretical interpretation of spectral sensitivity curves at long wavelengths. J Physiol 130:45–52

47. Göppert-Mayer M (1931) Über Elementarakte mit zwei Quantensprüngen. Ann Phys 9:273–294

48. Reperant J, Ward R, Miceli D et al (2006) The centrifugal visual system of vertebrates: a comparative analysis of its functional anatomical organization. Brain Res Rev 52:1–57

49. Werblin FS (1978) Transmission along and between rods in the tiger salamander retina. J Physiol 280:449–470

50. Kulkarni M, Schubert T, Baden T et al (2015) Imaging Ca^{2+} dynamics in cone photoreceptor axon terminals of the mouse retina. J Vis Exp:e52588

51. Kulkarni M, Trifunović D, Schubert T et al (2016) Calcium dynamics change in degenerating cone photoreceptors. Hum Mol Genet 25:3729–3740

52. Lin B, Masland RH (2006) Populations of wide-field amacrine cells in the mouse retina. J Comp Neurol 499:797–809

53. Peichl L, González-Soriano J (1994) Morphological types of horizontal cell in rodent retinae: a comparison of rat, mouse, gerbil and guinea pig. Vis Neurosci 11:501–517

54. Yin L, Masella B, Dalkara D et al (2014) Imaging light responses of foveal ganglion cells in the living macaque eye. J Neurosci 34:6596–6605

55. Palczewska G, Dong Z, Golczak M et al (2014) Non-invasive two-photon microscopy imaging of mouse retina and retinal pigment epithelium through the pupil of the eye. Nat Med 20:785–789

56. Antinucci P, Hindges R (2016) A crystal-clear zebrafish for in vivo imaging. Sci Rep 6:29490

57. Brainard DH (1997) The psychophysics toolbox. Spat Vis 10:433–436

58. Breuninger T, Puller C, Haverkamp S et al (2011) Chromatic bipolar cell pathways in the mouse retina. J Neurosci 31:6504–6517

59. Baden T, Schubert T, Chang L et al (2013) A tale of two retinal domains: near-optimal sampling of achromatic contrasts in natural scenes through asymmetric photoreceptor distribution. Neuron 80:1206–1217

60. Ivanova E, Toychiev AH, Yee CW et al (2013) Optimized protocol for retinal wholemount preparation for imaging and immunohistochemistry. J Vis Exp:e51018

61. Arey LB (1916) The movements in the visual cells and retinal pigment of the lower vertebrates. J Comp Neurol 26:121–201

62. Jacobs GH, Neitz J, Deegan JF (1991) Retinal receptors in rodents maximally sensitive to ultraviolet light. Nature 353:655–656

63. Wang J-S, Kefalov VJ (2011) The cone-specific visual cycle. Prog Retin Eye Res 30:115–128

64. Li PH, Field GD, Greschner M et al (2014) Retinal representation of the elementary visual signal. Neuron 81:130–139

65. Newman EA, Bartosch R (1999) An eyecup preparation for the rat and mouse. J Neurosci Meth 93:169–175

66. Boycott BB, Wässle H (1974) The morphological types of ganglion cells of the domestic cat's retina. J Physiol 240:397–419

67. Goodchild AK, Ghosh KK, Martin PR (1996) Comparison of photoreceptor spatial density and ganglion cell morphology in the retina of human, macaque monkey, cat, and the marmoset Callithrix jacchus. J Comp Neurol 366:55–75

68. Peichl L (2005) Diversity of mammalian photoreceptor properties: adaptations to habitat and lifestyle? Anat Rec Part A Discov Mol Cell Evol Biol 287:1001–1012

69. Szél A, Röhlich P, Caffé ARR et al (1992) Unique topographic separation of two spectral classes of cones in the mouse retina. J Comp Neurol 325:327–342

70. Bleckert A, Schwartz GW, Turner MH et al (2014) Visual space is represented by non-matching topographies of distinct mouse retinal ganglion cell types. Curr Biol 24:310–315

71. Rousso DL, Qiao M, Kagan RD et al (2016) Two pairs of ON and OFF retinal ganglion cells are defined by intersectional patterns of transcription factor expression. Cell Rep 15:1930–1944

72. Zhang Y, Kim I-J, Sanes JR et al (2012) The most numerous ganglion cell type of the mouse retina is a selective feature detector. Proc Natl Acad Sci USA 109:E2391–E2398

73. Joesch M, Meister M (2016) A neuronal circuit for colour vision based on rod–cone opponency. Nature 532:236–239

74. Wei W, Elstrott J, Feller MB (2010) Two-photon targeted recording of GFP-expressing neurons for light responses and live-cell imaging in the mouse retina. Nat Prot 5:1347–1352

75. Sterratt DC, Lyngholm D, Willshaw DJ et al (2013) Standard anatomical and visual space for the mouse retina: Computational reconstruction and transformation of flattened retinae with the Retistruct package. PLoS Comput Biol 9(2):e1002921

76. Sabbah S, Gemmer JA, Bhatia-Lin A et al (2017) A retinal code for motion along the gravitational and body axes. Nature 546:492–497

77. Schlichtenbrede FC, Mittmann W, Rensch F et al (2009) Toxicity assessment of intravitreal triamcinolone and bevacizumab in a retinal explant mouse model using two-photon microscopy. Invest Ophthalmol Vis Sci 50:5880–5887

78. Lin MZ, Schnitzer MJ (2016) Genetically encoded indicators of neuronal activity. Nat Neurosci 19:1142–1153

79. Bethge P, Carta S, Lorenzo DA et al (2017) An R-CaMP1.07 reporter mouse for cell-type-specific expression of a sensitive red fluorescent calcium indicator. PLoS One 12:e0179460

80. Davidson BL, Breakefield XO (2003) Neurological diseases: viral vectors for gene delivery to the nervous system. Nat Rev Neurosci 4:353–364

81. Vandenberghe LH, Auricchio A (2012) Novel adeno-associated viral vectors for retinal gene therapy. Gene Ther 19:162–168

82. Zhao Y, Araki S, Wu J et al (2011) An expanded palette of genetically encoded Ca^{2+} indicators. Science 96333:1888–1891

83. Dalkara D, Byrne LC, Klimczak RR et al (2013) In vivo-directed evolution of a new adeno-associated virus for therapeutic outer retinal gene delivery from the vitreous. Sci Transl Med 5:189ra76–189ra76

84. Lavis LD (2017) Teaching old dyes new tricks: biological probes built from fluoresceins and rhodamines. Annu Rev Biochem 86:825–843

85. Tada M, Takeuchi A, Hashizume M et al (2014) A highly sensitive fluorescent indicator dye for calcium imaging of neural activity in vitro and in vivo. Eur J Neurosci 39:1720–1728

86. Chang L, Breuninger T, Euler T (2013) Chromatic coding from cone-type unselective circuits in the mouse retina. Neuron 77:559–571

87. Wang YV, Weick M, Demb JB (2011) Spectral and temporal sensitivity of cone-mediated responses in mouse retinal ganglion cells. J Neurosci 31:7670–7681

88. Healy K, McNally L, Ruxton GD et al (2013) Metabolic rate and body size are linked with

perception of temporal information. Anim Behav 86:685–696

89. Lamb TD (1995) Photoreceptor spectral sensitivities: common shape in the long-wavelength region. Vision Res 35:3083–3091

90. Denk W, Svoboda K (1997) Photon upmanship: why multiphoton imaging is more than a gimmick. Neuron 18:351–357

91. Lyubarsky AL, Daniele LL, Pugh EN Jr (2004) From candelas to photoisomerizations in the mouse eye by rhodopsin bleaching in situ and the light-rearing dependence of the major components of the mouse ERG. Vision Res 44:3235–3251

92. Luo DG, Xue T, Yau KW (2008) How vision begins: an odyssey. Proc Natl Acad Sci U S A 105:9855–9862

93. Gray-Keller M, Denk W, Shraiman B et al (1999) Longitudinal spread of second messenger signals in isolated rod outer segments of lizards. J Physiol 519(Pt 3):679–692

94. Mank M, Reiff DF, Heim N et al (2006) A FRET-based calcium biosensor with fast signal kinetics and high fluorescence change. Biophys J 90:1790–1796

95. Zariwala HA, Borghuis BG, Hoogland TM et al (2012) A Cre-dependent GCaMP3 reporter mouse for neuronal imaging in vivo. J Neurosci 32:3131–3141

Chapter 11

Multiphoton Ca²⁺ Imaging of Astrocytes with Genetically Encoded Indicators Delivered by a Viral Approach

Rune Enger, Rolf Sprengel, Erlend A. Nagelhus, and Wannan Tang

Abstract

Decades of research have unraveled the complex functioning of neurons in the central nervous system. Our knowledge of the second main player of the brain – the non-excitable glial cells – clearly lags behind that of neurons. Pioneering work in the 1990s provided evidence that star-shaped glial cells – astrocytes – sense and modulate neuronal activity by intracellular Ca²⁺ signals. However, the precise roles of astrocytic Ca²⁺ signaling in brain physiology and pathophysiology are still highly controversial, largely due to technical limitations of previous Ca²⁺ imaging tools. With recent innovations in laser microscopy and engineering of molecular probes, the field of glioscience is undergoing a revolution. This chapter describes the application of multiphoton microscopy and genetically encoded fluorescent Ca²⁺ indicators to reveal astrocytic Ca²⁺ signals in acute brain slices and in vivo, both in anesthetized and in awake behaving animals.

Key words Astrocytes, Ca²⁺ signaling, Genetically encoded calcium indicator, GECI, GCaMP6, Glia, rAAV gene transduction, Two-photon microscopy

---

1 Introduction

There are large regional and interspecies differences in glia-neuron ratio, but overall, glial cells are considered to be as numerous as neurons in the mammalian brain [1, 2]. Astrocytes, the most prevalent type of glial cell in gray matter, have extensive arborizations and the diameter of their thinnest processes is only a few tens of nanometers. Typically, the territory of a single astrocyte exhibits a spherical shape and covers thousands of synapses [3]. Astrocytes also extend one or more specialized processes to the brain vasculature or the pia mater (Fig. 1). These so-called endfeet almost completely ensheathe the outer perimeter of the brain vasculature and form the glia limitans, a continuous layer along the brain surface [4].

It is well established that astrocytes serve as housekeepers that maintain the extracellular milieu of neurons. Specifically, they remove excess neurotransmitter and K⁺ released during neuronal activity, regulate the volume of the extracellular space, and facilitate

Espen Hartveit (ed.), *Multiphoton Microscopy*, Neuromethods, vol. 148,
https://doi.org/10.1007/978-1-4939-9702-2_11, © Springer Science+Business Media, LLC, part of Springer Nature 2019

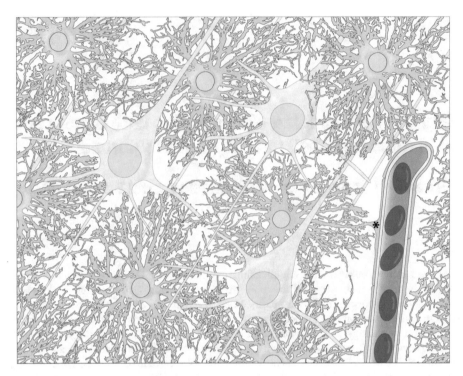

Fig. 1 Cartoon showing the morphology of cortical astrocytes. Astrocytes (green) are closely intermingled with neurons (light blue) in the gray matter. They have extensive arborizations and possess specialized end foot processes (asterisk) that ensheathe cerebral blood vessels

clearance of metabolic waste products [5–8]. Evidence also suggests that astrocytes provide metabolic support to neurons by supplying energy substrates and controlling local blood flow [9].

Astrocytes are not electrically excitable and can as such not communicate by action potentials. However, discoveries in the early 1990s pointed toward intracellular Ca^{2+} fluctuations being an important signaling mechanism that could convey information within an astrocyte and even spread to neighboring astrocytes through gap junctions [10]. Shortly thereafter, it was shown that astrocytic Ca^{2+} signals influenced neuronal activity [11, 12]. A series of subsequent studies provided evidence that astrocytes, by Ca^{2+}-dependent release of glutamate and other signaling molecules, coined gliotransmitters, contribute to the processing of neuronal signals. However, the concept of the tripartite synapse, with astrocytic processes surrounding pre- and postsynaptic neurons as functionally integrated elements of synapses, is still controversial [13, 14]. Disbelievers criticize that the studies have been performed in reduced (cell cultures) or immature systems (juvenile animals), with nonphysiological stimulation protocols, and that the kinetics of astrocytic Ca^{2+} signals are too slow to impact fast synaptic transmission. Moreover, most of the Ca^{2+} imaging studies have employed bulk-loading of organic Ca^{2+} indicators, a method with notable shortcomings. Firstly, bulk-loading, which involves cellular uptake of membrane-permeable, Ca^{2+}-insensitive acetoxy-methyl (AM) dye and subsequent action of intracellular esterases to generate membrane-impermeable, Ca^{2+}-sensitive dye, is quite ineffective in adult animals. Secondly, for poorly understood reasons, organic Ca^{2+} indicators mainly report on Ca^{2+} changes within the cell body and major branches, leaving out events in the numerous fine processes that constitute the majority of the astrocytic territory [15].

Evidently, key functions of astrocytes relate to measures not easily evaluated in reduced model systems or with conventional neuroscience methods. The recent developments in multiphoton laser scanning microscopy (MPLSM) and engineering of genetically encoded fluorescent indicators now enable us to visualize astrocytic actions with high spatiotemporal resolution in intact tissue. The following sections will outline what will need to be in place to study astrocytic Ca^{2+} signaling in vitro in acute brain slices as well as in vivo in anesthetized and awake mice with state-of-the-art genetically encoded Ca^{2+} indicators (GECIs). The acute brain slice model is a valuable addition to in vivo methods as it offers easy manipulation of cell activity with electrical stimulation, application of drugs, or alteration in the composition of the bath solution.

2 Materials

2.1 Choice of Probes and Promoters for Assessing Astrocytic Ca²⁺ Dynamics

After two decades of intense use of organic Ca^{2+} dyes to assess functions of astrocytes, the era of GECIs started. Several groups developed green fluorescent protein (GFP)-based GECIs using principles of fluorescence resonance energy transfer (FRET, [16–18]), chemiluminescence resonance energy transfer (CRET [19]), or conformational changes of single-fluorescent protein Ca^{2+}-sensor, such as GCaMP [20]. The GCaMP is an example of these early-developed GECIs and contains a circularly permuted enhanced GFP (cpEGFP), an N-terminal M13 fragment for calmodulin (CaM) binding, as well as a C-terminal CaM sequence. When Ca^{2+} binds to CaM, the interaction of Ca^{2+}-CaM-M13 induces a subsequent conformational change of cpEGFP, causing an increase in fluorescence intensity. Due to the high Ca^{2+} affinity of the GCaMP indicator, it has been widely applied for imaging studies of neural circuits and rapidly developed to more advanced generations (GCaMP2 to GCaMP7) providing improved signal-to-noise ratio and faster kinetics [21].

In 2010, it was reported that GCaMP2 and the membrane-tethered versions of GCaMP2 and GCAMP3 could be successfully applied to monitor Ca^{2+} changes in rat hippocampal astrocyte-neuron co-cultures by plasmid DNA transfection [18, 20]. Using confocal microscopy, spontaneous Ca^{2+} activity was detected in glial microdomains of cultured astrocytes [22, 23]. Subsequently, GCaMP3, GCaMP5, and GCAMP6 [24, 25] delivered by a recombinant adeno-associated virus (rAAV) approach (see below) showed spontaneous and evoked Ca^{2+} transients in astrocytes of adult mice, both in acute brain slices and in vivo [26–31]. Our studies were the first to visualize astrocytic Ca^{2+} signals by GECIs in combination with MPLSM [26–29].

Various GCaMP indicators have been tagged to different cellular compartments, such as the cell membrane and mitochondria. The first membrane-tethered Ca^{2+} indicator was Lck-GCaMP2 [23], using a strong N-terminal membrane-targeting dual acylation motif from Lck protein tyrosine kinase [32, 33] for obtaining a high plasma membrane expression and reporting on Ca^{2+} changes near the cell membrane. Since Lck-GCaMP5E (E specifies the type of GCaMP5, which like CCaMP6 comes in several variants) exhibits relatively smaller fluorescence changes in astrocyte profiles than cytosolic GCaMP5E [26], the membrane-tethered GCaMP indicators seem less suitable for revealing Ca^{2+} signals in astrocytic somata and main branches. A mitochondrial matrix-targeting sequence was used to achieve GCaMP5G/6s expression in mitochondria [34]. The mitochondrial-tagged GCaMP5G/6s (mito-GCaMP5G/6s) were targeted to either astrocytes or neurons and successfully revealed mitochondrial Ca^{2+} transients in primary cultures [34]. Mitochondrial-tagged GECIs may prove valuable for understanding the roles of mitochondrial Ca^{2+} dynamics in cellular metabolism and initiation of apoptotic cell death.

Expression of probes in specific cell types can be achieved by using cell-specific promoters. The Glial Fibrillary Acidic Protein (*GFAP*) promoter has been widely used to achieve rAAV transduction of astrocytes, even though subpopulations of neurons are also targeted [28]. There are at least two different versions of the *GFAP* promoter, a short version, gfaABC1D promoter (about 680 base pairs) [30, 31], and a long one (about 2200 base pairs) [26, 35]. In our hands, the 2200 base pair-long *GFAP* promoter exhibits significantly better specificity for astrocytes. Besides the rAAV approach, bacterial artificial chromosome (BAC) transgenic mouse lines expressing cyto-GCaMP6f/Lck-GCaMP6f have been generated with a Cre/ERT2-loxP-dependent strategy for MPLSM imaging of glial Ca^{2+} signaling [36]. The use of BAC lines, like other transgenic mouse lines, has the disadvantages of being time-consuming and expensive (high costs of animal housing) and will not be discussed further in this chapter.

2.2 Microscope

For imaging of acute brain slices, we use an Ultima multiphoton laser scanning microscope (Bruker/Prairie Technologies, Middleton, WI, USA) with a XLPLN 25× WMP 1.05NA water-immersion objective (Olympus) and a Chameleon Vision II (Coherent) laser.

For imaging of anesthetized and awake mice, we use a multiphoton microscope system assembled from components by Bruker/Prairie Technologies (model Ultima IV), Spectra-Physics

(Santa Clara, CA, USA) (laser model InSight DS), optical table and optomechanics by Standa Ltd. (Vilnius, Lithuania), optics by Bernhard Halle Nachfolger GmbH (Berlin, Germany), and electrooptical modulators by Qioptiq/Gasenger (Göttingen, Germany). We use a XLPLN 25× WMP 1.05NA, water-immersion objective for anesthetized acute preparations, and a Nikon 16× 0.8 NA water-immersion objective (model CFI75 LWD 16×W) for awake imaging due to a larger working distance of the latter (2 vs. 3 mm working distance).

All optical filters mentioned in the following description are by Chroma Technology Corporation (Bellows Falls, VT, USA). After having been reflected toward the detection unit by the main dichroic filter (type ZT473-488/594/NIRtpc), the signal light enters the system's four-channel detector house (only two channels are used for the work presented here), at the entrance of which, a type ZET473-488/594/NIRm filter is installed, shielding the photomultiplier tubes from the rest of reflective light from the laser beams. Inside the detector house, the light is split into two beams separated at 560 nm wavelength by the main signal light dichroic filter (T560lpxr). The "green" light (e.g., GCaMP6f and iGluSnFR fluorescence) is further guided by a secondary dichroic beam splitter (to enable separation from "blue" light) at 495 nm (T495lpxr) and filtered by a ET525/50m-2p band-pass filter, whereas the "red" light (e.g., jRCaMP1a and jRGECO1a fluorescence) is similarly guided by a secondary beam splitter (to enable separation from "amber" light) at 640 nm (T640lpxr) and subsequently filtered by a ET595/50m-2p band-pass filter. The photomultiplier tubes are Peltier-cooled units, model 7422PA-40 by Hamamatsu Photonics K.K. (Hamamatsu City, Japan).

Excitation wavelengths between 990 and 1020 nm are used in experiments with GCaMP5E/6f (at the higher end when in combination with jRCaMP1a or jRGECO1a).

3   Methods

3.1   Recombined Adeno-associated Virus Gene Transduction of Astrocytes

Numerous studies have employed recombinant AAV (rAAV) vectors to achieve gene transduction in experimental animals [37–39]. The use of rAAV has many advantages, including the apparent lack of pathogenicity, very mild immune responses, efficient transductions, and long-term, high-level gene expression. All of the AAV serotypes that have been isolated can infect various types of cells but show their natural tropism toward specific tissues and cells. In the brain, most AAV serotypes present neuronal tropism, while AAV serotypes 1, 2, 5, 6, 8, and 9 also transduce astrocytes [37, 40]. The wild-type AAV genome (ssDNA, about 4.8 kb) contains two open reading frames, Rep and Cap, flanked by two inverted terminal repeats (ITRs, the only cis element for the AAV production). The Rap and Cap are the genes encoding the viral Rep proteins (Rep 78,

Rep 68, Rep 52, and Rep 40, essential for AAV replication) and capsid proteins (VP1, VP2, and VP3; Fig. 2d). In addition to the Rep and Cap genes, AAV requires a helper plasmid containing essential genes from adenovirus (E4, E2a, and VA) mediating the AAV replication. In the rAAV system, the gene of interest (GOI) is placed between the ITRs in the rAAV encoding plasmid, and the Rep and Cap are supplied in trans by separate helper plasmids. After the virus production (Protocol 3.1.1), the GOI could be transduced via the packaged rAAV by stereotaxic injection (Protocol 3.2.1) into the desired brain regions according to specific coordinates. The GOI carried via rAAV vectors will typically reach its full expression 2–3 weeks after injection (Fig. 2a). Injected rAAVs should be of high purification quality (Fig. 2d, e), since cell debris and other contaminations may induce immune responses and tissue damage [41]. The genomic titer could be obtained by real-time PCR (RT-PCT, Fig. 2c) after the purification step.

3.1.1 Protocol:
Production and Purification
of rAAV

The most commonly used purification procedures during rAAV production include density gradient centrifugation (CsCl gradient and iodixanol gradient) [42] and affinity column chromatographic separation (heparin column and antibody affinity column) [43, 44]. The heparin column works best for rAAV serotype 2 (rAAV2) or any chimeric rAAVs containing serotype 2 [45]. Here we describe a single-step method with commercially available antibody affinity chromatography using AVB Sepharose column to obtain chimeric rAAV1/2, which transduces astrocytes with high efficiency [26, 28]. By using Human Embryonic Kidney 293 (HEK293) cell culture transfection [46], this method [47] allows us to obtain purified rAAVs with genomic titer of 1.0–8.0×10^{12} viral genomes/ml [26, 28, 29, 48, 49].

Detailed Workflow (Estimated Durations in Parenthesis)

1. Plasmid preparation (about 3 days). All plasmid DNA should be of high quality (optical density (260/280 nm) >1.80) and free of RNA contaminants.

   - AAV plasmids containing the GOI and cell-specific promoter flanked by AAV serotype 2 ITRs. The AAV plasmids can be grown in Sure, DH10B, or Stbl2 competent cells for conserving both ITRs. Three different restriction digests with SmaI, MscI, and AhdI of the purified AAV plasmid DNA are recommended for checking the presence of both ITR sequences in the plasmids.

     – pFdelta6, adenovirus helper plasmid (encoding E4, E2a, and VA). This plasmid should be grown in Stbl2 competent cells to avoid partial sequence deletions. Digest with HindIII to obtain bands of about 5.5 kb, 3 kb, 3 kb, 2.3 kb, and 1.5 kb.

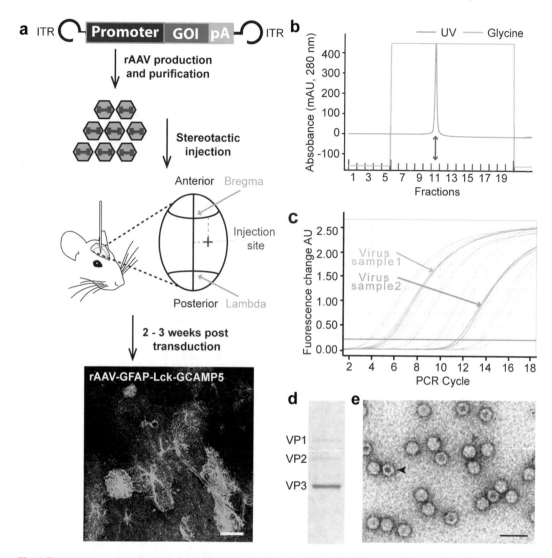

Fig. 2 Transduction of rAAV and virus purification. (**a**) Top, strategy of rAAV gene delivery with cell-type-specific promoter driving genes of interest. Middle, blue dashed lines and red cross represent the way of measuring and desired stereotaxic coordinates relative to bregma. Bottom, a green fluorescent MPLSM image of an acute hippocampal slice from a rAAV-*GFAP*-Lck-GCaMP5 transduced mouse. GOI gene of interest, pA poly A, ITR inverted terminal repeat. (**b**) Enlarged partial chromatogram of rAAV purification. The peak indicated by a double-headed arrow contains the purified rAAV eluted from the AVB antibody column as stated in the protocol (Sect. 3.1.1). Blue line represents the UV absorbance (in arbitrary units (AU)), light green line shows the elution phase of 50 mM glycine (pH 2.7). (**c**) The real-time PCR (RT-PCR) analysis (Light cycler, Prism 7000, Taq-man) to determine the genome-containing rAAV particles. The amplification curves of the six DNA standards in triplicates are given in gray. From left to right: $5.8 \times 10^{-4} - 5.8 \times 10^{-9}$ μg/ml. The amplification curves of two virus samples (triplets each) are shown in green. The final virus titer will be calculated accordingly. The intensity threshold to determine linear fluorescence increase is indicated by the blue line. (**d**) The rAAV1/2 capsid proteins VP1 (87 kDa), VP2 (72 kDa), and VP3 (62 kDa). Coomassie blue staining of purified rAAV particles. (**e**) Electron microscope picture of purified rAAV1/2 particles. Red arrowhead indicates a particle packed with viral DNA; black arrowhead shows a donut-shaped empty particle without viral DNA. Scale bars, 20 μm (**a**), 50 nm (**e**)

- pRV1, containing the AAV2 Rep and Cap sequences. This plasmid can be grown in Top10 and DH5alpha competent cells. Digest with XbaI to obtain DNA fragments of about 7.5 kb and 3.9 kb.

- pH21, containing the AAV1 Rep and Cap sequences. This plasmid can be grown in Top10 and DH5alpha competent cells. Digest with SacI to obtain DNA fragments of about 4.0 kb, 2.8 kb, and 0.5 kb.

2. Preparation of HEK293 cells (1 day).

   - Expand HEK293 cells (2–2.5×10^6) on 94 mm cell culture dishes in 10 ml standard Dulbecco's modified Eagle medium (DMEM) with low glucose-containing 10% fetal calf serum and 100 U/ml penicillin and 100 µg/ml streptomycin.

   - Culture the HEK293 cells until cells reach a confluency of about 50–60% (about 18–24 hours).

3. Transfection of viral plasmids (about 30 minutes for the transfection and 60–72 hours for the incubation).

   - The following materials are for a single batch of virus production (10 × 94 mm cell culture dish):

     - 25 µg AAV plasmid

     - 90 µg pFdelta6 plasmid

     - 25 µg pRV1 plasmid

     - 25 µg pH21 plasmid

     - 625 µl 2 M $CaCl_2$

     - 5 ml 2× BES-buffered saline (BBS, containing 50 mM BES (N,N-bis[2-hydroxyethyl]-2-aminoethanesulfonic acid), 280 mM NaCl, 1.5 mM Na_2HPO_4)

     - Double-distilled water (ddH_2O)

     - HEK293 cells (confluency 50–60%)

   - Prewarm ddH_2O, $CaCl_2$, and 2× BBS to 37 °C.

   - Mix all four abovementioned plasmids, 625 µl 2 M $CaCl_2$ and ddH_2O in a total volume of 5 ml.

   - Vortex and add 500 µl 2× BBS, mix.

   - Wait for 3 minutes, and then spread 1 ml mixture dropwise on each plate without changing medium.

   - Incubate the transfected HEK293 cells in a 3% CO_2 incubator for 20–24 hours.

   - Change medium (10 ml for each plate) and incubate with 5% CO_2 for another 36–48 hours.

4. Harvesting of rAAV (2 hours).

- 60 to 72 hours after transfection, collect culture medium, and gently scrap the cells together with the medium in two 50 ml Falcon tubes.

- Centrifuge the medium and cell collection at 1000 rpm at 4 °C for 10 minutes.

- Collect the medium supernatant and aliquot in four new 50 ml Falcon tubes (25 ml in each tube), and lysate the cell pellets in 10 ml lysis buffer (150 mM NaCl, 20 mM Tris, pH 8.0) in total for 10 minutes at RT.

- Centrifuge the cell lysates at 1000 rpm at 4 °C for 10 minutes. Collect the lysate supernatant and mix together with the medium supernatant.

- Add benzonase nuclease to the supernatant mixture to a final concentration of 50 units/ml, incubate at 37 °C for 1 hour.

- Remove cell debris by centrifuging at 3000 rpm for 10 minutes and transfer to new 50 ml Falcon tubes. At this stage, the sample can be stored either on ice for a short time period (no more than 4 hours) or at −80 °C for several weeks without a reduction of infectivity.

5. Purification of rAAV via fast protein liquid chromatography (FPLC) with HiTrap AVB Sepharose column.

   - Buffers and material: 20% ethanol, 1× PBS (137 mM NaCl, 2.7 mM KCl, 4.3 mM Na$_2$HPO$_4$ · 2H$_2$O, 1.4 mM KH$_2$PO$_4$) with pH 7.5, glycine 50 mM with pH 2.7, Tris-HCl 1 M with pH 8.0, HiTrap AVB Sepharose column (1 ml, GE Healthcare Life Sciences).

   - Equipment: Chromatography system of ÄKTAFPLC (GE Healthcare Life Sciences, discontinued, and replaced with ÄKTA pure).

   - System wash of the ÄKTAFPLC with 20% ethanol once. Insert the HiTrap AVB Sepharose column into ÄKTAFPLC, and equilibrate it with 10 ml 1× PBS (pH 7.5).

   - System wash with 1× PBS pH 7.5 (buffer inlet A) and 50 mM glycine pH 2.7 (buffer inlet B).

   - Filter the supernatant mixture (from step 4) with 0.22 μm Millex syringe filter (Merck Millipore) to prevent blocking of cellular debris in the ÄKTAFPLC.

   - Manually apply the filtered supernatant with a 60 ml syringe from the injection slot, and load with the ÄKTAFPLC program of manual loading onto the AVB column, avoiding air bubbles.

   - Wash the column with 20 ml 1× PBS pH 7.5, and then elute the rAAV from the AVB column with 50 mM glycine pH 2.7.

- Pick up the eluted fractions containing a peak shown on the monitor (Fig. 2b).

- Neutralize the pH of the picked fractions with 100 μl/ml 1 M Tris-HCl pH 8.0, each.

- Wash the fractions three times with 1× PBS pH 7.5 using Amicon Ultra-4 Centrifugal Filter units (Merck Millipore) to a final volume of about 200–300 μl.

- Filter the final purified rAAV with a 13 mm diameter syringe filter (0.22 μm pore size) under cell culture hood. Aliquot the purified rAAV and store at −80 °C until required.

- The real-time PCR could be performed with one aliquot to determine the genomic titer of the purified rAAV (Fig. 2c).

3.2 MPLSM of Astrocytic Ca²⁺ Signals in Acute Brain Slices

MPLSM of GECIs delivered by rAAV has made it possible to visualize the complexity of Ca^{2+} signals within astrocytes. Importantly, a plethora of Ca^{2+} signals has been described, including global transients (occurring more or less simultaneously in multiple astrocytes), localized transients (within a restricted compartment of the cell), and propagating waves [26, 30]. Interestingly, signal kinetics differ between astrocytic somata, fine processes, and end feet (Fig. 3).

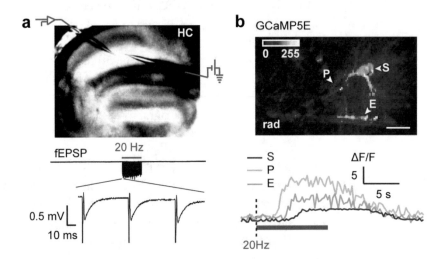

Fig. 3 Stimulation-evoked astrocytic Ca^{2+} signals in acute hippocampal slices. (**a**) Top, image of an acute hippocampal slice with stimulation and recording electrodes positioned in the CA3-CA1 Schaffer collateral/commissural fiber pathway. Bottom, example trace of field excitatory postsynaptic potential (fEPSP) before, during, and after a 20 Hz stimulation train. The fEPSP trace is expanded (scale bars apply to the expanded trace) during the stimulation period (indicated by a red bar). (**b**) Top, standard deviation (SD) fluorescence image of a rAAV-*GFAP*-GCaMP5E-transduced stratum radiatum (rad) astrocyte, with soma (S), process (P), and end foot (E) indicated with arrowheads. The image was obtained from an image series (4 Hz frame rate) taken during stimulation (20 Hz) of the Schaffer collateral/commissural fiber pathway. Scale bar, 10 μm. Bottom, color traces show the fluorescence response of the three astrocytic compartments indicated in image. Red bar, stimulation period. (Modified from Tang et al. [26] with permission from the Society for Neuroscience)

Accumulating evidence indicates that Ca^{2+} signals in different cellular compartments rely on diverse molecular mechanisms [26, 27, 34, 50].

The acute hippocampal slice is a versatile model for studying roles of astrocytes in well-characterized neuronal circuits. In particular, the CA3-CA1 Schaffer collateral/commissural fiber and mossy fiber systems are suitable for studying activity-evoked astrocytic responses. Application of drugs to acute slices offers the potential to delineate mechanisms underlying the complexity of astrocytic Ca^{2+} signals. MPLSM of acute brain slice preparations has demonstrated that astrocytes show both spontaneous and neuronal activity-evoked Ca^{2+} signals [26, 50], though their functional importance remains to be understood.

On the experimental day, mice are euthanized with 5% isoflurane in room air, and the brain is removed. Typically, transverse slices (400 µm) are cut with a vibroslicer in artificial cerebrospinal fluid (ACSF), pH 7.3, at 4 °C, bubbled with 95% O_2 and 5% CO_2, containing (in mM) 124 NaCl, 2 KCl, 1.25 KH_2PO_4, 2 $MgSO_4$, 2 $CaCl_2$, 26 $NaHCO_3$, and 12 glucose. Both in the resting and recording chambers, slices are continuously exposed to humidified gas (95% O_2 and 5% CO_2) at 28–32 °C and perfused with ACSF (pH 7.3).

Before inserting the stimulating and recording electrodes into the brain slice, a region with adequate expression of the fluorescent indicator(s) should be identified by MPLSM. We usually position the two electrodes 150–200 µm from each other. When stable synaptic field potentials have been present for at least 10 minutes, MPLSM synchronized with stimulation protocols can be applied according to experimental purpose and design (Fig. 3). Pharmacological reagents can be diluted in the ACSF and perfused into the recording chamber.

3.2.1 Protocol: Stereotaxic Injection of rAAV for Acute Brain Slice Experiments

We usually inject rAAV with GECI gene construct into a specific brain region [51] 2–3 weeks before the imaging session. This interval allows sufficient expression of GECI in target cells, without evidence of cell damage. However, the optimal expression time varies, depending on the experimental conditions, genes of interest, and injected dosage. The surgical procedures should be performed with great care to avoid brain damage. Note that high pressure applied during the rAAV injection step easily results in tissue death and astrogliosis [31].

Equipment and Materials

- Adult mice, aged about 8–20 weeks. All experiments must be done in accordance with local guidelines, regulations, and permission by the relevant local authorities.

- Purified rAAV. When the rAAV vectors are not well purified, immune responses and tissue death may occur acutely and chronically after injection. Wastes containing rAAV should be handled according to relevant guidelines.

- Ethanol, 70% for disinfection.

- Anesthetics and analgesics. Ketamine (100 mg/kg), xylazine (10 mg/kg), buprenorphine (0.15 mg/kg), and meloxicam (1 mg/kg).

- Eye cream/ointment (Ophtha A/S) and sterile saline (0.9% NaCl).

- Small animal stereotaxic apparatus (Kopf), temperature-controlled heating pad (World Precision Instruments), and surgical microscope (Leica).

- Surgical tools, including scissors, fine forceps, scalpel (Fine Science Tools), and sutures.

- Laboratory scale (Mettler Toledo), small animal hair shaver (Aesculap), dental drill with small drill bits (INTRAcompact 25CHC, KaVo Dental), and syringes (BD Medical).

- Micropipette puller (P 2000, Sutter) and micropipettes (5 μl, Sigma-Aldrich, BR708707-1000EA). After pulling the micropipettes, cut the tip with scissors manually to an opening of about 8–10 μm (inner diameter).

Procedure

1. Disinfection of surgical area and tools. Before starting, the surgical area should be disinfected with 70% ethanol, and surgical tools should be sterilized and dry before use.

2. Animal preparation and anesthesia. 5–10 minutes after the anesthesia injection, shave the fur around the injection site, and fix the animal on the stereotaxic frame with ear bars with moderate pressure and nose clamp. Gently pull the tongue out and place it aside. Cover both eyes with eye cream/ointment and keep the body temperature about 37 °C with the heating-pad.

3. Disinfect the skin with 70% ethanol or 2% iodine in ethanol, and then cut the skin above the injection site with scissors, and pull the skin aside to expose the skull. Gently clean the area around bregma and lambda, and level the bregma and lambda at the same plane. From this step on, a surgical microscope will be needed for the precise operation.

4. According to the coordinates of the target injection area (−2 mm, ±1.5 mm for bilateral hippocampal injection; Fig. 2a), measure the position of x and y coordinates.

5. Thin the skull over the target area using a dental drill until the skull turns almost transparent, and then use fine sharp forceps to tip through the bone and make a tiny opening. Do not drill through the skull, as it may cause injury to the brain parenchyma.

6. Fill the pulled micropipette by applying negative pressure using a tubing attached to the end of the micropipette and a 50 ml syringe connecting to the tubing. The volume of the rAAV could be calculated by unit indicator lines on the micropipette (1 μl/unit, each unit is 15 mm in length, and every 1 mm length represents about 70 nl volume of rAAV). Place the rAAV-filled micropipette with a holder connected to the arm of the stereotaxic frame above the injection site, attach the tubing to the end of the micropipette, and connect the tubing with the 50 ml syringe for applying positive pressure of injection. Under the surgical microscope, carefully insert the micropipette into the tiny opening and start to calculate z coordinates once the tip of the micropipette is touching the dura. For hippocampal injection, typically a two-step injection is performed at two discrete positions along the pipette axis with z coordinate depths of 1.2 and 1.5 mm, and at each position, about 280 nl rAAVs is injected.

7. After having lowered the tip of the micropipette to the desired depth, slowly apply pressure with the connected syringe, and manually control the speed of the injection under the surgical microscope no faster than 50 nl/minute. When finished, unplug the syringe immediately from the injection tubing. Wait at least 2 minutes before withdrawing the micropipette.

8. Suture the skin and clean the sutured area with sterile saline (0.9% NaCl).

9. Recovery from anesthesia and application of analgesic. Keep the animal warm until it fully recovers. Once the animal recovers, subcutaneously inject buprenorphine (0.15 mg/kg body weight), and return the animal to the home cage. Subcutaneously inject additional analgesic (meloxicam, 1 mg/kg body weight) on the first and second days after the operation. Monitor the animal for at least 3 days, observing any signs, such as wound scratching and inflammation of the surgical wound.

3.3 MPLSM of Astrocytic Ca^{2+} Signals In Vivo

A range of studies has illustrated complex roles of astrocytes in physiology and in disease models. Cases in point are the Ca^{2+} responses of cortical astrocytes to whisker stimulation [52] or visual stimuli [53]. Most of these studies were carried out before the introduction of high signal-to-noise ratio GECIs. Consequently, they were performed in acute preparations using synthetic Ca^{2+} indicators delivered by cell-patching or bulk-loading, with the limitations already mentioned (see Introduction). The following sections illustrate MPLSM of astrocytic activity in anesthetized and awake mice in the cortical spreading depression (CSD) model of the migraine aura.

3.3.1 *Astrocytic Activity in Anesthetized Mice*

MPLSM of anesthetized mice enables studies of the brain with intact blood flow. The fact that the mouse is anesthetized enables relatively invasive measurements and manipulations, as well as providing an opportunity to study disease models with severe outcome (e.g., studies of stroke and traumatic brain injury). However, there are some serious caveats regarding general anesthesia and whether or not the tissue has retained relevant physiological parameters. First of all, all types of anesthetics for mice cause some degree of respiratory depression [54]. This may be partly corrected by supplying oxygen to the air they are breathing, and controlled by monitoring blood oxygen saturation. However, such monitoring provides no information on the levels of CO_2 in the organism, which is a crucial factor in interpretation of brain physiology [55, 56]. To counteract this hypoventilation, the mice should be mechanically ventilated with a small animal ventilator through a tracheostomy or an endotracheal tube. Furthermore, it is important to ensure that the mouse has a stable, physiological body temperature, by use of a temperature-controlled heating plate. If left unheated at room temperature, the body temperature of an anesthetized mouse drops to the low 30s within minutes. The trauma of lengthy surgeries is also a major challenge to the mouse physiology, and one should go to great lengths to minimize both the invasiveness and duration of the surgical procedures. If you combine the cardiovascular effects of anesthesia with the surgical trauma and inadequacies in ventilation and temperature control, it adds up to a substantial chance for nonphysiological parameters in your specimen. Keep in mind that mice are small animals with high metabolism and limited reserves, and hence only slight and short-lasting deviations from optimal conditions could be detrimental to the experiment's outcome.

The gold standard to assess whether your preparation retains physiological or near-physiological values is to take arterial blood gas samples before and after experiments for assessment of at least blood oxygen level, CO_2, lactate, pH, and hematocrit. Ideally, and if possible, one should also consider additional monitoring, like continuous monitoring of blood oxygen saturation, capnography of expiratory air (although this is technically challenging), laser Doppler measurements of brain blood flow, and arterial blood pressure measurements.

Procedure: MPLSM of Anesthetized Mice

Anesthesia is induced in a chamber containing 4% isoflurane in room air enriched with 20% pure oxygen and subsequently maintained by nose cone flowing 2% isoflurane with the same gas mixture. This level of oxygen enrichment is sufficient in our hands to maintain arterial normoxia and saves oxygen gas. The induction chamber is cleared of isoflurane by an isoflurane scavenger system (article number 34-0387, Fluovac System, Harvard apparatus). The surgery is performed on a perforated surgical tabletop with active downward suction to clear away excess isoflurane from the nose cone. Body temperature is kept at 37 °C by a temperature-controlled heating plate. Since the temperature recorded by rectal probes may be slightly off, due to slight deviation in rectal probe placement, we recommend only using the rectal probe to correctly calibrate the heating plate and then use the plate's internal reference as feedback mechanism. Note that overheating is a potentially devastating complication when using closed-loop rectal temperature probes to set the temperature. Buprenorphine (0.15 mg/kg) is injected intraperitoneally and the mice are left for 10 minutes before surgery commences. We then perform a tracheotomy and insert an endotracheal tube for artificial ventilation. Anesthesia is maintained with a mixture of isoflurane (2% for surgical procedures, 1.5% during imaging) in room air with a small animal ventilator (SAR-1000; CWE Inc.). The left femoral artery is then cannulated for analyses of arterial blood gas values and the ventilator settings are adjusted to maintain physiological values (pO_2 = 80–120 mm Hg, pCO_2 = 25–35 mm Hg, pH = 7.30–7.50). In brief, the procedure involves dissecting free a proximal section of the femoral artery, followed by distal ligation. When the artery is separated from the vein, a spatula is inserted to facilitate incision of the artery later. For the distal ligation, we use 8-0 silk suture. A vessel clamp is temporarily placed proximally, immediately distal to where the vessels exit the inguinal canal, before a small incision is made to the artery with a straight scalpel blade, severing the lateral third of the diameter of the artery, and a catheter made of PE10 tubing filled with 0.9% NaCl connected to a 500 µl 23-gauge syringe is inserted. To make the catheter, PE10 tubing is pulled slightly to reduce its diameter before being cut to the desired length. The tip is then beveled to facilitate insertion. We use vessel cannulation forceps to advance the catheter (e.g., S&T Vessel cannulation forceps, 11 cm, article number 00574-11). An intravenous catheter may also be inserted in the femoral vein to enable injection of dextran coupled to fluorophores for visualization of the brain vasculature, although slow injection of small volumes of labeled dextrans intra-arterially seems to be unproblematic. There are several ways of inserting a catheter into the femoral vein. One may apply a similar technique as described above for the artery, but this is a lengthy and error-prone procedure, as the vein vessel wall is extremely fragile. To enable quick and secure femoral vein access, we use a slightly blunted tip of a 23-gauge insulin needle inserted into PE10 tubing to cannulate the vessel approximately 5 mm distal to the inguinal canal. If the artery is already cannulated, the skin cut can be advanced distally to expose the insertion site on the vein. Some of the connective tissue covering the vessels is quickly removed. To enable direct insertion of the catheter tip (only the metal part goes into the vein), we hold the connective tissue immediately to the side of the vessel and directly insert the catheter tip. To secure the catheter from falling out, we apply tissue adhesive onto the insertion site (for instance, 3M VetBond Tissue adhesive; although regular cyanoacrylate glue could also be used).

We then create a 2 mm craniotomy with center coordinates anteroposterior −3.0 mm, lateral +2.5 mm, relative to bregma as previously described [4]. In short, a dental drill is used to carefully carve a circular groove in the skull with intermittent application of saline (0.9% NaCl) for cooling, removal of debris, and softening of the bone until only approximately 0.1 mm of the bone thickness is left. Frequent application of saline decreases the risk of accidentally drilling all the way through the skull. The skull is then soaked with saline for 10 minutes more to soften before the bone flap is removed by fine forceps. Subsequently, the dura can be removed with a vessel dilator (article number: 08-969-007, AgnTho's AB) under saline. Agarose (0.8%) in 0.9% NaCl or artificial cerebrospinal fluid is applied at about 37 °C to the brain's surface and stabilized with a circular coverslip (3 mm diameter), the edge of which is fastened to the skull with a small amount of cyanoacrylate glue. To enable insertion of glass electrodes and delivery of drugs, the coverslip can be placed slightly off-center. For induction of CSD, we make a small secondary craniotomy approximately 4 mm rostral to the imaging window to allow epidural application of KCl (2 μl, 1 M) for induction of CSD waves. The small droplet of KCl is left in place for the duration of a single imaging trial (approximately 20 minutes). The rostral craniotomy is subsequently rinsed in saline and left bathed in saline until the next CSD wave is elicited. In experiments with K+-sensitive microelectrodes (described below), two electrodes are warranted (alternatively one could use double-barreled electrodes): one for K+ measurements and one to record changes in the DC potential. To facilitate placement of two electrodes, one may omit the coverslip altogether and use 1.1% agarose in saline or artificial cerebrospinal fluid instead, although this type of preparation is more prone to brain herniation through the craniotomy.

Application: Imaging Ionic Shifts Associated with CSD

We published one of the earliest reports using genetically encoded indicators to visualize astrocyte activity in intact mice [28] (Fig. 4). With superior spatiotemporal resolution, we imaged neuron-astrocyte interactions in CSD and disclosed the order of events at the CSD wave front.

CSD is a cellular phenomenon first described in 1944 by Aristides Leão [57]. It comprises a slow wave (3–5 mm/minute) of massive depolarization of brain tissue, followed by a refractory period where neurons are non-excitable [58, 59]. The phenomenon has been shown to be responsible for the perceptual disturbances of the migraine aura [60]. Similar phenomena have been described in the wake of traumatic brain injury, ischemic stroke,

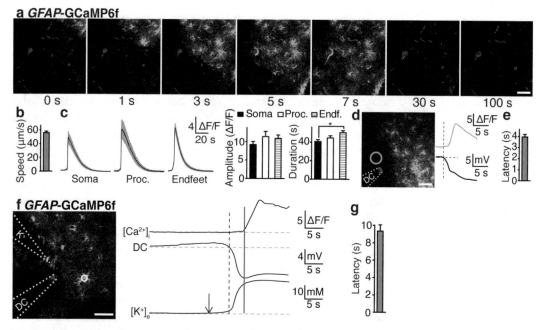

Fig. 4 Astrocytic Ca²⁺ changes during cortical spreading depression (CSD) in anesthetized mice. (**a**) Image series showing astrocytic GCaMP6f fluorescence (green) during passage of a CSD wave (time from Ca²⁺ wave arrival in upper right corner of field of view is indicated). Blood vessels are outlined with Texas red. (**b**) Speed of the astrocytic Ca²⁺ wave. (**c**) Average fluorescence traces (with 95% confidence interval shown in gray), amplitude, and duration (soma, $n = 44$ cells, 17 waves, 9 mice; large processes, $n = 14$ cells, 11 waves, 6 mice; end feet, $n = 45$ cells, 20 waves, 9 mice) of Ca²⁺ transients in astrocytic compartments. (**d**) (Left) Image of GCaMP6f fluorescence close to the tip of the DC potential electrode (dotted lines). Blue circle indicates sampled area, aligned to the wave front as it reaches the tip of the DC electrode. (Right) Astrocytic Ca²⁺ trace aligned to the DC potential. (**e**) Latency between negative DC deflection and astrocytic [Ca²⁺]ᵢ increase. (**f**) Relationship between extracellular K⁺ concentration ([K⁺]ₑ), DC potential, and astrocytic [Ca²⁺]ᵢ. (Left) Dotted lines indicate the placement of the DC electrode and the K⁺ sensitive microelectrode. The white circle denotes the astroglial region-of-interest that was sampled for the timing analyses. (Right) Latency between 0.25 mM [K⁺]ₑ rise (arrow over K⁺ trace) and increase in fluorescence (red vertical line). Dashed line indicates start of the negative DC potential shift. (**g**) Latency between [K⁺]ₑ increase and astrocytic [Ca²⁺]ᵢ increase. Scale bars, 50 μm. $*P < 0.05$; error bars, SEM. (Modified from Enger et al. [28] with permission from Oxford University Press)

and subarachnoid hemorrhage [61]. During the passage of a CSD wave, there are dramatic changes in the affected tissue. For instance, extracellular K⁺ can increase up to more than 60 mM; extracellular glutamate increases manyfold; at the same time as extracellular Na⁺, Ca²⁺ and Cl⁻ ions move intracellularly [62]. EEG is silenced, and there are multiphasic blood flow changes [63]. Experimentally, CSD may be triggered in a multitude of ways. The most commonly used and robust method involves application of KCl to the brain tissue, but application of glutamate, mechanical stimuli, and electrical stimulation may trigger the phenomenon as well [59]. We used topical application of K⁺ in a separate craniotomy, distant to the imaging window, to elicit CSD.

In our study, we combined astrocytic and neuronal Ca^{2+} imaging with extracellular DC potential measurements and K^+ measurements and were able to visualize astrocytic and neuronal activity at the CSD wave front. Furthermore, we used the genetically encoded glutamate indicator iGluSnFR, expressed in the membranes of neurons, to detect the onset of glutamate increase in the extracellular space during these events. Combining these technologies, by using the shifts in the DC potential, we were able to show that astrocyte Ca^{2+} elevations do not constitute the leading event in CSD propagation, but rather that the initial event is a slow increase in extracellular K^+, followed by an increase of neuronal intracellular Ca^{2+} concentration ($[Ca^{2+}]_i$), extracellular glutamate increase, and finally, an increase in astrocytic $[Ca^{2+}]_i$ [28].

Crucial to interpretation of the results of our study was the level of confidence we had with respect to the brain tissue being healthy and representative. Arterial blood gas measurements were taken to assure physiological values. On the one hand, extreme care was taken during surgery to not cause excessive tissue damage. On the other hand, the duration of surgery needed to be kept at a minimum, as physiological values rapidly decline during general anesthesia and there is a window of only a few hours where the mice usually maintain acceptable physiological values.

3.3.2 Astrocytic Activity in Awake Mice

The abovementioned caveats related to the use of general anesthesia during experiments can to a large degree be circumvented by the use of awake, unanesthetized mice, as these are able to maintain their physiological parameters by themselves. Experiments in awake animals are also not confounded by direct effects of anesthetic agents on the imaged parameter. Notably, >90% of astrocytic Ca^{2+} signals seem to be depressed by general anesthesia, although in a dose-dependent and anesthesia-type-dependent manner [64].

A handful articles report on astrocytic Ca^{2+} activity in awake, behaving mice. What is becoming more and more evident from these studies is the strong influence of neuromodulatory pathways for coordinating widespread astrocytic Ca^{2+} activation [65]. Even though some earlier studies reported on awake-like Ca^{2+} activity patterns in astrocytes from acute preparations [66], the first proper characterization of such activity came by Srinivasan and coworkers in 2015 [67], where the spontaneous and startle-mediated Ca^{2+} signals were described in wild-type and *Itpr2*$^{-/-}$ mice. The latter mouse strain lacks the astrocyte-specific inositol-1,4,5-triphosphate receptor subtype 2 (IP3R2), which is a major intracellular source of Ca^{2+} in astrocytes. A large proportion of Ca^{2+} transients in astrocytes seems to be dependent on noradrenergic input, as blocking such receptors reduced astrocytic Ca^{2+} transients by up to 90% in another report by Nedergaard and coworkers [65].

In some regard, such experiments in awake mice are similar to those described above, but notably, in the absence of anesthesia during imaging sessions, the surgical preparations need to be of a chronic nature. Specifically, imaging needs to be done through a chronic cranial window (Figs. 5 and 6), which demands a high degree of care and focus on animal welfare both postoperatively and during experiments.

To enable imaging of awake mice, the mice need to be head-restrained and placed on a relevant stage for the type of experiment being conducted. Currently, there are several different types of stages that can be used for awake animal imaging. These are the three stages we are currently using:

1. Styrofoam balls floating on air, e.g., [66]. Here, the mouse is placed on a styrofoam ball of 20–30 cm diameter (Fig. 5), which is levitated by air. To promote adaptation, supporting rails extending from the head fixation and backward may be added to the stage. Movements of the ball are typically recorded by modified optical computer mice in contact with the styrofoam ball.

2. Tilted disc [68]. For barrel cortex imaging during whisker stimulation, this is our stage of choice, as this stage enables relatively easy whisker tracking. Furthermore, the disc is rather stable, and the mouse may comfortably sit without moving, which could be an advantage for such studies as whisking always occurs during locomotion.

3. Tubes for immobilization [8]. For certain protocols, immobilization in a tube is necessary. This approach seems to need longer acclimatization for the mice, but eventually, the mice are so comfortable that they may even fall asleep.

We have most extensively used the floating styrofoam ball as a default stage, as this in our hands requires little acclimatization. The mice are routinely acclimatized to head fixation over a period of 2–3 days, but a large proportion of the mice handle the head fixation rather well even on the first day of training. The styrofoam spherical treadmill has the most degrees of freedom of all stages for head-fixed mice. A negative factor with these devices is that they are inherently unstable, promoting locomotion. Consequently, when more careful control of behavior is warranted, we prefer a tilted spinning disc where the mouse may sit comfortably without moving. Another disadvantage of the spherical treadmill is that the airflow necessary for levitating the ball effectively spreads large amounts of allergens.

To put cellular activity of astrocytes in a relevant context in an awake, behaving mouse, careful monitoring of animal behavior is required. A minimum should include tracking the movement of the ball/disc/treadmill in combination with an infrared surveillance camera. For more specific behavioral tasks, like imag-

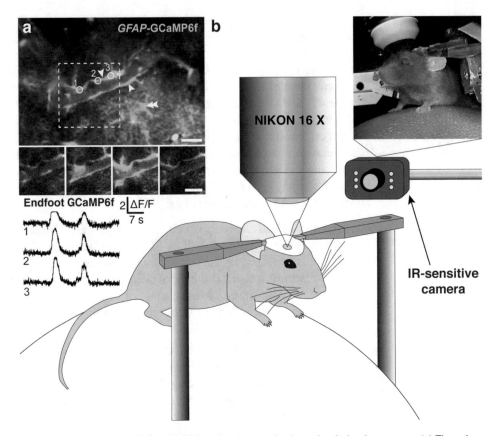

Fig. 5 Spontaneous astrocytic Ca^{2+} activity in a head-restrained awake, behaving mouse. (**a**) The micrograph depicts *GFAP*-GCaMP6f fluorescent astrocytes (green) and a capillary filled with Texas red-labeled dextran (70 kDa, red). Arrowheads denote astrocytic end feet; double arrowhead points to an astrocytic soma. Image sequence from boxed region shows spontaneous GCaMP6f fluorescence changes in end feet along the capillary. Fluorescent traces from regions-of-interest (end feet) 1, 2, and 3 below. Scale bars, 20 μm. (**b**) The mouse moves freely on a spherical treadmill and is monitored with an infrared camera while fluorescence is collected through MPLSM

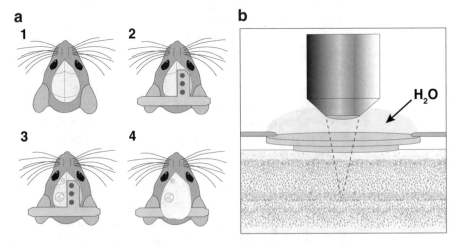

Fig. 6 Surgical procedure for chronic craniotomies for visualization of the cerebral cortex (see Sect. 3.3.2)

ing of the barrel cortex while stimulating whiskers, or imaging of the visual cortex during visual stimulation, additional monitoring is naturally warranted. We use custom-written LabView software to synchronize acquisition of the different data types, and save behavioral data.

Procedure: MPLSM of the Cerebral Cortex of Awake Mice during CSD

Anesthesia and preemptive analgesia are administered as described above. Furthermore, bupivacaine (5 mg/kg) is injected subcutaneously above the cranium. The skin is disinfected by 75% ethanol and 2% iodine in ethanol. Subsequently, a boat-shaped piece of skin overlying the cranium is cut away, and the skull is exposed and cleaned (Fig. 6a, 1). Grooves are then cut by scalpel in a chessboard pattern into the periosteum to enable strong adhesion of the cyanoacrylate glue that is subsequently applied to all exposed areas except where the craniotomies are created. After the cyanoacrylate glue is applied, a custom-made titanium head bar is mounted as in Fig. 6a, 2. It is of crucial importance that the skin edge is devoid of hair, slightly everted, and securely fastened by glue. A 2.5 mm craniotomy with center coordinates anteroposterior −3.0 mm, lateral +2.5 mm relative to bregma is created as described above (Fig. 6a, 3). Coordinates were chosen to maximize the distance from the craniotomy used for eliciting CSD to the imaging window, as well as to enable visualization of the visual cortex, which at least in humans is thought to have a propensity to CSD. In addition, the surrounding perimeter of the craniotomy is thinned and flattened. The skull is then soaked in saline for 10 minutes to soften before the bone flap is removed. Virus encoding fluorescent indicators is injected at three different surface positions, at approximately 350 μm in depth from the surface, distributed to stay clear of vasculature and to infect the central parts of the exposed brain surface. A window made of two custom-made laser-cut circular coverslips of 2.5 and 3.5 mm, respectively, glued together by ultraviolet-cured glue [69], is then centered in the craniotomy, so that the glass plug very slightly depresses the dura (Fig. 6b). The window is subsequently fastened by cyanoacrylate "gel" glue (Power Flex Super Glue, Loctite, Westlake, OH, USA) to the surrounding bone.

Furthermore, similarly to above, a small secondary craniotomy is made approximately 4 mm rostral to the imaging window to allow epidural application of KCl (2 μl, 1 M) for induction of CSD waves. This frontal craniotomy is covered by KWIK-SIL silicone elastomeric adhesive (World Precision Instruments). All exposed areas except the craniotomies are covered with dental cement (Fig. 6a, 4). Mice are given buprenorphine (0.15 mg/kg) subcutaneously once, 8–12 hours postoperatively (in addition to the preemptive dose). Meloxicam (1 mg/kg) is administered subcutaneously immediately postoperatively and once every 24 hours for 2 days postoperatively. If the surgery is performed correctly and care has been taken to maintain physiological parameters and a sufficient, but not excessive level of anesthesia during surgery, the

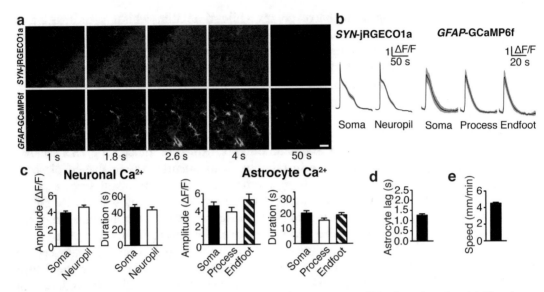

Fig. 7 Dual-color imaging of neuronal and astrocytic Ca^{2+} dynamics in CSD of awake mice. (**a**) Time-lapse image series of the neuronal jRGECO1a (red) and astrocytic GCaMP6f (green) fluorescence during passage of CSD. Scale bar, 25 µm. (**b**) Average fluorescence traces of neuronal jRGECO1a signals and astrocytic GCaMP6f signals. (**c**) Maximum amplitudes and durations of the fluorescent responses in astrocytic and neuronal compartments. (**d**) Time lag between the astrocytic and neuronal Ca^{2+} increase. (**e**) CSD wave propagation speed. Error bars, SEM. (Modified from Enger et al. (2017) [29] with permission from Oxford University Press)

mice should be mobile within minutes after the end of surgery, start eating and drinking within 1–2 hours, and exhibit normal exploratory behavior and grooming from the first postoperative day. For a period of at least 7 days postoperatively, the mice should be left to recover.

Application: Dual-Color Imaging of Ca^{2+} Signals in Neurons and Astrocytes of Awake Mice during CSD

We were the first to simultaneously visualize Ca^{2+} signals in neurons and astrocytes by combining green- and red-shifted genetically encoded Ca^{2+} sensors in awake behaving mice [29] (Fig. 7). We made chronic cranial windows as outlined above and were able to monitor astrocytic and neuronal activity or neuronal activity and extracellular glutamate dynamics simultaneously. As outlined above, we studied the events occurring during waves of CSD. We were able to demonstrate that key parameters are significantly different when studying awake mice, compared to studies in anesthetized mice. For instance, the propagation speed of CSD waves was at least 50% faster in the awake state. Similarly, the duration of Ca^{2+} increases in both neurons and astrocytes were of less than half the duration in the awake state. This strongly underscores the importance of studying unperturbed systems. We further found that mice devoid of the water channel aquaporin-4 (AQP4) displayed altered extracellular glutamate dynamics, and that a considerable amount of the CSD-evoked astrocytic Ca^{2+} responses remained in mice devoid of IP3R2.

3.4 Data Acquisition and Analyses

MPLSM results in large amounts of image data in addition to data from monitoring mouse behavior if imaging awake behaving mice. Since the time scale of astrocytic responses is rather prolonged, imaging series of spontaneous activity in astrocytes should be at least 5–10 minutes. For awake animal imaging, even though a very high acquisition rate rarely is needed to capture astrocytic signals, high frame rates are warranted to compensate for movement artifacts. In our experience, for awake, moving head-fixed mice, one should aim for frame rates of at least 15 Hz, preferably 30 Hz. A 5–10 minutes trial with two imaging channels and frames of 512×512 pixels, stored as 16-bit TIFF files, could easily amount to several gigabytes of image data. Hence, it is important to have a good pipeline for exporting data from the image acquisition computer for storage and processing. A second major step is to correct these time series for movement artifacts. In fact, movement artifacts always occur when mice are moving. Although they can affect all dimensions, they are most predominant in the x- and y-directions [70]. Movement correction tends to be computationally very demanding and should be performed as part of a batch-processing pipeline. We perform this offline on a dedicated computer. We have tried several movement correction algorithms, but are currently exclusively using the SIMA movement correction software [71]. After alignment, fluorescent activity needs to be extracted. Image segmentation is a particularly challenging task for astrocytes, as they have an intricate morphology. We are currently not able to automatically segment the various sub-regions of the astrocytes, in a reliable fashion, and for the time being, we manually segment the videos in regions of interest in a custom-written MATLAB program. Fluorescence intensities are extracted and normalized as $\Delta F/F_0$ values. Baseline is defined differently based on the experiment type in question. For instance, for imaging trials of CSD, the baseline is determined during the time preceding the wave appearing in the field of view. In image trials with spontaneous Ca^{2+} activity and conceptually no obvious baseline period, we determine the baseline by analyzing the distribution of fluorescent values throughout the time series in that particular ROI, apply a median filter, calculate the probability density estimate by use of the MATLAB function *ksdensity()*, find the maximum of the resultant distribution, and use this as F_0.

4 Outlook

The revolution in Ca^{2+} imaging tools has already unraveled that astrocytic Ca^{2+} signals are far more diverse and complex than hitherto appreciated. Notably, we have learned that some Ca^{2+} transients occur within small subcellular compartments, without involving the cell body or other parts of the astrocytic territory. Since the vast majority of papers dealing with astrocytic Ca^{2+} signals restricted their analysis to somatic Ca^{2+} transients in immature animals, GECIs and MPLSM should be used to re-examine the putative roles of astrocytic Ca^{2+} signals in brain physiology and pathophysiology. However, application of the new technology requires strategies for targeting GECIs to specific types of brain cells. Utilization of more selective promoters, such as the promoter for aldehyde dehydrogenase 1 family member L1 (Ald1l1) when studying astrocytes [36], will reduce bias due to unselective GECI expression. Suppression of astrocytic Ca^{2+} signals by anesthetics calls for studies in the awake animal, which, however, offers fewer opportunities for manipulation of cell activity. All-optical approaches for manipulation and readout are expected to overcome this hurdle [72] but require the use of optogenetic actuators and activity sensors that are cross-talk free, i.e., without spectral overlap, or distributed in compartments that are sufficiently far apart from each other to be separately stimulated and imaged by MPLSM. The toolkit of probes and sensors will expand and increase the utility. Notably, dual-color Ca^{2+} imaging with GECIs of different wavelengths will help us elucidate neuronal-glial interactions.

The plethora of different astrocytic Ca^{2+} signals in the awake animal also poses a challenge for the field. How are these signals best defined, characterized, and analyzed? Unbiased, standardized, and automatic descriptions of astrocytic Ca^{2+} are warranted. As several groups are aiming for this, it will hopefully be a reality within the not-too-distant future. Furthermore, movement correction and correlation of diverse neuronal and astrocytic Ca^{2+} activity patterns necessitate powerful computers and advanced statistics. It is also important to keep in mind that astrocytic Ca^{2+} fluctuations are extremely compartmentalized, occurring in tiny 3D volumes. A recent paper argues for 3D imaging [73], which further increases the level of complexity in data processing and analyses. The technical revolution will continue and rapidly expand our knowledge on astrocytes in the intact brain.

Acknowledgments

This work was supported by grants from South and Eastern Norway Regional Health Authority (2016070); the Research Council of Norway (grants 240476, 249988, and 262552); the European Union's Seventh Framework Programme for research, technological development, and demonstration under grant agreement no. 601055; and the Letten Foundation.

References

1. Sherwood CC, Stimpson CD, Raghanti MA, Wildman DE, Uddin M, Grossman LI, Goodman M, Redmond JC, Bonar CJ, Erwin JM, Hof PR (2006) Evolution of increased glia-neuron ratios in the human frontal cortex. Proc Natl Acad Sci U S A 103:13606–13611

2. Azevedo FAC, Carvalho LRB, Grinberg LT, Farfel JM, Ferretti REL, Leite REP, Jacob Filho W, Lent R, Herculano-Houzel S (2009) Equal numbers of neuronal and nonneuronal cells make the human brain an isometrically scaled-up primate brain. J Comp Neurol 513:532–541

3. Rusakov DA (2015) Disentangling calcium-driven astrocyte physiology. Nat Rev Neurosci 16:226–233

4. Takano T, Tian G-F, Peng W, Lou N, Libionka W, Han X, Nedergaard M (2006) Astrocyte-mediated control of cerebral blood flow. Nat Neurosci 9:260–267

5. Khakh BS, Sofroniew MV (2015) Diversity of astrocyte functions and phenotypes in neural circuits. Nat Neurosci 18:942–952

6. Kofuji P, Newman EA (2004) Potassium buffering in the central nervous system. Neuroscience 129:1045–1056

7. Nagelhus EA, Ottersen OP (2013) Physiological roles of aquaporin-4 in brain. Physiol Rev 93:1543–1562

8. Xie L, Kang H, Xu Q, Chen MJ, Liao Y, Thiyagarajan M, O'Donnell J, Christensen DJ, Nicholson C, Iliff JJ et al (2013) Sleep drives metabolite clearance from the adult brain. Science 342:373–377

9. Pellerin L, Magistretti PJ (2012) Sweet sixteen for ANLS. J Cereb Blood Flow Metab 32:1152–1166

10. Cornell-Bell AH, Finkbeiner SM, Cooper MS, Smith SJ (1990) Glutamate induces calcium waves in cultured astrocytes: long-range glial signaling. Science 247:470–473

11. Nedergaard M (1994) Direct signaling from astrocytes to neurons in cultures of mammalian brain cells. Science 263:1768–1771

12. Parpura V, Basarsky TA, Liu F, Jeftinija K, Jeftinija S, Haydon PG (1994) Glutamate-mediated astrocyte-neuron signalling. Nature 369:744–747

13. Haydon PG, Carmignoto G (2006) Astrocyte control of synaptic transmission and neurovascular coupling. Physiol Rev 86:1009–1031

14. Hamilton NB, Attwell D (2010) Do astrocytes really exocytose neurotransmitters? Nat Rev Neurosci 11:227–238

15. Tong X, Shigetomi E, Looger LL, Khakh BS (2013) Genetically encoded calcium indicators and astrocyte calcium microdomains. Neuroscientist 19(3):274–291

16. Heim R, Tsien RY (1996) Engineering green fluorescent protein for improved brightness, longer wavelengths and fluorescence resonance energy transfer. Curr Biol 6:178–182

17. Miyawaki A, Llopis J, Heim R, McCaffery JM, Adams JA, Ikura M, Tsien RY (1997) Fluorescent indicators for Ca²⁺ based on green fluorescent proteins and calmodulin. Nature 388:882–887

18. Tallini YN, Ohkura M, Choi B-R, Ji G, Imoto K, Doran R, Lee J, Plan P, Wilson J, Xin H-B, Sanbe A, Gulick J, Mathai J, Robbins J, Salama G, Nakai J, Kotlikoff MI (2006) Imaging cellular signals in the heart in vivo: Cardiac expression of the high-signal Ca²⁺ indicator GCaMP2. Proc Natl Acad Sci U S A 103:4753–4758

19. Baubet V, Le Mouellic H, Campbell AK, Lucas-Meunier E, Fossier P, Brulet P, Brûlet P (2000) Chimeric green fluorescent protein-aequorin as bioluminescent Ca²⁺ reporters at the single-cell level. Proc Natl Acad Sci U S A 97:7260–7265

20. Nakai J, Ohkura M, Imoto K (2001) A high signal-to-noise Ca²⁺ probe composed of a single green fluorescent protein. Nat Biotechnol 19:137–141

21. Tian L, Hires SA, Mao T, Huber D, Chiappe ME, Chalasani SH, Petreanu L, Akerboom J, McKinney SA, Schreiter ER, Bargmann CI, Jayaraman V, Svoboda K, Looger LL (2009) Imaging neural activity in worms, flies and mice with improved GCaMP calcium indicators. Nat Methods 6:875–881

22. Shigetomi E, Kracun S, Khakh BS (2010) Monitoring astrocyte calcium microdomains with improved membrane targeted GCaMP reporters. Neuron Glia Biol 6:183–191

23. Shigetomi E, Kracun S, Sofroniew MV, Khakh BS (2010) A genetically targeted optical sensor to monitor calcium signals in astrocyte processes. Nat Neurosci 13:759–766

24. Akerboom J, Chen T-W, Wardill TJ, Tian L, Marvin JS, Mutlu S, Calderón NC, Esposti F, Borghuis BG, Sun XR et al (2012) Optimization of a GCaMP calcium indicator for neural activity imaging. J Neurosci 32:13819–13840

25. Chen T-W, Wardill TJ, Sun Y, Pulver SR, Renninger SL, Baohan A, Schreiter ER, Kerr RA, Orger MB, Jayaraman V et al (2013) Ultrasensitive fluorescent proteins for imaging neuronal activity. Nature 499:295–300

26. Tang W, Szokol K, Jensen V, Enger R, Trivedi CA, Hvalby O, Helm PJ, Looger LL, Sprengel

R, Nagelhus EA (2015) Stimulation-evoked Ca^{2+} signals in astrocytic processes at hippocampal CA3-CA1 synapses of adult mice are modulated by glutamate and ATP. J Neurosci 35:3016–3021

27. Szokol K, Heuser K, Tang W, Jensen V, Enger R, Bedner P, Steinhäuser C, Taubøll E, Ottersen OP, Nagelhus EA (2015) Augmentation of Ca^{2+} signaling in astrocytic endfeet in the latent phase of temporal lobe epilepsy. Front Cell Neurosci 9:49

28. Enger R, Tang W, Vindedal GF, Jensen V, Helm PJ, Sprengel R, Looger LL, Nagelhus EA (2015) Dynamics of ionic shifts in cortical spreading depression. Cereb Cortex 25:4469–4476

29. Enger R, Dukefoss DB, Tang W, Pettersen KH, Bjørnstad DM, Helm PJ, Jensen V, Sprengel R, Vervaeke K, Ottersen OP, Nagelhus EA (2017) Deletion of aquaporin-4 curtails extracellular glutamate elevation in cortical spreading depression in awake mice. Cereb Cortex 27:24–33

30. Haustein MD, Kracun S, Lu X-H, Shih T, Jackson-Weaver O, Tong X, Xu J, Yang XW, O'Dell TJ, Marvin JS et al (2014) Conditions and constraints for astrocyte calcium signaling in the hippocampal mossy fiber pathway. Neuron 82:413–429

31. Shigetomi E, Bushong EA, Haustein MD, Tong X, Jackson-Weaver O, Kracun S, Xu J, Sofroniew MV, Ellisman MH, Khakh BS (2013) Imaging calcium microdomains within entire astrocyte territories and endfeet with GCaMPs expressed using adeno-associated viruses. J Gen Physiol 141:633–647

32. Zlatkine P, Mehul B, Magee AI (1997) Retargeting of cytosolic proteins to the plasma membrane by the Lck protein tyrosine kinase dual acylation motif. J Cell Sci 110:673–679

33. Benediktsson AM, Schachtele SJ, Green SH, Dailey ME (2005) Ballistic labeling and dynamic imaging of astrocytes in organotypic hippocampal slice cultures. J Neurosci Meth 141:41–53

34. Li H, Wang X, Zhang N, Gottipati MK, Parpura V, Ding S (2014) Imaging of mitochondrial Ca^{2+} dynamics in astrocytes using cell-specific mitochondria-targeted GCaMP5G/6s: Mitochondrial Ca^{2+} uptake and cytosolic Ca^{2+} availability via the endoplasmic reticulum store. Cell Calcium 56:457–466

35. Hirrlinger J, Scheller A, Hirrlinger PG, Kellert B, Tang W, Wehr MC, Goebbels S, Reichenbach A, Sprengel R, Rossner M, Kirchhoff F (2009) Split-Cre complementation indicates coincident activity of different genes in vivo. PLoS One 4(1):e4286

36. Srinivasan R, Lu TY, Chai H, Xu J, Huang BS, Golshani P, Coppola G, Khakh BS (2016) New transgenic mouse lines for selectively targeting astrocytes and studying calcium signals in astrocyte processes in situ and in vivo. Neuron 92:1181–1195

37. Zincarelli C, Soltys S, Rengo G, Rabinowitz JE (2008) Analysis of AAV serotypes 1-9 mediated gene expression and tropism in mice after systemic injection. Mol Ther 16:1073–1080

38. Shevtsova Z, Malik JM, Michel U, Bahr M, Kugler S (2005) Promoters and serotypes: targeting of adeno-associated virus vectors for gene transfer in the rat central nervous system in vitro and in vivo. Exp Physiol 90:53–59

39. Watakabe A, Ohtsuka M, Kinoshita M, Takaji M, Isa K, Mizukami H, Ozawa K, Isa T, Yamamori T (2015) Comparative analyses of adeno-associated viral vector serotypes 1, 2, 5, 8 and 9 in marmoset, mouse and macaque cerebral cortex. Neurosci Res 93:144–157

40. Aschauer DF, Kreuz S, Rumpel S (2013) Analysis of transduction efficiency, tropism and axonal transport of AAV serotypes 1, 2, 5, 6, 8 and 9 in the mouse brain. PLoS One 8(9):e76310

41. Wright JF, Zelenaia O (2011) Vector characterization methods for quality control testing of recombinant adeno-associated viruses. Methods Mol Biol 737:247–278

42. Strobel B, Miller FD, Rist W, Lamla T (2015) Comparative analysis of cesium chloride- and iodixanol-based purification of recombinant adeno-associated virus (AAV) vectors for preclinical applications. Hum Gene Ther Methods 112:1–29

43. Burova E, Ioffe E (2005) Chromatographic purification of recombinant adenoviral and adeno-associated viral vectors: methods and implications. Gene Ther 12(Suppl 1):S5–S17

44. Smith RH, Levy JR, Kotin RM (2009) A simplified baculovirus-AAV expression vector system coupled with one-step affinity purification yields high-titer rAAV stocks from insect cells. Mol Ther 17:1888–1896

45. Mietzsch M, Broecker F, Reinhardt A, Seeberger PH, Heilbronn R, Imperiale MJ (2014) Differential adeno-associated virus serotype-specific interaction patterns with synthetic heparins and other glycans. J Virol 88(5):2991–3003

46. Chen CA, Okayama H (1988) Calcium phosphate-mediated gene tranfer: a highly efficient transfection system for stably transforming cells with plasmid DNA. Biotechniques 6:632–638

47. McClure C, Cole KLH, Wulff P, Klugmann M, Murray AJ (2011) Production and titering of recombinant adeno-associated viral vectors. J Vis Exp (57):e3348

48. Tang W, Ehrlich I, Wolff SBE, Michalski A-M, Wölfl S, Hasan MT, Lüthi A, Sprengel R

(2009) Faithful expression of multiple proteins via 2A-peptide self-processing: a versatile and reliable method for manipulating brain circuits. J Neurosci 29:8621–8629

49. Berkel S, Tang W, Treviäo M, Vogt M, Obenhaus HA, Gass P, Scherer SW, Sprengel R, Schratt G, Rappold GA (2012) Inherited and de novo SHANK2 variants associated with autism spectrum disorder impair neuronal morphogenesis and physiology. Hum Mol Genet 21:344–357

50. Agarwal A, Wu PH, Hughes EG, Fukaya M, Tischfield MA, Langseth AJ, Wirtz D, Bergles DE (2017) Transient opening of the mitochondrial permeability transition pore induces microdomain calcium transients in astrocyte processes. Neuron 93:587–605.e7

51. Cetin A, Komai S, Eliava M, Seeburg PH, Osten P (2007) Stereotaxic gene delivery in the rodent brain. Nat Prot 1(6):3166

52. Wang X, Lou N, Xu Q, Tian G-F, Peng WG, Han X, Kang J, Takano T, Nedergaard M (2006) Astrocytic Ca²⁺ signaling evoked by sensory stimulation in vivo. Nat Neurosci 9:816–823

53. Schummers J, Yu H, Sur M (2008) Tuned responses of astrocytes and their influence on hemodynamic signals in the visual cortex. Science 320:1638–1643

54. Schwarte LA, Zuurbier CJ, Ince C (2000) Mechanical ventilation of mice. Basic Res Cardiol 95:510–520

55. Krnjević K, Randić M, Siesjö BK (1965) Cortical CO₂ tension and neuronal excitability. J Physiol 176:105–122

56. Tolner EA, Hochman DW, Hassinen P, Otáhal J, Gaily E, Haglund MM, Kubová H, Schuchmann S, Vanhatalo S, Kaila K (2011) Five percent CO₂ is a potent, fast-acting inhalation anticonvulsant. Epilepsia 52:104–114

57. Leão AAP (1944) Spreading depression of activity in the cerebral cortex. J Neurophysiol 7:359–390

58. Pietrobon D, Moskowitz MA (2014) Chaos and commotion in the wake of cortical spreading depression and spreading depolarizations. Nat Rev Neurosci 15:379–393

59. Somjen GG (2001) Mechanisms of spreading depression and hypoxic spreading depression-like depolarization. Physiol Rev 81:1065–1096

60. Takano T, Nedergaard M (2009) Deciphering migraine. J Clin Invest 119:16–19

61. Dreier JP (2011) The role of spreading depression, spreading depolarization and spreading ischemia in neurological disease. Nat Med 17:439–447

62. Kraig RP, Nicholson C (1978) Extracellular ionic variations during spreading depression. Neuroscience 3:1045–1059

63. Takano T, Oberheim N, Cotrina ML, Nedergaard M (2009) Astrocytes and ischemic injury. Stroke 40:S8–S12

64. Thrane AS, Rangroo Thrane V, Zeppenfeld D, Lou N, Xu Q, Nagelhus EA, Nedergaard M (2012) General anesthesia selectively disrupts astrocyte calcium signaling in the awake mouse cortex. Proc Natl Acad Sci U S A 109: 18974–18979

65. Ding F, O'Donnell J, Thrane AS, Zeppenfeld D, Kang H, Xie L, Wang F, Nedergaard M (2013) α1-Adrenergic receptors mediate coordinated Ca²⁺ signaling of cortical astrocytes in awake, behaving mice. Cell Calcium 54:387–394

66. Nimmerjahn A, Mukamel EA, Schnitzer MJ (2009) Motor behavior activates Bergmann glial networks. Neuron 62:400–412

67. Srinivasan R, Huang BS, Venugopal S, Johnston AD, Chai H, Zeng H, Golshani P, Khakh BS (2015) Ca²⁺ signaling in astrocytes from Ip3r2−/− mice in brain slices and during startle responses in vivo. Nat Neurosci 18:708–717

68. Danskin B, Denman D, Valley M, Ollerenshaw D, Williams D, Groblewski P, Reid C, Olsen S, Waters J (2015) Optogenetics in mice performing a visual discrimination task: Measurement and suppression of retinal activation and the resulting behavioral artifact. PLoS One 10(12):e0144760

69. Huber D, Gutnisky DA, Peron S, O'Connor DH, Wiegert JS, Tian L, Oertner TG, Looger LL, Svoboda K (2012) Multiple dynamic representations in the motor cortex during sensorimotor learning. Nature 484:473–478

70. Dombeck DA, Harvey CD, Tian L, Looger LL, Tank DW (2010) Functional imaging of hippocampal place cells at cellular resolution during virtual navigation. Nat Neurosci 13:1433–1440

71. Kaifosh P, Zaremba JD, Danielson NB, Losonczy A (2014) SIMA: Python software for analysis of dynamic fluorescence imaging data. Front Neuroinform 8:80

72. Emiliani V, Cohen AE, Deisseroth K, Hausser M (2015) All-optical interrogation of neural circuits. J Neurosci 35:13917–13926

73. Bindocci E, Savtchouk I, Liaudet N, Becker D, Carriero G, Volterra A (2017) Three-dimensional Ca²⁺ imaging advances understanding of astrocyte biology. Science 356(6339):eaai8185

Chapter 12

Two-Photon Imaging of Dendritic Calcium Dynamics In Vivo

Lucy M. Palmer

Abstract

Developments in microscopes and indicators over the past decade have paved the way for recording neural dynamics in vivo. Previously inaccessible small neuronal processes, such as dendrites and axons, can now be probed in vivo using two-photon microscopy, revealing their important role in information processing during behaviour. To perform such experiments, various tools, techniques and considerations are required. In this chapter, the procedures for recording dendritic calcium dynamics in vivo are detailed in a step-by-step manner. The various sources of calcium contributing to the recorded transients are discussed, and details are given on how to identify them from the characteristics of the recorded fluorescence transients. These procedural details and considerations regarding two-photon calcium imaging of dendritic activity are put in context of the importance of such recordings for understanding neural function during behaviour.

Key words Two-photon, Dendrites, In vivo, Behaviour, Procedures, Calcium imaging, Awake

1 Introduction

Dendrites are treelike neural projections that receive the majority of synaptic inputs onto a neuron. First reported by Deiters in 1865 [1], the role of dendrites in the processing of synaptic input has been heavily debated. Their small size (<1 μm diameter) hindered direct recordings, and therefore, their role in neural processing remained one of the mysteries of the neuronal structure. Historically, much of what we know about active dendritic processing has come from brain slice preparations. Although crucial to understanding their computational capabilities and integration rules, this preparation severs the most important feature of dendrites, encoding synaptic input patterns. Therefore, probing dendritic function in the intact preparation is vital to unravelling their role during information processing and behaviour. Measuring from such small structures in the intact brain is difficult. However, this is now possible with the advent of modern recording techniques, including multiphoton imaging.

Espen Hartveit (ed.), *Multiphoton Microscopy*, Neuromethods, vol. 148,
https://doi.org/10.1007/978-1-4939-9702-2_12, © Springer Science+Business Media, LLC, part of Springer Nature 2019

In the past few decades, two-photon fluorescence excitation [2] has enabled dendrites filled with fluorophores to be visualized deep into tissue (up to 800 μm deep), revealing the importance of dendrites in both receiving and transforming synaptic input in vivo. This and many more considerations regarding multiphoton imaging of dendritic calcium dynamics in vivo are explored in detail here. In this chapter, the importance of imaging calcium in dendrites is discussed, and the various sources of calcium influx through both passive and active processes are explained. The various calcium indicators available are listed, as well as the various considerations which must be taken into account when imaging dendritic activity in vivo. Lastly, the procedures for recording dendritic calcium dynamics in vivo are detailed in a step-by-step manner.

1.1 Why Image Calcium in Dendrites?

Calcium is a very important ion which is involved in many signalling cascades (for a review, see [3]). Sources of calcium influx in neurons are numerable, including calcium-permeable α-amino-3-hydroxy-5-methyl-4-isoxazolepropionic acid (AMPA) and N-methyl-D-aspartate (NMDA) glutamate-type receptors, voltage-gated calcium channels (VGCC), transient receptor potential type C (TRPC) channels and nicotinic acetylcholine receptors (nAChR). Furthermore, calcium can also be released from internal stores which is mediated by inositol tris-phosphate receptors (IP3R) and ryanodine receptors (RyR) (Fig. 1a). To monitor calcium activity in living cells, many different calcium indicators have been developed over the past half century. Historically, these calcium indicators were in the form of organic dyes and more recently, the development of genetic calcium indicators is beginning to rival the synthetic indicators in both calcium affinity and application [4] (discussed in detail below).

Due to their large signal-to-noise ratio, calcium indicators are currently one of the most popular methods to image neuronal activity. However, although calcium is involved in many signalling cascades at the molecular level, most in vivo imaging studies to date have not focused on the role of calcium as a second messenger, but instead use the measured changes in calcium fluorescence to infer neural activity. Generally speaking, in most studies, calcium imaging is used as a proxy for reporting voltage activity. However, it must be stressed that recorded dendritic calcium fluorescence signals are just changes in the calcium concentration within a neuronal compartment.

1.2 Sources of Calcium Signals in Dendrites

Interpreting the measured calcium responses can be problematic for the following reasons:

1. *Resolution.* Calcium indicators are generally not able to resolve single action potentials [4]. To date, synthetic indicators (such as Oregon Green BAPTA variants) have better signal-to-noise ratio than genetic indicators, although this may change as more effort is currently expended on generating better genetic indicators [4].

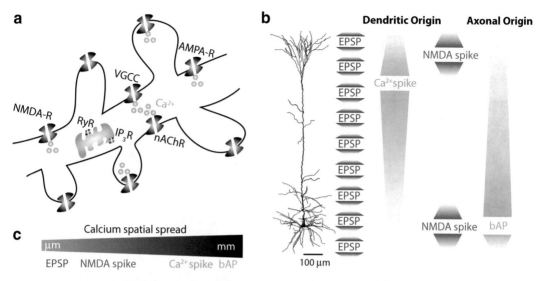

Fig. 1 Sources of calcium influx in dendrites. (**a**) Sources of calcium influx in dendrites are calcium-permeable α-amino-3-hydroxy-5-methyl-4-isoxazolepropionic acid (AMPA) and N-methyl-D-aspartate (NMDA) glutamate-type receptors, voltage-gated calcium channels (VGCC), nicotinic acetylcholine receptors (nAChR), and release of calcium from internal stores via activation of ryanodine receptors (RyR) and IP3 receptors (IP3R). Channel and receptor location are schematic. (**b**) At the cellular level, dendritic calcium can be due to both voltage events which originate in the axon (backpropagating action potential, bAP) or the dendrite (EPSP, Ca²⁺ spike, NMDA spike). (**c**) The voltage event underlying the recorded changes in dendritic calcium can be determined by assessing the location (**b**) and spatial spread of the calcium transient

2. *Buffering.* The buffering of endogenous calcium by calcium indicators (particularly those of high affinity) causes a decrease in the amplitude and an increase in the duration of the recorded calcium [5, 6].

3. *Non-linearity.* The linearity of the calcium indicator is heavily dependent on the frequency of action potentials. Above certain action potential firing frequencies, calcium indicators are non-linear due to their slow decay (large time constant of decay) [7]. Dendrites further complicate the linearity of calcium indicators as the extent of linearity is also dependent on the dendritic location [7].

4. *Identity.* Many voltage events which occur within dendrites cause calcium influx (Fig. 1b). Identifying the voltage event causing the recorded dendritic activity can be troublesome, especially in vivo, and is largely dependent on the sensitivity of the indicator and the recording conditions. Assessing the dendritic location and spatial distribution of the recorded calcium influx can highlight the potential source of calcium (Fig. 1b). Pharmacological manipulations can further illustrate the receptors/channels gating the calcium influx, adding to the weight of evidence.

    The main dendritic voltage events which have been recorded in vivo, including backpropagating action potentials, calcium-dependent dendritic spike/plateau potentials, NMDA-dependent dendritic spikes and EPSPs, are listed in Table 1 and described in detail below.

Table 1
Identification of neural voltage events leading to calcium transients. Based on the spatial spread, branch amplitude, inter-event amplitude and distal dendrite invasion of the recorded calcium transients, identification of the voltage event (backpropagating action potential (bAP), Ca²⁺ spike, NMDA spike, EPSP) can be putatively inferred

	Spatial spread	Branch amplitude	Inter-event amplitude	Distal dendrite invasion
bAP	Global	Uniform	Non-uniform	No, single bAP Yes, burst
Ca²⁺ spike	Global	Uniform	Non-uniform	Yes
NMDA spike	Local (30 μm)	Non-uniform	Uniform	No
EPSP	Local (1 μm)	Non-uniform	Uniform	No

1.2.1 Backpropagating Action Potentials

Sodium action potentials are initiated in the axon in most neurons in vitro [8–10]. Although there is no direct evidence for the site of action potential initiation in vivo, it is generally assumed to be the same as in vitro [11, 12]. Since the early 1990s, it has been clear that action potentials not only travel down the axon but can also invade dendrites [13–24]. These backpropagating action potentials (bAPs) have been shown to interact with incoming synaptic input which is seen as one of the cellular mechanisms behind learning and memory (for a review, see [25]) and coincidence detection [6, 7, 26–31]. bAPs were first recorded in vivo using two-photon calcium imaging and intracellular electrophysiology [32], and they have since been shown to reliably invade the basal dendrites [33] but not the distal apical tuft dendrites [20, 33, 34] of cortical pyramidal neurons in vivo. For in vivo recordings, action potential backpropagation raises issues of interpretation since unfortunately, the calcium signature of bAPs in the dendrites can also be confused with other dendritic signals (namely, calcium-dependent dendritic spike/plateau potentials).

Identifying dendritic calcium transients as bAPs in vivo:

1. Global signal which invades the entire dendritic arbour

2. Relatively uniform amplitude throughout a dendritic branch

3. Non-uniform amplitude between different events (as the number and frequency of action potentials contributing to the transient may differ)

4. Do not typically invade distal dendrites efficiently (e.g., single bAPs do not evoke a calcium signal in the distal dendrites of layer 5 pyramidal neurons)

1.2.2 Calcium-Dependent Dendritic Spikes/Plateau Potentials

Local input onto the distal apical dendrite can activate calcium conductances that generate all-or-none calcium-dependent dendritic spikes/plateau potentials [35, 36]. These supralinear events are mediated primarily by L-type calcium channels [37] and have a very powerful effect on the firing of the neuron, switching the firing pattern to burst-firing mode [38]. Furthermore, the invasion of distal dendrites by calcium-dependent dendritic spikes/plateau potentials may support coincidence detection [39, 40], which is regulated by various ion channels [40, 41]. Calcium-dependent dendritic spikes/plateau potentials are vital for sensory perception. Using genetic calcium indicators, Xu et al. [42] illustrated the occurrence of calcium plateau potentials in the distal dendrites of layer 2/3 pyramidal neurons in the barrel cortex during active whisking. Takahashi et al. [43] further illustrated that these calcium plateau potentials are causally linked to whisker deflection detection, and they have been shown to cause synaptic plasticity [44, 45].

Identifying dendritic calcium transients as calcium-dependent dendritic spikes/plateau potentials in vivo:

1. Global signal which invades the entire dendritic arbour
2. Relatively uniform amplitude throughout a dendritic branch
3. Relatively uniform amplitude between different events
4. Typically invade distal dendrites efficiently

1.2.3 NMDA-Dependent Dendritic Spikes

The thin (sub-micron) dendrites in the basal and apical tuft can generate so-called NMDA spikes in response to synaptic input [39, 46, 47]. NMDA spikes, as the name implies, are generated by the regenerative activation of NMDA-receptor channels (after relief from voltage-dependent Mg^{2+} block) resulting in a non-linear voltage response to highly localized (<30 μm) synaptic input [34]. NMDA spikes do not have as great an impact on the firing rate as calcium spikes, as multiple NMDA spikes in distal dendrites of cortical pyramidal neurons are required to alter somatic firing [34, 39]. However, they increase the electrical coupling of fine dendrites to the soma [48], allowing distal inputs to overcome their electrotonic disadvantage and have a greater influence on somatic potentials than proximal inputs [34, 49]. Consequently, NMDA spikes are important during sensory processing and behaviour. Using a combination of techniques, NMDA spikes have been shown to encode sensory information [34] and angular tuning [50] in the tuft dendrites of layer 2/3 cortical neurons and basal dendrites of layer 4 stellate neurons, respectively.

Identifying dendritic calcium transients as NMDA spikes in vivo:

1. Local signal (30 μm spatial spread; [34])
2. Non-uniform amplitude throughout dendritic branch
3. Relatively uniform amplitude between different events

1.2.4 Excitatory Postsynaptic Potentials (EPSP)

Dendrites can receive hundreds/thousands of synaptic inputs at any given time. Since the resulting calcium influx is small during an EPSP compared to other dendritic voltage events (e.g., dendritic spikes), it is often hard to detect EPSPs in vivo. However the development of increasingly sensitive probes [4, 51] has recently enabled spine dynamics to be investigated in response to various sensory inputs. Many studies looking at spine activity in vivo have addressed the issue of whether synaptic inputs during certain paradigms are 'clustered' in the dendritic tree. Spine calcium activity has been subsequently assessed in many physiological systems with conflicting results, with some studies supporting [52–55] and others not supporting [56, 57] functional synaptic clustering. Since the dendritic calcium signal resulting from synaptic input is highly dependent on the spine neck resistance [58–61], most synaptic input onto a single spine does not result in a measurable calcium influx in the parent dendrite [4].

Identifying dendritic calcium transients as EPSPs in vivo:

1. Extremely spatially restricted signal

2. Relatively uniform amplitude between different events

1.2.5 Other Sources of Cellular Calcium

Calcium changes that are generated by calcium entry through voltage- and ligand-gated channels are the best characterized. However, the release of calcium from intracellular stores can also result in increased dendritic calcium although these events will not be discussed in detail here.

2 Materials

2.1 Calcium Indicators Used for Dendritic Imaging

The loading of calcium indicators into neurons is a crucial first step for imaging the dynamics of dendritic activity in vivo. When choosing the right indicator, various factors must be considered including the preparation, the question (single neuron or population imaging) and the application (cell permeant/impermeant). Details of the synthetic and genetic calcium indicators which are currently typically used when imaging dendritic activity in vivo are described below.

2.1.1 Synthetic (Organic) Calcium Indicators

Synthetic calcium indicators were first used over half a century ago [62, 63] and have since been continuously developed. The most popular synthetic calcium indicators were developed from the hybridization of calcium-selective chelators (BAPTA or EGTA) with aromatic chromophores in the laboratory of Roger Tsien [64]. Members of the Oregon Green BAPTA dye family are typically the synthetic calcium indicators of choice for dendritic imaging due to their high resting fluorescence, large signal-to-noise ratio and easy delivery. Depending on the membrane permeability of the dye, it can be loaded either extra- or intracellularly to investigate important dendritic processes in vivo [33, 34, 56, 65–68]. More recently, other synthetic dyes are now available which have higher signal-to-noise ratios [69].

2.1.2 Genetically Engineered Calcium Indicators (GECIs)

Potentially one of the greatest advances in the past decade which has propelled the field of dendritic computation in vivo has been the development of genetically encoded calcium indicators (GECIs). While the early GECIs were limited due to their slow response kinetics and low signal-to-noise ratios [70], highly sensitive GECIs have now been developed which can detect single action potentials in vivo [4]. Although synthetic indicators are generally still 'better' at detecting single action potentials than GECIs [4], GECIs generally have stable expression and therefore enable calcium activity to be recorded over long periods of time. Varieties of the single-fluorophore GECI, GCaMP, have been used extensively to measure dendritic activity in vivo in the cortex [34, 40, 42–45, 55, 71, 72], hippocampus [73] and olfactory bulb [74].

3 Methods

3.1 Procedures for Recording Dendritic Calcium Dynamics In Vivo

The following procedures outline the important steps required for calcium imaging in vivo including details about the (1) virus injection surgery (if not using transgenic mice), (2) head-post and chronic window implantation, (3) habituation for awake recordings and (4) dendritic calcium imaging. The details below are based on mouse models and all procedures must adhere to local animal welfare regulations.

3.1.1 Virus Injection

Instead of using transgenic animals expressing calcium indicators, viral transduction protocols can be used to express GECIs in brain regions of interest

1. Anesthetize the mouse with either 1.5–3.0% isoflurane (vol/vol, in O_2) delivered via nose mask or an intraperitoneal injection of ketamine (0.1 mg/g body weight) and xylazine (0.01 mg/g body weight).

2. Anesthesia is maintained throughout the surgical procedure and depth of anesthesia is assessed frequently by checking vital signs. If required, anesthetic top-ups are given (increase isoflurane concentration or administer 10% of initial volume of ketamine/xylazine dose).

3. Place the mouse on a heat pad (maintained at ~37° C) via rectal probe.

4. Cover both eyes with eye ointment (petrolatum ophthalmic ointment).

5. Inject lidocaine hydrochloride subcutaneously above the skull (~50 µl; 1% in saline).

6. Stabilize the head in a stereotaxic frame.

7. Remove the hair from the top of the head.

8. Clean the scalp with three alternating washes of 70% ethanol and betadine.

9. Using sterile surgical tools, make a small incision above the brain region of interest (~1 cm long).

10. Gently drill a small hole (~1 mm diameter) with a dental drill (1/4″ drill bit) in the skull above the brain region of interest, ensuring not to break through and damage the underlying brain. To prevent warming of the underlying brain, flush with sterilized saline regularly.

11. Place the virus injection glass pipette (5–10 μm tip diameter) into the hole, intently watching as the pipette tip 'punches through' the dura. This is crucial so you are able to zero the z coordinate to the surface of the brain.

12. Move the injection pipette forward to the required depth at a slow-medium pace (10 μm per second).

13. Slowly inject the virus at a given depth (typically, 30–70 nl for adeno-associated viruses). Wait >5 minutes to allow the injected virus to settle before bringing the injection pipette to the surface and out of the brain. For dendritic imaging, it is typically advantageous to sparsely label neurons to limit dendritic overlap, limit background fluorescence and better enable the imaged dendritic branch to be traced back to the soma.

14. Suture the incision and place betadine on the sutured wound.

15. Administer pain relief, meloxicam (1–3 mg/kg body weight), 30 min prior to cessation of anesthesia.

16. Place mouse back into the home cage with heat mat and monitor recovery.

17. Wait 3–5 weeks for virus expression (according to particular virus injected).

3.1.2 Head-Post and Chronic Window Implantation

To enable chronic dendritic imaging, a head-post and chronic window must be implanted above the brain region of interest. This step must be performed whether the calcium indicator was genetic or synthetic.

1. Perform steps 1–7 described above.

2. Using sterile surgical tools, remove all the skin from above the skull.

3. Clear exposed skull initially with scalpel and then with bone scraper. Note: ensure no membranes remain as otherwise the head-bar will not be able to be affixed to the skull.

4. Apply a thin layer of cyanoacrylate glue to the edge of the exposed skin. Ensure there is absolutely no fluid/blood coming from the exposed skin.

5. Secure the head-post to the skull, initially with a thin layer of cyanoacrylate glue and then with strong dental cement (e.g., C&B Metabond, Parkell). Alternatively, UV curable glue can be used (Optibond, Kerr).

6. Place mouse in the head-post frame.

7. For 3 mm chronic window, place 3 mm diameter round coverslip on skull with brain region of interest in the middle. If possible, try to avoid skull sutures. Sketch around the coverslip with the tip of standard #5 forceps as a guide to the craniotomy dimensions.

8. Slowly drill along the sketched craniotomy outline with a dental drill (1/4″ drill bit), ensuring not to break through and damage the underlying brain. To prevent warming of the underlying brain, flush with sterilized saline regularly. Continue drilling through the skull until the underlying cortical surface vasculature is visible (especially when the bone is damp with saline). Do not break through the skull with the drill.

9. Use the tip of standard #5 forceps to make a small incision in the remaining thinned skull bone outlined by the dental drill (i.e., remaining laminar bone). To do this, place the forceps into the trabecular bone at one side of the circular craniotomy. Grab the craniotomy 'island' (i.e., bone in the middle of the drilled craniotomy) and slowly lift it, being careful not to damage the underlying dura and brain surface. The craniotomy island should come off easily if the drilling was sufficiently through to the trabecular bone. Place Ringer on the craniotomy as soon as the dura is exposed.

10. Ultimately the dura should not bleed, but if it does, wash the craniotomy with Ringer until bleeding slows to a stop. If bleeding does not subside with time, then either dry the dura to cease blood flow within the blood vessels or use Gelfoam (wettened) on the dura surface to stop the bleeding.

11. Sterilize a 3 mm #1 circular coverslip by bathing in 70% ethanol. Remove all fluid from the skull, and ensure it is dry before placing the cover glass into the craniotomy so it is flush with the skull (Fig. 2). Use cyanoacrylate glue to secure the coverslip in place, as close to the brain as possible (within 100 μm). Additionally, to ensure stability of recordings, a 5 mm coverslip can be glued onto the 3 mm coverslip (using optical adhesive; Norland optical adhesive) or a thin layer of melted agar (1.5–2% in modified Ringer) can be placed between the coverslip and dura.

12. Carefully cover the edge of the cyanoacrylate/coverslip with regular dental cement, and extend this over the entire exposed skull surface.

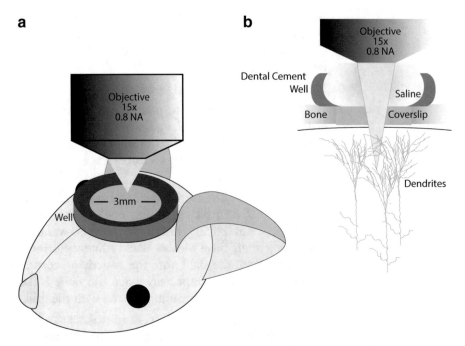

Fig. 2 In vivo calcium imaging preparation. (**a**) Schematic illustrating the typical preparation for two-photon imaging of dendritic activity in rodents in vivo. A chronic window is inserted in the skull, and a dental cement well is fabricated on top of the head. (**b**) Dendrites filled with fluorophore can be visualized through the chronic window using two-photon microscope through an objective with a high NA and long focal length

13. If your head-post is simply a 'post', build a well (bucket) for submerging the imaging objective using conventional cheaper dental cement (e.g., jet repair, Lang Dental). This involves step-by-step application of the dental cement, slowly building the height of the well's wall (thickness, 1–2 mm). The ultimate dimensions of the well depend on the objective used, but for most objectives, it should be roughly 10 mm in diameter and 5 mm high (Fig. 2).

14. Administer pain relief, Meloxicam (1–3 mg/kg body weight), via intraperitoneal injection, 30 minutes prior to cessation of anesthesia.

15. Cover the coverslip with inert silicone (e.g., kwik-cast, WPI) for protection, and place the mouse back into the home cage with heat mat, and monitor recovery. Administer pain relief as per local ethical requirements (typically daily for up to 5 days).

16. Allow the animal to recover for 7–10 days before imaging commences.

3.1.3 Habituation for Calcium Imaging Experiments in Awake, Behaving Animals

Gently train the implanted mice to tolerate having head movement restrained (i.e., being head fixed) by preventing head movement (in a step-wise manner). This involves the following steps which can take up to a week:

1. To provide extra motivation and reward, mice can be water deprived on a 5/2 schedule. Here, they are given 1–3 ml water at the experimental setup only (for 5 days) and allowed free access to water for 2 days. To provide even further motivation, sucrose (0.1 g/ml) can be added to the reward water. Mouse weight should be measured daily to ensure they do not lose greater than 10–20% body weight.

2. Handle the animals prior to commencing habituation.

3. In the home cage, manually hold the head-post to prevent the mouse from moving away. After 3 seconds, release the head-post. Repeat this every few minutes for increasing periods of time until the mouse is comfortable with being restrained by the head-post.

4. Put the head-post frame into the home cage, and fix the animal via the head-post to the frame in a stepwise manner, starting with a duration of 3 minutes (then 5, 10, 20 minutes). Immediately release the animal if it shows any signs of distress.

5. Put the head-post frame on the experimental apparatus, and again, fix the animal via the head-post to the frame in a stepwise manner, starting at 3 minutes (then 5, 10, 20 minutes). Immediately release the animal if it shows any signs of distress. This will make the animal comfortable with head fixation at the setup.

3.1.4 Dendritic Calcium Imaging

Calcium imaging can be performed in either the anesthetized or awake state. In essence, the imaging protocol is the same.

1. Ensure all the recording devices are synchronized [75].

2. If performing experiments in the awake state, start timer, and ensure imaging session is no longer than the time ethically permitted (typically <2 hours as mice lose motivation).

3. Place the animal on a heat pad in the head-post frame, and put under the microscope.

4. Remove the inert protective silicon, place Ringer in the well and submerge the objective. Try to orient the coverslip perpendicular to the optical axis of the microscope.

5. Under careful visual control, locate neurons expressing the calcium indicator (Fig. 3). Be watchful of the laser power used (typically limit laser power to <30 mW at the back focal plane of the objective). Remember, every photon counts and therefore always limit the duration of excitation light to the bare minimum. Light exposure can be systematically reduced by precise temporal control using an electro-optic modulator (Pockels cell, Conoptics) to block the laser light during (1) the 'flyback' of the mirrors and/or (2) the edges of the imaged region (during resonant scanning).

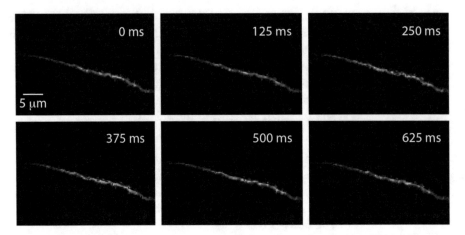

Fig. 3 Example of dendritic calcium imaging in vivo. Neocortical layer 2/3 pyramidal neuron dendrites were filled with the synthetic calcium indicator Oregon Green 488 BAPTA-1 via a patch pipette. Spontaneous calcium influx was measured as an increase in fluorescence in a dendrite located within the upper cortical layers using two-photon microscopy. Each panel illustrates the changes in calcium fluorescence at various time points, reported here from the start (0 ms) to the end (625 ms) of the dendrite-specific change in calcium fluorescence

6. Ensure excessive laser intensity is not being used which can cause photobleaching and photodamage. If the baseline fluorescence of the imaged dendrite is decreasing over time, then decrease the laser intensity until fluorescence at baseline (F_0) is stable.

7. Always know the depth you are acquiring images from – this equates to the dendritic subregion. For example, basal dendrites of layer 2/3 pyramidal neurons are located approximately 300 μm from the pia, whereas tuft dendrites are only 50–100 μm deep.

8. Typically, dendrites which are traversing in the horizontal plane are of greatest interest as information regarding the spatial spread can be addressed (detailed above).

9. It can be tempting to image from a region of bright dendrites, but these are typically unhealthy as high resting fluorescence is a good indicator of excessive calcium influx. Many high signal-to-noise calcium indicators have extremely low resting fluorescence (e.g., GCaMP).

10. Depending on the question at hand, be mindful of the field of view and μm per pixel imaged. Dendrites are typically only 1 μm wide, and according to Nyquist sampling, you would need at least 2 pixels per μm.

11. After each region imaged, ensure you acquire a high-quality z-stack, again, taking Nyquist sampling into account. This will enable you to trace the imaged dendrites back to their soma, which will give further information about the dendrites imaged and the cell type. Also zoom out and take another high quality

z-stack, to gain a representation of the neuronal population surrounding the imaged dendrite of interest. It may also enable further identification.

12. Analyse images with the software of choice (e.g., ImageJ/ FIJI). If possible, perform some preliminary analysis on the first few trials to ensure all the correct imaging parameters are met (no photobleaching/movement artifact etc.)

3.2 Considerations When Imaging Dendritic Activity In Vivo

3.2.1 Imaging Depth

One of the main restrictions of functional imaging is the tissue penetration of the excitation light. To date, most calcium imaging of dendritic activity in vivo has been conducted using two-photon microscopy. Since most commonly used calcium indicators are maximally excited by the simultaneous absorption of two photons of red light, only calcium signals from neurons/dendrites that are located up to a depth of 400–600 μm can be recorded. However, advances in optics and indicators have attempted to push the depth of functional imaging, with varying success. Since there is an inverse relationship between wavelength and scattering/absorption, longer wavelengths can penetrate deeper into tissue. This simple fact has led to the development of red-shifted fluorophores which are excited at longer wavelengths than the commonly used calcium indicators.

Recently, red GECIs have been developed which are based on red-shifted fluorescent proteins. Since they are excited by 1040 nm or 1100 nm light, red GECIs not only facilitate deep-tissue imaging, but they also enable dual-colour imaging with GFP-based reporters and the use of optogenetics with calcium imaging (for a review, see [76]). Unfortunately, since red-shifted fluorophores are significantly more lipophilic than fluorescein-like dyes [77], they accumulate in intracellular compartments. Therefore, although promising, their fluorescence signal is currently not comparable to the commonly used blue-excited indicators.

3.2.2 Imaging from Optically Sectioned (Transected) Dendrites

Due to the morphological polarization of most principal neurons in the cortex and hippocampus, the optical sectioning by conventional two-photon imaging typically results in recording calcium activity from transverse sections of dendrites in vivo. This, of course, limits (1) the number of pixels recorded from, (2) the spatial distribution of the recorded signal (hence making identification hard) and (3) simultaneous recordings along a dendritic axis. Various techniques have been developed to bypass the limitations imposed by the optical sectioning of two-photon microscopy by imaging volumes of tissue. Conventional two-photon microscopy employs a pair of nonresonant galvanometric scanning mirrors which restrict frame acquisition rates to less than 10 Hz speeds which are too slow for volumetric imaging of functional calcium activity over large fields of view. The use of acousto-optical deflectors (AODs), which steer laser beams by using the diffraction of light by sound waves of different frequencies, can achieve imaging rates fast enough to image volumes [78, 79]. Another approach to image tissue volumes at functionally sufficient temporal resolution is by mechanical movement of the microscope objective. Here, the objective lens is set in continuous up-and-down motion using a piezoelectric actuator and the laser is scanned through the tissue using a pair of galvanometric mirrors [80]. This has enabled the calcium dynamics to be recorded from hundreds of neurons across a 250 × 250 × 250 μm volume at rates of up to 10 Hz [80, 81]. The speed of tissue volume imaging can be further improved by employing a user-defined scanning trajectory [82, 83] or imaging plane [73]. Using a laser beam with an axially elongated focus (Bessel beam; 15–400 μm axial full width at half maximum), 3D volumes of tissue can be measured at rates of up to 30 Hz [69]. Although this sacrifices axial resolution, dendrites and even spines can still be resolved. Video-rate volumetric functional imaging has been performed in various preparations including the mouse, zebrafish and fruit fly [69].

3.2.3   Motion Control

Brain movement is a large consideration when imaging from small neuronal structures such as dendrites. The brain naturally pulsates due to the pulsing of blood through vessels. This is exacerbated by the craniotomy procedure, which can introduce greater area for the brain to move as a whole. There are physical procedures that can be performed in an attempt to minimize the amount of movement under the craniotomy. Downward pressure can be applied to physically dampen brain motion by (1) agar placed on the craniotomy between the brain and imaging window and (2) double coverslips preparation. Post hoc correction of motion in the x- and y-direction can be performed, within reasonable limits, and many motion-control software packages are available online.

References

1. Deiters O (1865) Untersuchungen über Gehirn und Rückenmark des Menschen und der Säugethiere. Vieweg, Braunschweig

2. Denk W, Strickler JH, Webb WW (1990) Two-photon laser scanning fluorescence microscopy. Science 248:73–76

3. Berridge MJ (1998) Neuronal calcium signaling. Neuron 21:13–26

4. Chen TW, Wardill TJ, Sun Y, Pulver SR, Renninger SL, Baohan A, Schreiter ER, Kerr RA, Orger MB, Jayaraman V, Looger LL, Svoboda K, Kim DS (2013) Ultrasensitive fluorescent proteins for imaging neuronal activity. Nature 499:295–300

5. Helmchen F, Imoto K, Sakmann B (1996) Ca^{2+} buffering and action potential-evoked Ca^{2+} signaling in dendrites of pyramidal neurons. Biophys J 70:1069–1081

6. Markram H, Lubke J, Frotscher M, Sakmann B (1997) Regulation of synaptic efficacy by coincidence of postsynaptic APs and EPSPs. Science 275:213–215

7. Letzkus JJ, Kampa BM, Stuart GJ (2006) Learning rules for spike timing-dependent plasticity depend on dendritic synapse location. J Neurosci 26:10420–10429

8. Stuart G, Schiller J, Sakmann B (1997) Action potential initiation and propagation in rat neocortical pyramidal neurons. J Physiol 505:617–632

9. Clark BA, Monsivais P, Branco T, London M, Häusser M (2005) The site of action potential

initiation in cerebellar Purkinje neurons. Nat Neurosci 8:137–139

10. Palmer LM, Stuart GJ (2006) Site of action potential initiation in layer 5 pyramidal neurons. J Neurosci 26:1854–1863

11. Destexhe A, Lang EJ, Pare D (1998) Somato-dendritic interactions underlying action potential generation in neocortical pyramidal cells in vivo. In: Bower J (ed) Computational neuroscience: trends in research. Plenum Press, New York, pp 167–172

12. Kole MH, Stuart GJ (2012) Signal processing in the axon initial segment. Neuron 73:235–247

13. Stuart GJ, Sakmann B (1994) Active propagation of somatic action potentials into neocortical pyramidal cell dendrites. Nature 367:69–72

14. Spruston N, Schiller Y, Stuart G, Sakmann B (1995) Activity-dependent action potential invasion and calcium influx into hippocampal CA1 dendrites. Science 268:297–300

15. Häusser M, Stuart G, Racca C, Sakmann B (1995) Axonal initiation and active dendritic propagation of action potentials in substantia nigra neurons. Neuron 15:637–647

16. Larkum ME, Rioult MG, Lüscher HR (1996) Propagation of action potentials in the dendrites of neurons from rat spinal cord slice cultures. J Neurophysiol 75:154–170

17. Bischofberger J, Jonas P (1997) Action potential propagation into the presynaptic dendrites of rat mitral cells. J Physiol 504:359–365

18. Williams SR, Stuart GJ (2000) Action potential backpropagation and somato-dendritic distribution of ion channels in thalamocortical neurons. J Neurosci 20:1307–1317

19. Lemon N, Turner RW (2000) Conditional spike backpropagation generates burst discharge in a sensory neuron. J Neurophysiol 84:1519–1530

20. Waters J, Larkum M, Sakmann B, Helmchen F (2003) Supralinear Ca²⁺ influx into dendritic tufts of layer 2/3 neocortical pyramidal neurons in vitro and in vivo. J Neurosci 23:8558–8567

21. Larkum ME, Watanabe S, Lasser-Ross N, Rhodes P, Ross WN (2008) Dendritic properties of turtle pyramidal neurons. J Neurophysiol 99:683–694

22. Bathellier B, Margrie TW, Larkum ME (2009) Properties of piriform cortex pyramidal cell dendrites: implications for olfactory circuit design. J Neurosci 29:12641–12652

23. Ledergerber D, Larkum ME (2010) Properties of layer 6 pyramidal neuron apical dendrites. J Neurosci 30:13031–13044

24. Casale AE, McCormick DA (2011) Active action potential propagation but not initiation in thalamic interneuron dendrites. J Neurosci 31:18289–18302

25. Feldman DE, Brecht M (2005) Map plasticity in somatosensory cortex. Science 310:810–815

26. Egger V, Feldmeyer D, Sakmann B (1999) Coincidence detection and changes of synaptic efficacy in spiny stellate neurons in rat barrel cortex. Nat Neurosci 2:1098–1105

27. Feldman DE (2000) Timing-based LTP and LTD at vertical inputs to layer II/III pyramidal cells in rat barrel cortex. Neuron 27:45–56

28. Sjostrom PJ, Turrigiano GG, Nelson SB (2001) Rate, timing, and cooperativity jointly determine cortical synaptic plasticity. Neuron 32:1149–1164

29. Holmgren CD, Zilberter Y (2001) Coincident spiking activity induces long-term changes in inhibition of neocortical pyramidal cells. J Neurosci 21:8270–8277

30. Froemke RC, Dan Y (2002) Spike-timing-dependent synaptic modification induced by natural spike trains. Nature 416:433–438

31. Stuart GJ, Häusser M (2001) Dendritic coincidence detection of EPSPs and action potentials. Nat Neurosci 4:63–71

32. Svoboda K, Denk W, Kleinfeld D, Tank DW (1997) In vivo dendritic calcium dynamics in neocortical pyramidal neurons. Nature 385:161–165

33. Hill DN, Varga Z, Jia H, Sakmann B, Konnerth A (2013) Multibranch activity in basal and tuft dendrites during firing of layer 5 cortical neurons in vivo. Proc Natl Acad Sci U S A 110:13618–13623

34. Palmer LM, Shai AS, Reeve JE, Anderson HL, Paulsen O, Larkum ME (2014) NMDA spikes enhance action potential generation during sensory input. Nat Neurosci 17:383–390

35. Larkum ME, Zhu JJ, Sakmann B (1999) A new cellular mechanism for coupling inputs arriving at different cortical layers. Nature 398:338–341

36. Schiller J, Schiller Y, Stuart G, Sakmann B (1997) Calcium action potentials restricted to distal apical dendrites of rat neocortical pyramidal neurons. J Physiol 505:605–616

37. Perez-Garci E, Larkum ME, Nevian T (2013) Inhibition of dendritic Ca²⁺ spikes by GABA_B receptors in cortical pyramidal neurons is mediated by a direct $G_{i/o}$-$\beta\gamma$-subunit interaction with Ca_v1 channels. J Physiol 591:1599–1612

38. Larkum ME, Zhu JJ (2002) Signaling of layer 1 and whisker-evoked Ca²⁺ and Na⁺ action potentials in distal and terminal dendrites of

rat neocortical pyramidal neurons in vitro and in vivo. J Neurosci 22:6991–7005

39. Larkum ME, Nevian T, Sandler M, Polsky A, Schiller J (2009) Synaptic integration in tuft dendrites of layer 5 pyramidal neurons: a new unifying principle. Science 325:756–760

40. Harnett MT, Xu NL, Magee JC, Williams SR (2013) Potassium channels control the interaction between active dendritic integration compartments in layer 5 cortical pyramidal neurons. Neuron 79:516–529

41. Harnett MT, Magee JC, Williams SR (2015) Distribution and function of HCN channels in the apical dendritic tuft of neocortical pyramidal neurons. J Neurosci 35:1024–1037

42. Xu NL, Harnett MT, Williams SR, Huber D, O'Connor DH, Svoboda K, Magee JC (2012) Nonlinear dendritic integration of sensory and motor input during an active sensing task. Nature 492:247–251

43. Takahashi N, Oertner TG, Hegemann P, Larkum ME (2016) Active cortical dendrites modulate perception. Science 354:1587–1590

44. Gambino F, Pages S, Kehayas V, Baptista D, Tatti R, Carleton A, Holtmaat A (2014) Sensory-evoked LTP driven by dendritic plateau potentials in vivo. Nature 515:116–119

45. Cichon J, Gan WB (2015) Branch-specific dendritic Ca²⁺ spikes cause persistent synaptic plasticity. Nature 520:180–185

46. Major G, Polsky A, Denk W, Schiller J, Tank DW (2008) Spatiotemporally graded NMDA spike/plateau potentials in basal dendrites of neocortical pyramidal neurons. J Neurophysiol 99:2584–2601

47. Polsky A, Mel BW, Schiller J (2004) Computational subunits in thin dendrites of pyramidal cells. Nat Neurosci 7:621–627

48. Nevian T, Larkum ME, Polsky A, Schiller J (2007) Properties of basal dendrites of layer 5 pyramidal neurons: a direct patch-clamp recording study. Nat Neurosci 10:206–214

49. Branco T, Häusser M (2011) Synaptic integration gradients in single cortical pyramidal cell dendrites. Neuron 69:885–892

50. Lavzin M, Rapoport S, Polsky A, Garion L, Schiller J (2012) Nonlinear dendritic processing determines angular tuning of barrel cortex neurons in vivo. Nature 490:397–401

51. Mao T, O'Connor DH, Scheuss V, Nakai J, Svoboda K (2008) Characterization and subcellular targeting of GCaMP-type genetically-encoded calcium indicators. PLoS One 3:e1796

52. Bollmann JH, Engert F (2009) Subcellular topography of visually driven dendritic activity in the vertebrate visual system. Neuron 61:895–905

53. Takahashi N, Kitamura K, Matsuo N, Mayford M, Kano M, Matsuki N, Ikegaya Y (2012) Locally synchronized synaptic inputs. Science 335:353–356

54. Winnubst J, Cheyne JE, Niculescu D, Lohmann C (2015) Spontaneous activity drives local synaptic plasticity in vivo. Neuron 87:399–410

55. Wilson DE, Whitney DE, Scholl B, Fitzpatrick D (2016) Orientation selectivity and the functional clustering of synaptic inputs in primary visual cortex. Nat Neurosci 19:1003–1009

56. Jia H, Rochefort NL, Chen X, Konnerth A (2010) Dendritic organization of sensory input to cortical neurons in vivo. Nature 464:1307–1312

57. Chen X, Leischner U, Rochefort NL, Nelken I, Konnerth A (2011) Functional mapping of single spines in cortical neurons in vivo. Nature 475:501–505

58. Koch C, Zador A (1993) The function of dendritic spines: devices subserving biochemical rather than electrical compartmentalization. J Neurosci 13:413–422

59. Yuste R, Denk W (1995) Dendritic spines as basic functional units of neuronal integration. Nature 375:682–684

60. Palmer LM, Stuart GJ (2009) Membrane potential changes in dendritic spines during action potentials and synaptic input. J Neurosci 29:6897–6903

61. Harnett MT, Makara JK, Spruston N, Kath WL, Magee JC (2012) Synaptic amplification by dendritic spines enhances input cooperativity. Nature 491:599–602

62. Shimomura O, Johnson FH, Saiga Y (1962) Extraction, purification and properties of aequorin, a bioluminescent protein from the luminous hydromedusan, Aequorea. J Cell Comp Physiol 59:223–239

63. Ashley CC, Ridgway EB (1968) Simultaneous recording of membrane potential, calcium transient and tension in single muscle fibers. Nature 219:1168–1169

64. Tsien RY (1980) New calcium indicators and buffers with high selectivity against magnesium and protons: design, synthesis, and properties of prototype structures. Biochemistry 19:2396–2404

65. Varga Z, Jia H, Sakmann B, Konnerth A (2011) Dendritic coding of multiple sensory inputs in single cortical neurons in vivo. Proc Natl Acad Sci U S A 108:15420–15425

66. Chen X, Rochefort NL, Sakmann B, Konnerth A (2013) Reactivation of the same synapses

during spontaneous up states and sensory stimuli. Cell Rep 4:31–39

67. Palmer LM, Schulz JM, Murphy SC, Ledergerber D, Murayama M, Larkum ME (2012) The cellular basis of GABA_B-mediated interhemispheric inhibition. Science 335:989–993

68. Grienberger C, Chen X, Konnerth A (2014) NMDA receptor-dependent multidendrite Ca^{2+} spikes required for hippocampal burst firing in vivo. Neuron 81:1274–1281

69. Lu R, Sun W, Liang Y, Kerlin A, Bierfeld J, Seelig JD, Wilson DE, Scholl B, Mohar B, Tanimoto M, Koyama M, Fitzpatrick D, Orger MB, Ji N (2017) Video-rate volumetric functional imaging of the brain at synaptic resolution. Nat Neurosci 20:620–628

70. Looger LL, Griesbeck O (2012) Genetically encoded neural activity indicators. Curr Opin Neurobiol 22:18–23

71. Mittmann W, Wallace DJ, Czubayko U, Herb JT, Schaefer AT, Looger LL, Denk W, Kerr JN (2011) Two-photon calcium imaging of evoked activity from L5 somatosensory neurons in vivo. Nat Neurosci 14:1089–1093

72. Gentet LJ, Kremer Y, Taniguchi H, Huang ZJ, Staiger JF, Petersen CC (2012) Unique functional properties of somatostatin-expressing GABAergic neurons in mouse barrel cortex. Nat Neurosci 15:607–612

73. Sheffield ME, Dombeck DA (2014) Calcium transient prevalence across the dendritic arbour predicts place field properties. Nature 517:200–204

74. Sachidhanandam S, Sreenivasan V, Kyriakatos A, Kremer Y, Petersen CC (2013) Membrane potential correlates of sensory perception in mouse barrel cortex. Nat Neurosci 16:1671–1677

75. Micallef AH, Takahashi N, Larkum ME, Palmer LM (2017) A reward-based behavioral platform to measure neural activity during head-fixed behavior. Front Cell Neurosci 11:156

76. Dana H, Mohar B, Sun Y, Narayan S, Gordus A, Hasseman JP, Tsegaye G, Holt GT, Hu A, Walpita D, Patel R, Macklin JJ, Bargmann CI, Ahrens MB, Schreiter ER, Jayaraman V, Looger LL, Svoboda K, Kim DS (2016) Sensitive red protein calcium indicators for imaging neural activity. Elife 5:e12727. https://doi.org/10.7554/eLife.12727

77. Collot M, Wilms CD, Bentkhayet A, Marcaggi P, Couchman K, Charpak S, Dieudonne S, Hausser M, Feltz A, Mallet JM (2015) CaRuby-Nano: a novel high affinity calcium probe for dual color imaging. Elife 4:05808. https://doi.org/10.7554/eLife.05808

78. Reddy GD, Saggau P (2005) Fast three-dimensional laser scanning scheme using acousto-optic deflectors. J Biomed Opt 10:064038. https://doi.org/10.1117/1.2141504

79. Reddy GD, Kelleher K, Fink R, Saggau P (2008) Three-dimensional random access multiphoton microscopy for functional imaging of neuronal activity. Nat Neurosci 11:713–720

80. Göbel W, Helmchen F (2007) New angles on neuronal dendrites in vivo. J Neurophysiol 98:3770–3779

81. Göbel W, Helmchen F (2007) In vivo calcium imaging of neural network function. Physiology (Bethesda) 22:358–365

82. Katona G, Kaszas A, Turi GF, Hajos N, Tamas G, Vizi ES, Rozsa B (2011) Roller Coaster Scanning reveals spontaneous triggering of dendritic spikes in CA1 interneurons. Proc Natl Acad Sci U S A 108:2148–2153

83. Katona G, Szalay G, Maak P, Kaszas A, Veress M, Hillier D, Chiovini B, Vizi ES, Roska B, Rozsa B (2012) Fast two-photon in vivo imaging with three-dimensional random-access scanning in large tissue volumes. Nat Methods 9:201–208

Chapter 13

Voltage Imaging with ANNINE Dyes and Two-Photon Microscopy

Christopher J. Roome and Bernd Kuhn

Abstract

Voltage imaging is a tried and tested tool for revealing changes in neuronal membrane voltage at high temporal and spatial resolution in vitro and in vivo. However, single-photon in vivo voltage imaging using cameras and synthetic dyes does not allow depth resolution in highly scattering tissue such as the mammalian brain and risks introducing artifacts due to phototoxicity and bleaching. In contrast, voltage imaging with synthetic electrochromic dyes and two-photon excitation near the red spectral edge of absorption circumvents these challenges, allowing depth-resolved measurement of voltage changes at high spatial and temporal resolution in scattering tissue with negligible phototoxicity and bleaching.

Here, we describe how to image voltage using two-photon microscopy and the voltage-sensitive dyes ANNINE-6 and ANNINE-6plus. The key advantages of these dyes are that the voltage response is linear, the temporal resolution is essentially limited by the imaging speed, and phototoxicity and bleaching can be neglected when an excitation wavelength close to the red spectral edge of absorption is chosen. We report how to image membrane voltage of dissociated cells in culture. We provide protocols for imaging average membrane voltage in vitro (in brain slices) and in anesthetized and awake animals. Finally, we describe the labeling of single neurons in vivo and how to measure supra- and sub-threshold voltage changes in their dendrites in the awake animal. Voltage can be imaged from internally labeled cells for at least 2 weeks after a single electroporation. Dendritic voltage imaging can be combined with electrical recording, calcium imaging, and/or pharmacology.

Key words Voltage-sensitive dye, VSD, Voltage imaging, ANNINE, Membrane potential, Two-photon, In vitro, In vivo

1 Introduction

Electrical recording has been the key technique for studying neuronal activity, offering accurate voltage and current measurements at high temporal resolution. However, electrical recordings typically suffer from limited spatial resolution. To overcome this problem, in the very early days of functional imaging, L.B. Cohen, B.M. Salzberg, and colleagues screened roughly 1000 synthetic dyes to assess their voltage sensitivity [1, 2]. Based on screening

Espen Hartveit (ed.), *Multiphoton Microscopy*, Neuromethods, vol. 148,
https://doi.org/10.1007/978-1-4939-9702-2_13, © Springer Science+Business Media, LLC, part of Springer Nature 2019

results and/or a mechanistic understanding of fluorescence modulation by membrane voltage, L.M. Loew, A. Grinvald, P. Fromherz, and colleagues developed new and better voltage-sensitive dyes (VSDs) [3–8]. VSDs have since been used with high-speed, low-noise cameras to image modular brain activity in vitro and in vivo [9, 10], voltage changes of single neurons in cell cultures and slice preparations [11–15], and membrane voltage in many other applications [16]. For a recent primer on voltage imaging, see [17].

Genetically encoded voltage indicators have also been developed over the last two decades, each with their own advantages and challenges, but promise to become useful tools in the near future [18, 19].

Despite its long history, voltage imaging has remained challenging and has consequently been overshadowed by rapidly developing calcium imaging techniques. This is because voltage signals are typically 10–100 times faster than calcium signals, and thus more difficult to detect. Furthermore, membrane voltage change is spatially restricted to the plasma membrane, whereas calcium concentration changes are detected throughout the intracellular volume. Finally, the signal amplitude of VSDs are much smaller (typically $1/10$ or $1/100$) than that of calcium indicators.

Phototoxicity and bleaching of VSDs have also been serious drawbacks when voltage imaging from delicate neuronal structures. Therefore, most neuroscientists opt for calcium imaging as a proxy for neuronal activity, despite being limited to supra-threshold events.

Nonetheless, voltage imaging with synthetic voltage dyes offers important advantages over calcium indicators and genetically encoded voltage indicators. Among the wide variety of synthetic voltage indicators, in particular electrochromic VSDs possess several interesting features that render them ideally suited for measuring neuronal membrane potentials. Electrochromic dyes shift their spectrum, or, in other words, change their color due to the electric field over the membrane. This spectral shift is based on a physical effect which does not involve movement of the dye or any type of binding. Therefore, the spectral shift is almost instantaneous, and so voltage imaging with synthetic electrochromic dyes has a temporal resolution essentially limited by the imaging speed. In contrast, many genetically encoded voltage indicators have a slower time constant. Additionally, voltage signals measured with synthetic electrochromic dyes are proportional to the membrane voltage. In contrast, calcium signals represent the convolution of the calcium concentration and the binding characteristics of the molecular probe, which introduce delays and nonlinearities. In general, voltage imaging with electrochromic dyes not only reports supra-threshold activity as calcium imaging does, but also sub-threshold activity and hyperpolarization, both of which are essential for understanding the dynamics of dendritic integration and neuronal networks.

Here, we focus on protocols for voltage imaging with two-photon microscopy. This combination allows depth-resolved voltage imaging in vitro and in vivo. It also reduces phototoxicity and bleaching if the correct excitation wavelength is chosen. We describe cell culture experiments, brain slice experiments, and in vivo experiments with bulk-loaded tissue and single-cell labeling.

Additionally, we focus on ANNINE dyes, i.e., ANNINE-6 and ANNINE-6plus, which were developed and designed based on elaborate physical chemistry studies and theoretical models in the laboratory of P. Fromherz at the University of Ulm, Germany, and later at the Max Planck Institute of Biochemistry, Martinsried, Germany [5, 6, 20–26]. ANNINE dyes are purely electrochromic, i.e., they exhibit spectral shifts only due to the electric field over the cell membrane but not due to any other effects [27].

1.1 Mechanism of Voltage Sensitivity of ANNINE Dyes

ANNINE dyes, like other electrochromic VSDs, have a hydrophilic headgroup with electric charges and a hydrophobic tail with no electric charges (Fig. 1a). The headgroup of ANNINE-6 has a positive and a negative charge, while the headgroup of ANNINE-6plus has two positive charges. Due to this design, ANNINE molecules bind to lipid membranes (Fig. 1b).

A second important design feature is that the chromophore, consisting of a ring system and nitrogen atoms, is asymmetric: On one side of the chromophore, the nitrogen is in the ring, on the other side it is outside the ring. Due to this asymmetry, the charges in the chromophore are not symmetrically distributed. Interestingly, if the chromophore is excited by absorption of a photon, this charge distribution changes. An electron which was in the ground state on one side of the chromophore moves during the excitation process toward the other side (Fig. 1c). When the dye emits a photon (fluoresces), the electron returns to the original position.

In the case that the dye molecule is bound to a membrane, the charge shift within the chromophore will interact with the electric field over the membrane. For example, if the negatively charged electron is moved during the absorption process with the electric field over the membrane (the direction of the electric field is defined from plus to minus), then more energy will be needed for the absorption process to occur. This results in a shift of the absorption spectrum to higher energy, i.e., toward shorter wavelength or toward blue. During the emission process, the charge moves back to the original position. But now the charge moves against the external electric field and, therefore, gains energy. Therefore, the emission spectrum is also shifted to shorter wavelengths, and if the electric field over the membrane with the dye molecules changes, the spectral shift will change accordingly. Conversely, if the dye

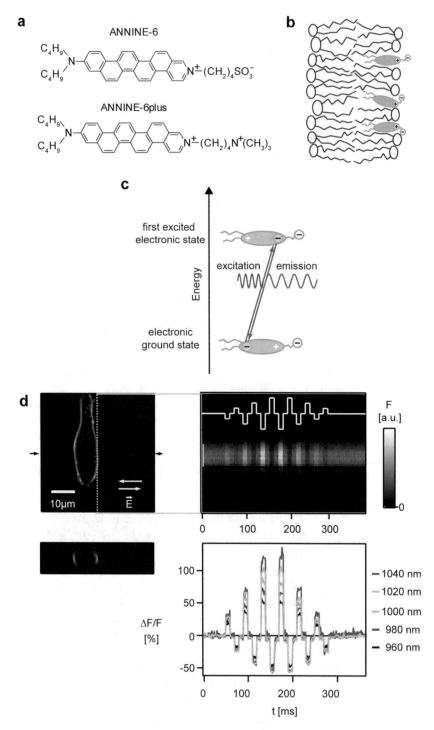

Fig. 1 ANNINE-6 and ANNINE-6plus; basic properties. (**a**) Structure of ANNINE-6 and ANNINE-6plus. Both have a charged and therefore hydrophilic head group (right) and an uncharged hydrophobic tail (left). ANNINE-6plus is more water-soluble than ANNINE-6 due to two positive charges compared to a positive and a negative charge. (**b**) Sketch of a VSD molecule bound to a lipid membrane. The hydrophilic head groups align with the lipid head groups. The hydrophobic tail and chromophore are surrounded by hydrophobic hydrocarbon chains

molecules are bound to the opposite side of the membrane, a spectral shift will occur in the opposite direction.

Neurons change their membrane potential if they are electrically active. For example, during an action potential, the membrane potential changes from -70 mV at rest to $+20$ mV. Therefore, if the membrane of a neuron is labeled with a VSD, the neuronal voltage change will shift the spectrum of the voltage-sensitive dyes with a linear relation.

The spectral shift must be converted into an intensity change that can be detected with a photomultiplier tube (PMT). Importantly, no intensity change due to a membrane voltage change can be detected if the dye is excited close to the maximum of the excitation spectrum and photons of the full emission spectrum are collected by the PMT. Surprisingly (at first sight), and most importantly, the highest sensitivity (relative fluorescence change per voltage change) of electrochromic dyes, and specifically of ANNINE dyes, is found close to the red spectral edge of absorption [28]. This is true for both one-photon and two-photon excitation (Fig. 1d). As the probability of absorption close to the spectral edge is low, relatively high laser power is necessary for voltage imaging. Fortunately, the required laser power is still low enough not to cause any observable damage.

1.2 Voltage Imaging with Two-Photon Excitation

Voltage imaging is always a fight against noise. Optical noise follows the Poisson statistics, so if n photons are detected on average, the expected noise is $\sqrt{n}$. The optical voltage signal (measured in % intensity change) must be larger than the relative noise level ($\sqrt{n}/n$, measured in % intensity fluctuation).

As voltage signals are typically in the range of 0.1–10% intensity change, voltage imaging has been in the domain of camera imaging, which enables collection of millions of photons, thereby achieving low relative noise levels. Signals as small as 0.3% relative fluorescence change can be detected in single trials, if more than 1,000,000 photons are detected (relative noise 0.1%) with a signal-to-noise ratio (SNR) of 3. Two- or multi-photon microscopy

Fig. 1 (continued) of the lipid molecules. (**c**) Charge shift within a VSD during the absorption and emission process. During the absorption process, the center of charge shifts. This shift occurs due to the asymmetry of the chromophore. (**d**) Relative fluorescence change of ANNINE-6 increases toward the red spectral edge of two-photon absorption. A HEK293 cell labeled with ANNINE-6 is shown in the x-y plane and the x-z plane at the location indicated by arrows (left). A two-photon line scan was taken along the membrane (yellow dotted line). External electric fields $\vec{E}$ with different amplitudes and direction (white trace, switching directions indicated by white arrows) were applied while scanning along the membrane. The spatio-temporal map shows bright and dark stripes correlating with the applied external electric fields (right). Intensity gray scale is given in arbitrary units [a.u.]. Line scan band (indicated by yellow) was spatially averaged to show relative fluorescence changes corresponding to six different (3 positive, 3 negative) membrane voltage changes (bottom). Relative fluorescence changes in response to the same external electric fields increase with increasing excitation wavelength, i.e., closer to the red spectral edge of the absorption spectrum. (Modified with permission from Elsevier [28])

typically operates in a regime of tens to hundreds of photons. For example, 100 photons result in a relative noise of about 10%. Therefore, so far, only a few studies have been published using two-photon microscopy for voltage imaging [28–33]. A related technique, voltage imaging by second harmonic generation, is similarly challenging [17, 34].

One way to increase the number of generated photons is to increase the volume of the excitation point spread function; this, however, decreases the spatial resolution. The volume of the point spread function can be increased by underfilling the back focal plane of the objective by introducing a telescope into the laser path, which narrows the beam two- or three-fold [35]. An alternative is to use a Bessel module consisting of a spatial light modulator, three lenses, and an annular mask to generate an axially elongated Bessel focus [36]. With this technique, the axial excitation can be increased up to 400 μm.

Another consideration is to convert the linearly polarized excitation light into circularly polarized light by placing a $\lambda/4$ plate with the primary axis turned 45° against the direction of the laser polarization into the excitation light path. Thereby, all dye molecules perpendicular to the laser beam axis can be excited and not just those parallel to the original polarization of the laser [37].

Two-photon microscopy is typically slow due to the necessity of scanning the sample. Scanning speed can be increased by line scans and bi-directional scanning. With a galvo scanner, a temporal resolution of up to 2 kHz can be reached by bidirectional line scan. Resonant scanners additionally increase the speed about 15-fold. Several other fast-scanning two-photon microscopes have been developed and will become important for voltage imaging [38–40]. If no spatial resolution is required, parking the laser beam is an interesting and relatively simple approach to measure voltage at a single location at very high temporal resolution [31].

Another important advantage of two-photon microscopy is that it allows simultaneous excitation of different chromophores due to the wide two-photon absorption spectra. Therefore, for example, voltage and calcium can be imaged simultaneously [32, 37]. It is also possible to combine two-photon voltage imaging with other indicators or with membrane-bound fluorophores to aid in movement correction.

1.3 Two-Photon Voltage Imaging of Cell Cultures

In low-density cell cultures, cell membranes are exposed as single sheets. After applying VSD to the bath, voltage imaging can be performed with one-photon excitation (see for example [41]) or two-photon excitation from identified structures (Fig. 1d). If cell cultures get confluent, voltage imaging becomes more difficult as fluorescence signals of different cells might get mixed.

1.4 Two-Photon Voltage Imaging of Bulk-Loaded Tissue

Brain tissue is tightly packed with intermingling cell membranes, as can be seen in any electron microscope image. By applying a VSD non-specifically to the extracellular space, all cell surfaces get labeled. This type of labeling is called bulk-loading. It is impossible to resolve single neurons with one-photon excitation due to the optical resolution limit and scattering. Even depth-resolved imaging is not possible. Two-photon microscopy, however, allows depth-resolved measurement of average membrane potentials and in special cases even signals from single neurons.

Bulk-loading can be done by exposing the surface of the brain to a dye solution after removal of the dura [9], by injection of the dye via pipette during an acute experiment [30], by injection of dye via pipette through a PDMS (silicone) membrane covering the craniotomy [42], or, similarly, through a chronic cranial window with an access port [43].

In tissue bulk-loaded with VSD, all cell surfaces of dendrites, axons, somata, and glial cells are labeled. Every membrane-bound dye molecule contributes to the intensity of a measurement, but only those that undergo a membrane voltage change contribute to the signal. Therefore, the signal of voltage imaging is proportional to the ratio of membrane surface participating in a voltage change to the overall membrane surface in the measured volume and to the amplitude of the voltage change.

Typically, axons show large voltage changes (action potentials), but only for very short times (in mammals, in vivo, less than 1 ms) and they contribute a large fraction to the overall neuropil membrane surface. The surface of a single axon is relatively small however, and during firing of an action potential, the transient depolarization is typically followed by a slower hyperpolarization. Additionally, as action potentials are typically not perfectly synchronized, depolarizing and hyperpolarizing signals will cancel out in an average measurement. Therefore, the axonal signal can only be seen under special conditions, as for example, in electrically stimulated parallel fibers of cerebellar slices [44].

Dendrites also contribute substantially to the overall neuropil membrane surface, especially spiny dendrites. Dendritic postsynaptic potentials can synchronize (e.g., during a sensory input) and are typically slow (several milliseconds) compared to action potentials. Typical voltage changes of postsynaptic potentials in dendrites are 10s of mV. Consequently, their large surface area and slow, synchronized depolarization makes them the main source of the voltage signal in bulk-loaded tissue [45].

Somata barely contribute to the overall membrane potential signal as their surface area is neglectable compared to axons and dendrites. But, surprisingly, it was shown that scanning along the rim of Purkinje neuron somata in cerebellar brain slices yields a signal with a SNR good enough to detect simple spikes in bulk-loaded tissue [32]. However, this might be limited to Purkinje neurons and there are so far no reports of any other cell types where this would be possible.

Glia cells, which contribute significantly to the overall membrane surface, are not expected to participate in fast voltage changes, and, therefore, only dilute the signal.

Statistics on membrane surfaces, i.e., the surface ratio between dendrites, axons, and astrocytes (see for example [46]) help to estimate the expected voltage signal [30]. In most voltage imaging studies, it is assumed that the voltage signal originates from the dendrites only.

Two-photon microscopy allows the measurement of average membrane voltage in response to sensory stimulation, typically after averaging, and average membrane potential oscillations, typically after temporal smoothing [30].

1.5 Two-Photon Voltage Imaging of Internally Labeled Neurons

To overcome the problem of signal mixing in bulk-loaded tissue, it is possible to load single neurons internally via a whole-cell recording pipette in slices. This technique was pioneered by S.D. Antic and D.P. Zecevic with electrochromic dyes which have a good water-solubility [11, 15]. Cameras can be used for imaging voltage signals from dendrites, somata, and axons. Alternatively, two-photon microscopy can be used for imaging by parking the beam for spot measurements [31] or by scanning short lines [29].

Loading neurons with VSD is challenging because, on one hand, the dye should bind to the membrane and, on the other hand, it should be soluble during the loading process. As ANNINE-6 is not water-soluble, the less hydrophobic ANNINE-6plus was developed [23]. ANNINE-6plus shares with ANNINE-6 the same chromophore, spectral characteristics, and it is also purely electrochromic [23]. Only the water solubility and membrane affinity are different. ANNINE-6plus saturates at about 1 mM in pure water. However, it precipitates quickly in saline. In general, ANNINE-6plus has a much lower solubility than the JPW dyes [11, 15], which are typically used for internal labeling. In our hands, all attempts to fill neurons with ANNINE-6plus in aqueous solutions failed. However, ANNINE-6plus is soluble in ethanol and it saturates at about 6 mM. We were able to fill Purkinje neurons reliably by electroporation with ANNINE-6plus (3 mM) in ethanol. In preliminary experiments, we were also able to label cortical pyramidal neurons with ANNINE-6plus. Purkinje neurons survive this procedure well and the labeling technique subsequently allows reliable supra- and sub-threshold voltage imaging in vivo from dendrites for many minutes without bleaching and phototoxic effects [37]. The labeling lasts for at least 2 weeks but, of course, the highest SNR of voltage imaging is found during the first few days after electroporation [37].

In general, there is the concern that VSD bound to intracellular organelle membranes, for example, of the endoplasmic reticulum, might dilute the voltage signal by contributing to the fluorescence, but not to the voltage signal [47]. Therefore, the calibration of the relative fluorescence change to voltage should be considered an approximation. However, for distal Purkinje dendrites, the voltage signals correspond well with dendritic patch-clamp experiments under similar conditions. For example, spikelets of dendritic complex spikes in distal dendrites of Purkinje neurons were estimated to have voltage amplitudes of 34 ± 7 mV [37] with ANNINE-6plus imaging, while patch-clamp experiments under similar conditions resulted in amplitudes of 35 ± 4 mV [48].

2 Materials

Chronic cranial window surgery as described in [43, 49, 50].
Glass window with access port as described in [43]. For additional advice, see Sect. 3.2.

2.1 Two-Photon Microscope Setup

In vivo two-photon microscope setup (e.g., MOM, Sutter Instrument).

Ti:sapphire femtosecond laser (e.g., Coherent or Spectra-Physics); for ANNINE-6 and ANNINE-6plus, the wavelength range of 950–1040 nm is important; for voltage imaging, we mainly excite at 1020 nm.

GaAsP PMTs for high-detection quantum yield in combination with low sensitivity above 900 nm (e.g., H10770PA-40, Hamamatsu).

$\lambda/4$ plate in the wavelength range of 1020 nm (e.g., WPQW-IR-4M, Sigma-koki).

Telescope (2× or 3×) for narrowing the beam (e.g., BE02R/M, Thorlabs).

Emission filter longpass 540 nm or bandpass 540–750 nm (e.g., FF01-650/200-25, Semrock); note: if GaAsP PMTs are used, no emission filter is needed for excitation wavelengths above 1000 nm.

2.2 Microscope Testing

Fluorescent microbeads (e.g., 19507-5, Fluoresbrite Polychromatic Red Microspheres 0.5 μm, Polyscience)

Agarose Type III-A (A9792, Sigma-Aldrich)

2.3 Pipettes for Bulk-Loading, Single-Cell Loading, and Electrophysiological Recording

Bulk-loading: Quartz or borosilicate tubing (e.g., outer diameter 1 mm, inner diameter 0.7 mm; Sutter Instrument)

Single-cell loading: Borosilicate tubing (e.g., outer diameter 1 mm, inner diameter 0.58 mm; Sutter Instrument)

Electrophysiological recording:

Quartz electrodes (inner diameter 0.7 mm; Sutter Instrument)

Alexa Fluor 594 Hydrazide (A10438, Invitrogen Molecular Probes)

Agarose (A9793, Sigma-Aldrich)

Pipette puller for quartz pipettes (P-2000, Sutter Instrument) or borosilicate tubing (P-2000, P-97, Sutter Instrument)

Beveller (e.g., BV-10, Sutter Instrument)

2.4 Injection Apparatus for Bulk- and Single-Cell Loading

Micromanipulator (e.g., MP-285, Sutter Instrument) with pipette holder with port for pressure application (e.g., W.P.I. or HEKA); tower for micromanipulator (e.g., Sutter Instrument)

Bulk-loading: Beveled quartz or borosilicate pipettes (10–15 μm opening)

Single-cell loading: Borosilicate pipettes (1 μm tip diameter, 7–10 MΩ impedance measured with saline)

Tubing (typically silicone tubing with 1 mm inner diameter and thick wall) (e.g., from As One, www.as-1.co.jp)

Syringe (fitting to the tubing) preferably with Luer taper (e.g., from As One or W.P.I.)

T-connector (fitting to the tubing) preferably with Luer taper (e.g., from As One or W.P.I.)

3-way (or 4-way) stopcock preferably with Luer taper (e.g., from As One or W.P.I.)

Pressure meter (e.g., PM015R, W.P.I.)

For single-cell loading: Stimulus isolator (ISO-Flex, A.M.P.I.) for electroporation

2.5 Preparation of ANNINE-6 and ANNINE-6plus Stock Solution for Bulk- and Single-Cell Loading

ANNINE-6 (MW 592.8) or ANNINE-6plus (MW 717.6) (Dr. Hinner & Dr. Hübener Sensitive Farbstoffe GBR, Munich, Germany, http://www.sensitivefarbstoffe.de/; info@sensitivefarbstoffe.de. Alternatively, contact bkuhn@oist.jp)

Desiccator

Vortexer

0.22 μm small volume filter (e.g., UFC30GV00, Ultrafree-MC-GV Centrifugal filter 0.22 μm, Millipore)

Small centrifuge

For bulk-loading:

DMSO (e.g., Nacalai tesque); DMSO is hygroscopic; to keep DMSO water-free, add molecular sieves 3 Å (e.g., 23355-44, Nacalai tesque) to the DMSO bottle

Pluronic F-127 (e.g., AAT Bioquest)

Heater to keep tubes with dye at up to 70 °C

For single-cell loading: Ethanol 99.5%

2.6 Solutions

Ringer's solution (135 mM NaCl, 5.4 mM KCl, 1.0 mM $MgCl_2$, 1.8 mM $CaCl_2$, 5 mM Hepes (free acid), in H_2O, adjust to pH 7.2 with NaOH)

Saline (150 mM NaCl in H_2O)

3 Methods

3.1 Procedure for Voltage Imaging with ANNINE-6 in Cell Cultures

Voltage imaging from dissociated cells in culture is the easiest voltage imaging task and allows imaging of subcellular, cellular, and network voltage changes. The cell membranes are exposed and, therefore, labeling and imaging is relatively easy. In addition, signal amplitudes are high compared to bulk-loaded tissue.

For labeling cell cultures, prepare an ANNINE-6 or ANNINE-6plus stock solution by dissolving ANNINE-6 (0.5 mM) or ANNINE-6plus (2.0 mM) in a solution of 20% Pluronic F-127 in DMSO. This stock solution can be kept for a long time (many months) at room temperature if protected from light.

For the labeling solution, mix ANNINE-6 or ANNINE-6plus stock solution with saline (0.9% NaCl in water) or Ringer's solution at a ratio of 1:50 or 1:100, respectively. It is important to prepare enough solution to cover the cell culture dish completely with labeling solution to a height of about 2 mm. The labeling solution should be prepared just before labeling the cell culture.

Then, the culture medium of the cell culture dish is removed (but kept in a separate dish) and the labeling solution is added to the cell culture. After 5 minutes, the labeling solution is removed, and the cell culture is carefully washed with dye-free saline or Ringer's solution. In most cases, washing is not necessary. Enough dye-free saline or Ringer's solution should be added so that the cell culture is well covered (typically a few millimeters). If the labeling is not bright enough, the procedure can be repeated, and/or the incubation time can be increased from 5 to 10 minutes.

Now, the cell cultures can be used for voltage imaging with two-photon microscopy or with standard confocal microscopy.

Under the two-photon microscope, only the outline of the cells should be visible. The linearly polarized laser beam will excite parallel-oriented dye molecules but no perpendicularly oriented molecules. Spherical cells will have bright poles and a dim equator.

This can be overcome by inserting a $\lambda/4$ plate into the excitation light path (see Sect. 3.4).

For cell cultures, it is also very convenient to use confocal microscopy for voltage imaging as scattering can be neglected. In most confocal microscopes, laser excitation at 488 nm, 502 nm, and/or 514 nm is available. For these excitation wavelengths, ANNINE-6 and ANNINE-6plus show high sensitivity. In combination with a 540 nm emission longpass filter and line scans along the membrane, voltage imaging can be performed easily.

3.2 Preparation of a Cover Glass Window with Access Port

Chronic cranial windows are widely used for in vivo imaging [49]. The craniotomy is sealed by a glass cover slip and does not allow subsequent access to the brain. Here, we describe how to prepare a glass window with access port. The access port allows dye application, drug application, and electrical recording.

The procedure was described before in [43] (Fig. 2a), but in the following paragraphs, we give some additional advice for preparing the access port. The windows with access port should be prepared before the surgery and can be stored for many weeks.

The glass window (typically round with 5 mm diameter) is held with an alligator clip using soft silicone tubing to cover the alligator clip teeth to protect and grip the glass (see Fig. 1 in [43]). It might be convenient to file down the alligator clip teeth to make the glass more accessible. To reduce the force of clamping, the crocodile clamps can be bent open. The alligator clip can then be mounted on a stand or stereotaxic frame, or even held by hand for drilling under a microscope. All drilling is done by hand, on dry glass, using a conical stone drill bit. The free hand is used to blow compressed air on the glass to clear away the glass dust while drilling. To begin making the hole, it is best to drill at higher speed holding the drill at an angle of about 45° to the glass to make an initial hole through the glass with a size of less than 0.5 mm and not necessarily round. It is important to be patient: the glass should always be sanded away rather than chipped. Then, the drill is held perpendicular to the glass, and the tip of the conical drill bit is inserted into the hole. While using lower drilling speed, the glass is grinded to increase the hole size and make a smooth edge. It is best to do this on both sides of the glass to make the hole symmetrical and with perpendicular edges. It is important to avoid creating an angle at the glass edge as this will distort imaging near the hole.

In general, there are several alternative methods for fabricating the hole in any shape and size, for example, sand blasting with a mask or laser-induced etching of fused silica glass. However, sand blasting, in our hands even with the finest grain size, resulted in rough edges. Laser-induced etching is limited to fused silica glass which is expensive but allows cutting of the access port and additionally the shape of the glass off the cover slip center.

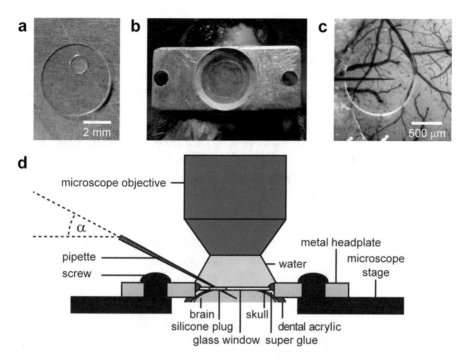

Fig. 2 Chronic cranial window with access port for dye loading, drug application, and electrophysiological recording. (**a**) Glass cover slip with a 1 mm silicone access port. (**b**) Chronic cranial window with an access port can be mounted just as a regular chronic window on top of the dura mater. (**c**) The access port allows repeated brain manipulations for weeks and keeps the imaging site under the glass stable. A dye-filled pipette coming from the left is positioned over a region having only a few blood vessels to avoid bleeding by puncturing vessels. The pipette is then lowered along its axis into the brain. After retraction of the pipette, the silicone seals the brain again. (**d**) Schematic of brain manipulation through a chronic cranial window with access port. A typical access angle α is 25° [43]

Typically, the access port will be off the center. The center of the glass window is reserved for optimal imaging (Fig. 2a).

After drilling the hole, PDMS (typically Sylgard 184) is prepared to fill the hole. To pipette Sylgard, a 1 ml pipette tip is cut with a blade to have an opening of about 4 mm. As it is difficult to pipette a precise volume of Sylgard, it is best to measure the weight and add catalyst accordingly at a ratio of 10:1. It is important to mix thoroughly. As the Sylgard/catalyst mixture is not viscous enough to fill the hole, it is necessary to heat it first, for example, on a glass coverslip over a hot glass bead sterilizer for a few minutes. The Sylgard/catalyst mixture should just be tacky enough to prevent it spreading over the glass window.

Using a 0.5 mm metal drill bit, a droplet of tacky Sylgard/catalyst mixture is picked up on the drill bit tip and applied to the glass hole by hand, by inserting the drill bit tip into the hole and carefully allowing the droplet to touch the hole edges. The silicone should only be attached to the inside edges of the hole, lying approximately flat with the glass, and not protrude far above or below the surface of the window. Then, the window with the filled hole is heated again, over a glass bead sterilizer, for example, to fully set the silicone in the hole. After the silicone is set, the window can be cleaned, for example, with detergent, and rinsed well with water. Finally, the window is dried with compressed air. The windows with access port should be prepared in advance and can be stored for several months.

Before mounting the window with access port, it is put into the alligator clip and sterilized above the hot glass bead sterilizer during surgery until it is used. It is important to keep the glass clean, sterile, and free of dust.

3.3 Chronic Cranial Window Surgery with Access Port for In Vivo Imaging

A chronic cranial window with access port is required for in vivo imaging of average membrane potential after bulk-loading and in vivo imaging of voltage from single cells after electroporation.

All animal experiments and genetic manipulations must be approved by local authorities following the respective rules and regulations.

As for any in vivo experiment, surgery is crucial for high-quality imaging. Several protocols are available [43, 49, 50]. Here, we briefly describe our procedure for somatosensory cortex. We anesthetize the mouse with 2–3% isoflurane in air or 100% oxygen in an anesthesia initiation box. Then, we inject dexamethasone (2 mg/kg i.m.) to reduce the immune response and carprofen (7.5 mg/kg i.p.) to reduce inflammation and pain. After shaving the head, we apply hair-removal cream "for sensitive skin" with cotton applicators. The cream dissolves all remaining hair within 2 minutes and should not be applied for too long time to avoid skin irritation. The skin over the future craniotomy is removed. For the craniotomy, a groove is drilled around the target region, leaving a bone patch about 4 mm in diameter intact. It is important to drill until the remaining bone is very thin and flexible, but not so thin as to break through the bone and damage the brain. Bone debris can be removed by careful application of compressed air. To check the thinned bone, slight pressure is applied to the central bone patch and the light reflected off the shiny thinned bone is observed to evaluate if the bone is thin enough. The most critical step of the surgery is to lift the bone. To do this in a controlled way, we glue a toothpick to the center of the bone patch with superglue or dental acrylic and then carefully lift the bone without putting pressure on the brain [43]. This technique dramatically increased our success rate of obtaining an intact dura after bone removal (almost 100% success rate for an experienced person).

Then, the window with an access port is mounted, and oriented such that a pipette, which must enter via the access port, can reach the imaging site under the center of the glass at the correct depth. The glass should fit flat on the brain or be pressing it down slightly to reduce movement artifacts during imaging. To do this,

it is necessary to remove all bone ridges around the future craniotomy, i.e., the bone patch, before lifting the bone. Also, no bone
debris should remain under the glass to avoid bone regrowth. The
window is sealed to the bone with superglue. A custom-made
headplate (Fig. 2b), light enough to be carried by the animal and
strong enough to allow stable head fixation, is fixed to the bone
with dental acrylic. As dental acrylic, we use MetaBond (also called
SuperBond in some countries). The advantage of this dental acrylic
is that it bonds very strongly to the skull while other dental acrylics
which we tried need a superglue coating of the bone as a mediator.
Unfortunately, MetaBond is expensive and only available through
dental supply companies (therefore, in some countries, a dental
license might be required to buy MetaBond). The wound around
the implant is closed by fixing the skin to the implant with superglue. An analgesic, like buprenorphine, should be injected before
the animal awakes. Several days of recovery should be allowed. If
the surgery was successful, a crisp blood vessel pattern can be seen
under the window (Fig. 2c).

Before the VSD injection, it is useful in many cases to map the
brain region with imaging of intrinsic signals [30, 51, 52].

3.4 Two-Photon Microscope for Voltage Imaging

In general, any regular two-photon microscope can be used for
voltage imaging. However, several specifications and simple modifications can help to increase the SNR (Fig. 3). Care must be taken
to follow all laser safety guidelines as infrared high-power femtosecond laser pulses can cause serious damage in the unprotected eye.

As a first modification, a $\lambda/4$ plate with the slow or fast axis
turned $45°$ against the polarization direction of the laser can be
inserted to convert linearly polarized excitation laser light into circularly polarized light (Fig. 3a). The slow or fast axis of the $\lambda/4$
plate is typically marked on the mounting of the plate. Mounting
of the $\lambda/4$ plate in a rotation mount allows easy adjustment of the
angle. The direction of laser polarization can be found in the specifications. After installation, for testing purpose only, a polarizer
placed behind the $\lambda/4$ plate should not be able to block the laser
beam completely regardless of its orientation. Now all dye molecules with their dipole moment in the plane perpendicular to the
light propagation direction, irrespective of how their dipole
moments are oriented in the plane, can be excited.

The two-photon point spread function of excitation is proportional to the numerical aperture (NA) of the objective, laterally by
$1/NA$ and axially by $1/NA^2$ [53]. By increasing the point spread
function as described earlier, more dye molecules can be excited;
therefore, the SNR can be improved at the expense of spatial resolution. To do this, the back focal plane of the objective can be
under-filled by narrowing the excitation laser beam diameter,
thereby reducing the effective NA of excitation. Underfilling the
back aperture allows using high-NA objectives, which are crucial

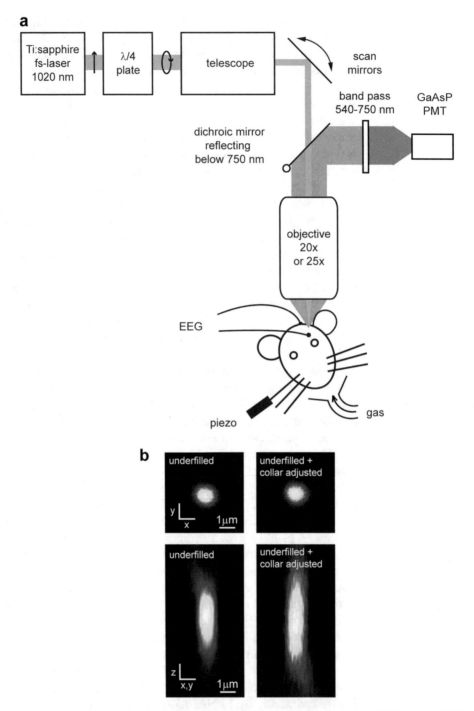

Fig. 3 Schematic drawing of a two-photon microscope for voltage imaging with ANNINE-6 or ANNINE-6plus. (**a**) To optimize voltage imaging, a $\lambda/4$ plate and a telescope are inserted into the excitation light path. The $\lambda/4$ plate converts the linearly polarized light into circularly polarized light. The telescope decreases the laser beam diameter and thereby underfills the back aperture of the objective. Underfilling the objective's back aperture increases the size of the point spread function of excitation, especially in z-direction, and thereby increases the excitation volume. Larger excitation volume results in more photons and therefore in a reduction of the

for efficient fluorescence detection, and at the same time allows excitation of dye molecules in an increased volume. Typically, the laser beam diameter is decreased to 1/2 or 1/3 with a telescope inserted into the light path between the laser and the microscope.

The point spread function can be measured by imaging sub-resolution fluorescent microbeads (typically 200 or 500 nm) in a 1.5% agarose gel. To prepare a sample, add 1.5% agarose to water and heat while stirring until the agarose dissolves and the foggy emulsion becomes a transparent solution. Then, add microbeads (dilute typically 1000–10,000 times), stir thoroughly, and add a droplet of the hot and liquid solution on a glass slide and cover the droplet with a glass cover slip. The agarose will solidify quickly. Now, the test sample can be used with water immersion. It should be possible to find mainly isolated beads. If the bead density is too high, it is necessary to prepare a new sample with higher dilution of the beads. To measure the point spread function, a bead is 3D reconstructed with high zoom (typically 5 or higher) and small z-step size (typically 0.5 μm or less).

Some objectives allow adjusting the focal spot with a collar to optimize the point spread function at different depths of imaging in the brain (e.g., XLPLN25XWMP2, 25× NA 1.05, Olympus). However, this feature can also be used to artificially widen the point spread function and thereby increase the excitation volume (Fig. 3b).

It is important to use PMTs with high quantum efficiency. GaAsP PMTs typically have a quantum yield 3–5× higher in the visible wavelength range than regular PMTs. As for all two-photon imaging, the PMT sensor size should be large to also collect scattered photons. GaAsP PMTs degrade with exposure to light. Therefore, it is useful to keep a record of laser output, PMT voltage setting, and measured intensity with a specific objective lens of a defined, reproducible sample (e.g., a fluorescein solution with specific molarity). If intensity levels deviate seriously, steps must be taken to find the problem. A spare set of PMTs is typically a good investment and allows for a quick comparison between old and new PMTs.

Emission filters should have as wide a range as possible (for ANNINE-6 or ANNINE-6plus a 540 nm longpass or a 540–750 nm bandpass filter), or can be omitted if the PMT is not sensitive to the

Fig. 3 (continued) relative noise $\sqrt{n}/n$. The objective collects the emitted photons from the sample. Objectives with high numerical aperture are recommended to collect as many photons as possible, again to reduce the relative noise. The dichroic beam splitter and a band pass filter select the wavelength range of ANNINE-6 or ANNINE-6plus emission with high voltage sensitivity. The GaAsP PMT detects the photons with high quantum yield. Modified with permission from National Academy of Sciences, U.S.A [30]. (**b**) The excitation volume can be increased by underfilling the back aperture of the objective (left) and, additionally, by turning the collar of the 25x, NA 1.05 objective (Olympus) to maximal imaging depth even when imaging close to the surface (right) [37]

excitation wavelength, as in the case of GaAsP detectors and excitation wavelengths above 1000 nm.

Laser stability is crucial for low noise experiments. To test laser stability and noise, a fluorescein solution (about 50 μM can be used, or fluorescent beads visualized in a 1.5% agarose matrix as described above) can be imaged. For example, when imaging a fluorescein solution or a sample with beads for several seconds, a time course of every pixel in the image can be analyzed. Typically, a wide range of mean intensities will be found as the intensity decreases toward the edges of the image or due to the intensity differences of the imaged structures. If the variance (SD^2) increases linearly with the mean intensity and the extrapolation for intensity = 0 results in variance = 0, then the microscope setup is shot-noise limited. The shot noise originates from the photon statistics which follows a Poisson distribution. If a different noise behavior is found, it is necessary to search for the noise source (e.g., unstable laser, unstable Pockels cell, electronic noise, etc.).

3.5 Injection of VSD into the Brain for Bulk-Loading

At first, an ANNINE-6 or ANNINE-6plus stock solution should be prepared.

Larger quantities of ANNINE dye can be dissolved in ethanol and then split into convenient aliquots. For example, 1 ml of ethanol is added to a flask containing 0.2 mg ANNINE-6, and 10 aliquots of 100 μl containing 20 μg ANNINE are filled into 500 μl microcentrifuge tubes. After drying the dye in a desiccator, the tubes can be stored at −20 °C protected from light for years.

ANNINE-6 is insoluble in water or saline without additional solvents. ANNINE-6plus dissolves in water to about 1 mM [23]; however, it does not dissolve well in saline. Therefore, a stock solution of ANNINE-6 and ANNINE-6plus is prepared in DMSO at a concentration of 0.5 and 2.0 mM, respectively. This corresponds to adding 66 μl of DMSO to a tube with 20 μg ANNINE-6, and 14 μl of DMSO to a tube with 20 μg ANNINE-6plus. Similarly, 20% Pluronic F-127 in DMSO can be used as a solvent. This stock solution can be kept for many months at room temperature protected from light.

Before use, the stock solution (typically a tube with 20–200 μl) is heated to 70 °C in a water bath or heating block for at least 15 minutes. In general, ANNINE-6 is more difficult to load into tissue at high concentrations, but it persists longer in the tissue than ANNINE-6plus.

To inject the dye into the brain, the PDMS membrane of the access port and the dura mater must be penetrated. This works best with a beveled sharp quartz pipette (Fig. 4a). The angle between the grinding plate and the pipette axis should be in the range of 25–45°; we use 45°. Borosilicate pipettes also work but are less stiff than quartz pipettes. A typical opening is 15 μm for the small diameter of the elliptic opening. To allow a volume measure-

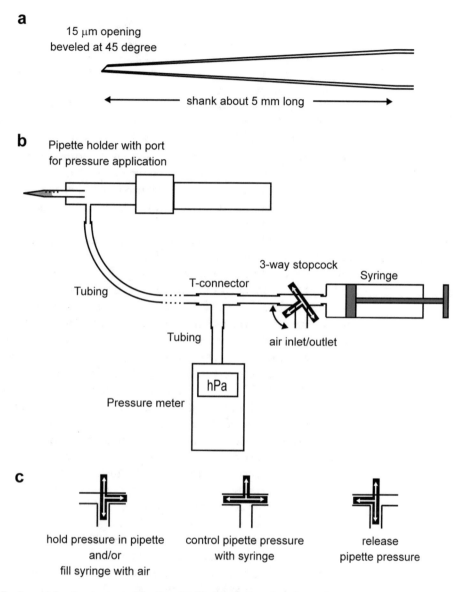

Fig. 4 Equipment for dye pressure injection. (**a**) Sketch of a beveled sharp pipette used for dye injection in vivo. (**b**) A 3-way stopcock and a pipette holder with pressure application port allow applying air pressure with a syringe. A manometer is used to measure the pressure during the dye loading. (**c**) Different settings for the 3-way stopcock in (**b**). The arrows on the 3-way stopcocks indicate the airflow directions during different procedures

ment, it is useful to draw marks on the pipette trunk to estimate later the volume of injected solution. It can be determined by observing the meniscus and by calculating the volume in the pipette per unit length. For example, when using capillaries with an inner diameter of 0.7 mm, a change of the meniscus by 1 mm corresponds to 385 nl. Typically, 500–2000 nl are filled into a pipette and injected into the brain.

Before the experiment, a pipette setup, typically used for patch clamp experiments, should be prepared that allows pressure application and pressure measurement (Fig. 4b). Typically, PVC and/or silicone tubing is used to connect to a 10 ml or 20 ml syringe via a T-connector to a pressure meter. Additionally, a 3-way stopcock should be inserted close to the syringe to control airflow into and out of the tubing and to keep pipette pressure constant by manual control during the injection (Fig. 4c). Before the animal experiment, the apparatus should be tested to determine whether the pipette can reach the access port of the cranial window, whether it fits under the objective that will be used, whether the tubing system holds pressure (test 70 hPa), and whether the markings on the pipette can be observed.

Now an anesthetized mouse can be mounted on the in vivo imaging stage. The heated ANNINE-6 or ANNINE-6plus stock solution gets diluted typically 1:50 or 1:20 with saline (0.9% NaCl in H_2O) and backfilled into the pipette. It is best to mix the solution just before filling the pipette, because eventually the dye will precipitate. Then, the pipette is mounted on the pipette holder and lowered onto the access port. The approach is observed under a low-magnification objective, and a region under the access port without major blood vessels is selected. When the pipette tip touches the PDMS (Sylgard) membrane, detected by visual observation, the pipette position is zeroed, and the pipette is slowly lowered along the axis of the pipette through the Sylgard membrane and the dura into the brain. This entry is typically observed by two-photon microscopy imaging of the fluorescent solution in the pipette, but it is recommended to combine it with wide-field imaging for better overview and clearer view of the blood vessel pattern. The pipette is moved slowly to allow the brain tissue to relax, typically a few minutes per 100 μm. The target location of the dye delivery is typically between 100 μm and 1000 μm below the pia, which can be judged with the z-focus shift of the two-photon microscope. This can be done under visual control with wide-field imaging and/or two-photon microscopy. When reaching the target location, it is best to wait 5–10 minutes so that the tissue can relax and close the gaps between pipette and tissue. Then, the pressure in the pipette is slowly increased until the position of the meniscus changes, as observed by eye or a magnifying glass. The injection should be done slowly and steadily without sharp peaks to keep the tissue as undisturbed as possible. Typically, 1000 nl are injected within 15 minutes (or slower) to label a volume with 400 μm diameter.

For labeling a larger area, it is advised to inject at several locations or at several positions along the pipette path through the tissue. For the latter, it is better to do the injections on the path into the tissue, rather than on the way back out, to avoid leakage of the dye to the surface along the injection track. After injection, it is best to wait a few minutes until the dye has bound to the membrane and the pressure in the brain has adjusted to avoid dye solution being squeezed out along the retraction path of the pipette. Then the pipette is slowly retracted and the Sylgard seals the access port again. Now, the animal can recover for a few hours (typically 2–12 hours) to allow DMSO to be washed out and the dye to be dispersed evenly.

3.6 Two-Photon Imaging of Average Membrane Voltage Changes In Vivo

After mounting the mouse again on the in vivo stage, either anesthetized or awake, the labeled brain region should be evaluated by two-photon microscopy. This is typically done with a wavelength around 950 nm. The absorption cross section of ANNINE-6 at 950 nm is higher than in the range of 1020–1040 nm (used for functional imaging), and therefore the image is brighter. There should be a clear contrast between the dark cell bodies and the bright neuropil. Bright puncta might be the result of precipitated dye. Also, bright capillaries might be observed. This becomes more obvious a few days after labeling.

Now two-photon voltage imaging can be performed with an excitation wavelength in the range of 1020–1040 nm and the appropriate emission filter. Line scans typically show the lowest noise and fluctuation levels and offer a temporal resolution of up to 2 kHz for bi-directional scans with regular galvo scanners. Laser power is typically in the range of 60 mW when imaging close to the surface (cortical layers 1 and 2) and up to 120 mW in deeper layers (cortical layers 3 and 4). Resonant scanners are about 15 times faster and allow box scans (512×16 or 512×32 pixels per image) at similar temporal resolution (1 kHz or 500 Hz).

One simple way to observe a voltage signal is to apply sensory stimulation. For example, after finding the position of a specific barrel in the somatosensory cortex of the mouse by imaging of intrinsic optical signals (Fig. 5a) and subsequent bulk-loading with VSD (Fig. 5b), the principle vibrissa of the barrel can be stimulated. The response can be imaged, and a stimulus-triggered average can be calculated (Fig. 5c). The signal increases with excitation near the red spectral edge of absorption (Fig. 5d). This proves that a voltage change was measured and that the observed intensity change is not a movement artifact.

Sensory responses can be repeatedly applied, which allows stimulus-triggered averaging. Thereby, small signals can be revealed that otherwise would be masked by noise. The software controlling the microscope can supply trigger outputs for synchronizing the stimulus to the imaging. We use ScanImage [54] and MScan (Sutter Instrument).

An alternative is to image sleep spindle activity (Fig. 5e). Sleep spindles are bursts of oscillatory brain activity which occur during stage 2 NREM sleep. Sleep spindles are also observed under anesthesia (e.g., 1.5% isoflurane in oxygen or air, body temperature at

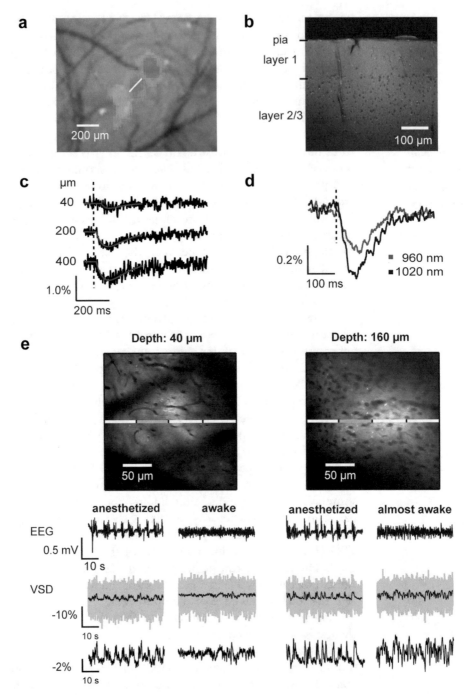

Fig. 5 Voltage imaging of average membrane potential of neuronal tissue in vivo. (**a**) Imaging of intrinsic signals reveals the location of primary responses in barrel cortex. In this montage, three vibrissa responses (green, magenta, blue) in relation to the blood vessel pattern are shown. Following the imaging of intrinsic signals, the brain region of interest was bulk-loaded with VSD. (**b**) Two-photon image of a barrel cortex brain slice prepared after in vivo bulk-loading with ANNINE-6 shows homogeneous labeling. Somata (here of layer 2/3) and blood vessels remain dark. (**c**) Two-photon line scans performed at the location indicated in (**a**) (white line) at different depths below the brain surface were spatially averaged. Vibrissa stimulation causes an average membrane

37 °C). They are easily recorded electrically with an EEG. To image sleep spindles, the brain region of interest is bulk-loaded with ANNINE-6 or ANNINE-6plus. Then, line scans can be recorded through the cortical tissue depth-resolved due to two-photon excitation (Fig. 5e, line scan position indicated in the images). To increase the SNR, imaging data can be spatio-temporally filtered. For example, Fig. 5e shows a trace where the yellow segment of the line scan was spatially averaged (yellow trace), and then temporally averaged by a box car filter. Additionally, a trace is shown below where the full line scan was spatially averaged and box car filtered. The SNR increases at the expense of spatial and temporal resolution. Voltage imaging of sleep spindles is very useful for testing and comparing imaging setups, surgeries, and dye labeling.

Brain oscillations up to at least 15 Hz can be recorded for hours (recording at 1 kHz and subsequent spatio-temporal filtering) without detectable phototoxicity or bleaching. However, for line scans, the stability of the preparation is very important as movement artifacts cannot easily be corrected by subsequent alignment during offline analysis. If sequences are going to be averaged, it should be done off-line. This allows sorting out trials with obvious movement artifacts. The stability of the preparation depends on the brain region, the surgery, and on the position of the mouse on a wheel or ball during the experiment.

Washout of ANNINE-6 is slow, allowing many minutes or even hours of recording. ANNINE-6plus is washed out faster than

---

Fig. 5 (continued) depolarization, which results in decrease of fluorescence if the dye is bound to the outer leaflet of the membrane and the dye is excited at the red spectral absorption edge. (**d**) Under the same experimental conditions, the excitation wavelength was changed. As expected from a purely electrochromic dye, excitation closer to the spectral edge results in a larger relative fluorescence change. Such an experiment confirms the mechanism of voltage sensing. (**e**) In vivo two-photon images of cerebral cortex layer 1 and layer 2/3. Blood vessels and somata appear as dark shadows, as the dye is bound to the outer leaflet of the cell membranes. Line scans allow measurement of local average membrane depolarization in anesthetized and awake animals, which correlates with the local electro-encephalogram (EEG). The yellow trace is the averaged trace of the line scan segment indicated above. The black, overlaid trace is the corresponding filtered trace (boxcar smoothing, 200 ms). The segmentation allows analyzing cross-correlations among neighboring brain segments. For example, cross-correlations among segments are higher in anesthetized than in awake animals [30]. The trace at the bottom shows the relative fluorescence change of the full line scan after filtering with 200 ms boxcar smoothing. EEG signals are biphasic, representing sources and sinks while the VSD signal shows average membrane depolarization (negative fluorescence change). As the neuronal activity de-correlates during the transition from anesthetized to awake state, the average voltage signal becomes noisy (160 μm below the dura mater, right VSD traces) and then flat (40 μm, right VSD traces) in the fully awake animal. Average membrane voltage measurements can be recorded for hours without photo bleaching. (Modified with permission from National Academy of Sciences, U.S.A [30])

ANNINE-6. ANNINE-6plus labeling can be considered stable for several hours.

3.7 Analysis of Average Membrane Potential Changes During the Experiment

Typically, voltage signals are not immediately visible because the relative fluorescence change is smaller and faster than changes observed when, for example, imaging with fluorescent calcium indicators. For a first analysis during the imaging session, it is convenient to use a program like ImageJ [55] or FIJI [56] to quickly filter and collapse/average line scans. This gives immediate feedback.

As an example, we show voltage imaging of sleep spindles. After averaging a part of the line scan (indicated by yellow, Fig. 5e), a noisy trace (yellow) is obtained. If the spatiotemporal line scan image is filtered temporally ("Gaussian Blur 3D" of ImageJ and FIJI) and then the line scan segment is averaged, the EEG-correlated membrane depolarization becomes visible (Fig. 5e, black trace overlaying the yellow trace).

3.8 Procedure for In Vitro Voltage Imaging After ANNINE-6 Labeling In Vivo

Also for slice experiments, it is recommended to inject ANNINE-6 in vivo. The slices can be cut after a few hours. This has the advantage that ANNINE-6 can equilibrate completely, DMSO is washed out, and no precious imaging time is wasted by waiting for adequate labeling to occur after adding dye in vitro. Importantly, in our hands, slice labeling after cutting slices was not successful because ANNINE-6 and ANNINE-6plus label dead surface cells brightly, while healthy deeper cells remain almost unlabeled.

In this case, surgery can be simplified. After removing hair and mounting the animal in a stereotaxic frame, the skin is cut over the target region and pushed aside to expose the bone. Then, a 1 mm hole is drilled into the bone. Care should be taken not to drill through the bone. The last layer of bone should be removed with a fine needle tip leaving the dura mater intact. Then, ANNINE-6 can be injected stereotaxically (for injection procedure, see above). For starting parameters, we recommend injecting about 500–1000 nl of 5% ANNINE-6/DMSO stock solution in saline (prepared after heating of the stock solution for at least 15 minutes at 70 °C). Injection speed should be low (typically 15 minutes for 1000 nl). If necessary, several injections can be made at different depths or in different locations. The injected volume can also be adapted. For sensitive tissue, ANNINE-6/DMSO concentration in saline should be reduced from 5% to 1%.

After injection and retraction of the pipette, the skin is closed again over the craniotomy and either sewed or glued together with superglue. Then the animal can recover. Labeling typically lasts a few days, but it is best to start the in vitro experiment (slicing) 2–12 hours after injection.

3.9 Two-Photon Imaging of Single-Cell Membrane Voltage Changes In Vivo

Measuring the average membrane potential of bulk-loaded tissue with two-photon microscopy is a relatively simple experiment and can be used for mapping depth-resolved postsynaptic potentials in vivo or in slice preparations. To understand the neuronal network on a cellular level, one can optically record from single neurons or small networks of neurons in vivo, favorably in awake animals. This can be achieved by filling single cells with a VSD, waiting until the dye diffuses and gets incorporated into the membrane, and then image with two-photon microscopy (e.g., while the awake mouse is sitting on a treadmill; Fig. 6a, b). For example, by line-scanning along dendrites, spatio-temporal dendritic voltage maps with sub-millisecond temporal and micrometer spatial resolution can be obtained from cerebellar Purkinje neurons revealing supra- and sub-threshold electrical activity in awake animals [37].

Interestingly, the spectral characteristics of ANNINE-6plus and GCaMP6 allow simultaneous voltage and calcium imaging [37]. Additionally, to record signals at dendritic and somatic locations simultaneously, an extracellular recording electrode can be inserted through the cranial window access port and positioned at the soma to record extracellular spikes, while optical imaging at the dendrites records dendritic activity [37].

In the following, we describe how to label and image dendrites of single Purkinje neurons in awake mice with millisecond temporal and several micrometer spatial resolution. It is important to know that Purkinje neurons have about three spines per micron at the distal dendrite [57]. This increases the dendritic membrane surface dramatically such that the spines contribute 80% to the total membrane surface [57]. This huge surface area makes Purkinje neurons a perfect target for voltage imaging. Additionally, the fan-like dendritic tree of Purkinje neurons allows simultaneous line-scanning across several dendritic branches.

At first, a chronic cranial window with access port ([43]; for further tips, see Sect. 3.2) is mounted over the cerebellum of a mouse (Fig. 6c). A few days after the window surgery, the anesthetized mouse is headfixed on the injection stage. An Adeno-Associated Viral vector (AAV), delivering the gene of GFP, GCaMP, or another green fluorescent chromophore, is injected through the access port into the granule layer of the cerebellum. FITC (50 μM) can also be included in the injection pipette solution to monitor the extent and location of the virus injection under the two-photon microscope. To specifically target Purkinje neurons, several promoters or promoter systems are available [37, 58]. When the chromophore expression is bright enough, typically after 5–10 days, the mouse can be used for targeted labeling of single or multiple Purkinje neurons.

An ANNINE-6plus solution in ethanol is prepared at 3 mM which is half the saturation concentration (Fig. 6d). The solution is shaken using a vortexer (10 minutes) or an ultrasound sonicator (5 minutes). Then, the solution is filtered with a 0.22 μm centrifugal filter.

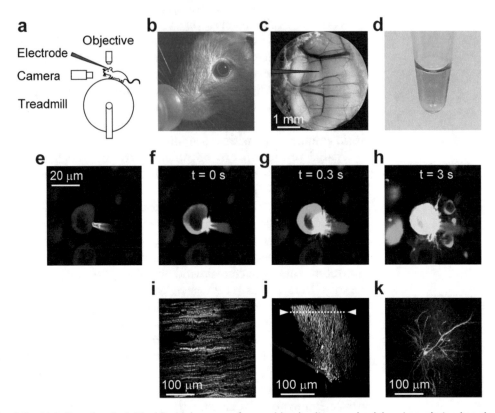

Fig. 6 Dual labeling of an individual Purkinje neuron for combined voltage and calcium two-photon imaging in awake mice. (**a**) Sketch of the setup with a mouse mounted on a treadmill under a two-photon microscope. An electrode is used to fill single Purkinje neurons by electroporation and to electrically record from their soma. (**b**) A behavioral camera allows detailed observation of the pupil, the vibrissa, and the face of the mouse. (**c**) A chronic cranial window with access port on the vermis of the cerebellum allows entering the brain with a pipette (schematically indicated). (**d**) ANNINE-6plus dissolved in pure ethanol at 3 mM concentration. (**e**) A patch pipette filled with ANNINE-6plus in ethanol is used to label a single GCaMP6f-expressing Purkinje neuron by electroporation in the anesthetized mouse. (**f–h**) A GCaMP6f-labeled Purkinje neuron filled at the soma with ANNINE-6plus by electroporation, at $t = 0$, 0.3, and 3 seconds, respectively; 24 hours after labeling a Purkinje neuron with ANNINE-6plus, the dye spread out evenly as can be seen (**i**) in the cross section of the Purkinje neuron dendrite as an overlay of the green channel (GCaMP6f) and the red channel (ANNINE-6plus). (**j**) Reconstruction of the Purkinje neuron in the red channel (ANNINE-6plus). (**k**) It is also possible to label cortical pyramidal neurons with ANNINE-6plus by electroporation. The image shows a z-projection of a layer 3 pyramidal neuron. (Modified from [37])

Mice with a chronic cranial window and access port over the cerebellum are anesthetized and head mounted. A borosilicate glass patch pipette with 1 μm tip diameter (7–10 MΩ in saline; test pipette resistance first to optimize the pipette pulling; always use fresh pipettes for electroporation) is filled with the ANNINE-6plus/ethanol solution (typically 5 μl; but just enough solution to have electrical contact between the tip of the pipette and the silver wire inside the pipette) and mounted in a pipette holder. The pipette holder has a chlorinated silver wire connected to a stimulus isolator. The port of the pipette holder for pressure application is connected to a pressure meter and a syringe. Neutral pressure is applied when entering the brain through the access port to prevent leakage of the dye/ethanol solution into tissue. The ethanol in the pipette prevents clogging of the tip. During the approach of a target cell, the pipette tip is visualized with the two-photon microscope. ANNINE-6plus dissolved in ethanol shows bright fluorescence at 1000 nm excitation wavelength and allows positioning of the pipette on a target Purkinje neuron labeled with, for example, GFP or GCaMP6 (Fig. 6e). A slight deformation of the soma should be observed to ensure close contact (Fig. 6e). For electroporation, a stimulus protocol of 50 negative current pulses (−30 µA), 1 ms in duration at 100 Hz, is delivered (Fig. 6f–h). Neutral pressure is applied to the patch pipette, and the pipette is retracted immediately after the cell is loaded and replaced for further single-neuron labeling. Typically, several Purkinje neurons can be filled per mouse on the same day. After loading the cells with dye, mice are returned to their cages to allow the dye to spread to distal dendrites and throughout the entire cell. ANNINE-6plus is highly lipophilic, so dye diffusion can take several hours.

After about 20 hours, a brightly labeled Purkinje neuron is selected for the imaging experiment (Fig. 6i, j).

In general, also other cells can be electroporated with this method. For example, we were able to label cortical pyramidal neurons (Fig. 6k). For a first trial, the electroporation parameters should be chosen close to the parameters suggested for Purkinje neurons. Then, pulse amplitude, length, number, and/or frequency can be optimized for best survival of the target cells.

If a labeled cell is selected for imaging, line scans can be performed along the dendritic tree (Fig. 7a). Laser power typically used for imaging dendritic voltage changes in Purkinje neurons at 1020 nm excitation wavelength is in the range of 40–80 mW. If GCaMP6 is used as a second label, large calcium transients of the dendritic complex spikes can be easily observed. The large calcium transients are associated with the dendritic voltage changes and are therefore reliable indicators of complex spikes. As the absorption cross section of GCaMP6 at 1020 nm is less than 10% of the peak at 900 nm, ANNINE-6plus and GCaMP6 can be excited simultaneously without bleaching one or the other. In other words, both chromophores are excited close to their red spectral edge of absorption.

By averaging the full line scan spatially, a noise level of less than 5% should be reached. Then, dendritic complex spikes composed of several 1–2 ms spikelets will be clearly visible with fluorescence changes of about 15%.

The labeling lasts for days or even weeks (Fig. 7b, c). However, there is a slow but steady reduction of intensity due to washout of ANNINE-6plus and therefore a reduction of the SNR. Such long-term experiments suggest that ANNINE-6plus has no serious

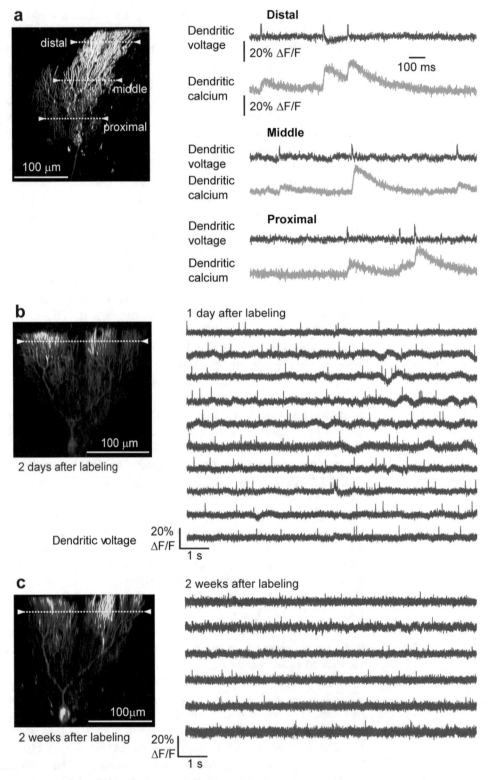

Fig. 7 Simultaneous voltage and calcium imaging of dendritic activity in a Purkinje neuron and long-term imaging. (**a**) Reconstruction of a Purkinje neuron 1 day after labeling with ANNINE-6plus. The GCaMP-labeled Purkinje neuron was filled with the dye by targeted electroporation. Averaged line scans along the distal and

pharmacological effects on the Purkinje neuron. Also, no bleaching or phototoxicity was observed even after several minutes of line scanning along the Purkinje neuron dendrite.

Using different excitation wavelengths confirms the spectral shift as the mechanism of voltage sensitivity as the relative fluorescence changes increase toward the red spectral edge of absorption (Fig. 8a). As the biological signal varies, multiple measurements are required to average out the biological variability and then to clearly show the wavelength-dependent increase in sensitivity (Fig. 8b).

The main advantage of voltage imaging is to obtain spatio-temporal information of voltage changes. With two-photon microscopy, line scans can be recorded with sufficient speed to resolve voltage signals with millisecond precision. For example, by scanning along the dendritic tree of Purkinje neurons, spatio-temporal maps of dendritic voltage (Fig. 9a) and calcium (Fig. 9b) can be imaged. Interestingly, local subthreshold events can be visualized, studied, and quantified in the voltage map (Fig. 9a). Spatio-temporal coherence of a signal (termed hotspot) indicates a voltage change, while a single hot pixel indicates noise.

Additionally, simultaneous extracellular electrical activity can be recorded (Fig. 9c) by using quartz electrodes (0.7 mm ID, Sutter Instrument) beveled at a blunt angle of 40 degrees, with a tip resistance of 7–10 MΩ. To visualize the electrode with the two-photon microscope, 50 µM Alexa Fluor 594 Hydrazide is added to the electrode solution (0.9% NaCl in water). This allows visually guided positioning of the electrode tip, for example, onto the soma of a neuron (Fig. 6j). To help reduce tip blockage and prevent dye leakage into the tissue, electrode tips should be filled with 1–2% agarose gel containing 50 µM Alexa and 0.9% NaCl in water. In this case, it is not necessary to apply pressure to the electrode, and it can be inserted directly through the silicone access port in the cranial window, with the ground electrode placed above the window under the objective with saline as immersion medium.

To keep some spatial information but to reduce noise, the line-scan can be binned into a few segments. For example, in the den-

Fig. 7 (continued) proximal dendrite at the position indicated (left) show rapid voltage depolarization events in the red channel (red traces) and long-lasting calcium transients in the green channel (green traces). The signals correspond to dendritic complex spikes. As ANNINE-6plus is applied internally, depolarization results in a positive fluorescence change when excited at the red spectral edge of absorption. Note, the signal direction is inverted compared to extracellular labeling (e.g., see Fig. 5c). (**b**) Reconstruction of a Purkinje neuron 2 days and (**c**) 2 weeks after electroporation with ANNINE-6plus. Spatially averaged, uncorrected recordings of dendritic voltage 1 day and 2 weeks after labeling (right) at location indicated (left) show reliable dendritic complex spike detection but no bleaching or phototoxicity. The SNR in the later recording decreased in comparison to the earlier one. All line scans were recorded at 2 kHz [37]

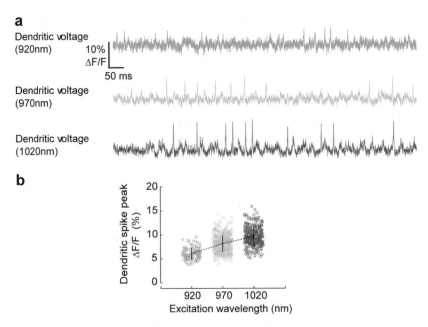

a

Dendritic voltage (920nm)

10% ΔF/F

50 ms

Dendritic voltage (970nm)

Dendritic voltage (1020nm)

b

Fig. 8 Relative fluorescence change is excitation wavelength-dependent, as shown before in HEK293 cell cultures (Fig. 1d). (**a**) Averaged line scan recordings from distal spiny dendrites of a Purkinje neuron in an awake mouse with excitation wavelength of 920, 970, and 1020 nm. (**b**) The peak amplitudes of dendritic complex spikes fluctuate due to biological variability for each excitation wavelength. The mean signal increases with excitation closer to the red spectral edge of absorption confirming the mechanism of voltage sensitivity, i.e., the spectral shift [37]

drites of Purkinje neurons, the number and timing of spikelets in spiny dendrites can be determined with sub-millisecond precision (Fig. 9d).

A very useful tool is in vivo pharmacology in combination with simultaneous voltage and calcium imaging [37]. Beveled quartz pipettes (10–15 μm opening) are filled with drug solution and then inserted through the access port into the brain. The drug is

Fig. 9 (continued) complex spikes at high SNR. (**b**) The corresponding dendritic calcium spatio-temporal map shows large transients for every dendritic complex spike. Note the spatial variability of the peak amplitude of calcium transients within single dendritic complex spikes and the amplitude variability between different dendritic complex spikes. (**c**) The access port allowed simultaneous extracellular electrical recordings from the soma while imaging voltage and calcium transients from the dendrites. Simple spikes (somatic Na^+ spikes) result in a current sink at the soma, while complex spikes (dendritic Ca^{2+} spikes) result in a dominant current source signal at the soma. (**d**) Different parts of the dendritic tree show a different number of spikelets during the same complex spike event. The number of spikelets correlates with the amplitude of the calcium transients in each part of the dendritic tree. Open arrowheads indicate spatially localized low activity and filled arrowheads high activity. Spatially localized dendritic spikelets during complex spikes correlate with a local boost in the dendritic calcium transient (small arrowheads). Note that the depolarization is caused by the influx of Ca^{2+}. Therefore, the voltage and calcium signal should have the same onset. The observed delay in onset of the calcium signal is the result of the dynamics of the calcium indicator GCaMP6f [37]

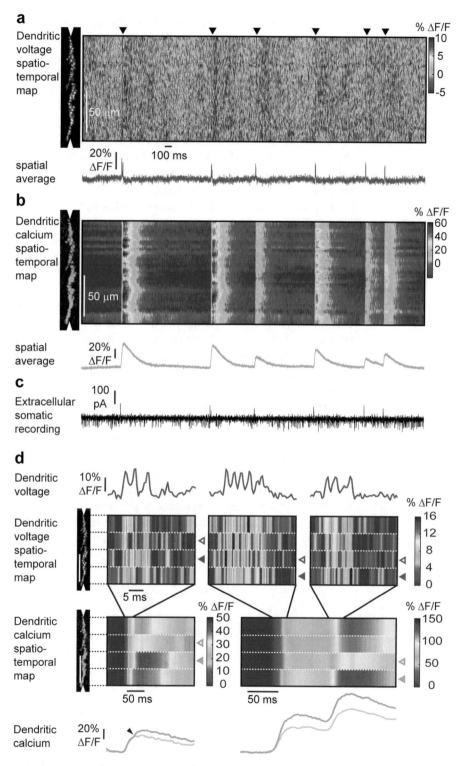

Fig. 9 Simultaneous voltage and calcium imaging of Purkinje neuron dendrites and electric recording from the soma in the awake mouse. (**a**) A line scan at 2 kHz was taken along the Purkinje neuron dendrite (left) to measure a voltage spatio-temporal map. Note red vertical lines (black triangles) indicating activation of the full dendritic tree during a complex spike, and background pattern with red hotspots indicating localized subthreshold electrical activity. The spatially averaged dendritic voltage trace shows dendritic

released by pressure application (typically <20 hPa) through the pressure port of the pipette holder or by iontophoresis.

3.10 Basic Offline Analysis of Voltage Imaging Data

Analysis of voltage imaging data can be separated into three consecutive steps: Movement correction, noise reduction by spatial and/or temporal averaging, and calculation of relative fluorescence change.

Movement artifacts should be eliminated as much as possible during the experiment. Movement correction should not be necessary for imaging data from cultures or slices. Unfortunately, in in vivo experiments, movement artifacts sometimes cannot be avoided. Some movement artifacts can be corrected or reduced for by shifting lines in line scan data or shifting images in movie data. The shift is determined by cross-correlating each line or image with a reference line or image. The reference line or image is calculated by averaging a segment of the recording with low movement artifacts.

As described in the Introduction, noise can be reduced by increasing the number of photons contributing to a signal. This can be done by adding neighboring pixels and thereby sacrificing spatial resolution (spatial smoothing), or it can be done by adding up data of the same pixel temporally and thereby sacrificing temporal resolution (temporal smoothing). This requires careful consideration of the expected spatial and temporal extent of the measured voltage signal. If not, important features might get filtered out. Typically, regions of interest which are expected to show the same signal are combined to one trace and then temporally averaged. Alternatively, a continuous spatio-temporal filter is applied.

Finally, the relative fluorescence change is calculated. The relative fluorescence change is independent of the brightness of the sample and therefore independent of the number of dye molecules. In general, it reflects the voltage change of the sample. It is calculated by subtracting the baseline fluorescence from the time-dependent fluorescence. The baseline can be a constant

$$\frac{\Delta F}{F}(t) = \frac{F(t) - F_{baseline}}{F_{baseline}}$$

or the temporally filtered $F(t)$ trace, $F_{baseline}(t)$

$$\frac{\Delta F}{F}(t) = \frac{F(t) - F_{baseline}(t)}{F_{baseline}(t)}$$

The temporal filtering of $F(t)$ acts as a low-pass filter, and therefore $\frac{\Delta F}{F}(t)$ will only show signals above the cut-off frequency of the temporal filter applied. The filter can be, for example, a sliding box filter.

If other chromophores contribute to the measured intensity or if VSD is bound to membranes that do not participate in a voltage change, then the voltage signal will be "diluted" and therefore be reduced. If autofluorescence or spectral overlap of a second chromophore is known, it can be subtracted as background $F_{background}$.

$$\frac{\Delta F}{F}(t) = \frac{\left(F(t) - F_{background}\right) - \left(F_{baseline}(t) - F_{background}\right)}{F_{baseline}(t) - F_{background}} = \frac{F(t) - F_{baseline}(t)}{F_{baseline}(t) - F_{background}}$$

The relative fluorescence change can then be converted into a voltage change according to Table 1 which results in an estimate of the average membrane voltage change in the spatio-temporal voxel defined by the imaging parameters and the filtering.

If two different chromophores are imaged simultaneously, spectral unmixing might become necessary due to spectral overlap [37]. For an introduction to spectral unmixing, see, for example, [59].

3.11 Calibration of Relative Fluorescence Change to Voltage

In sparse cell cultures, it is possible to calibrate the optical voltage signal to membrane voltage change. The optically measured voltage signal must be compared with an electrical measurement. After labeling the cell culture, patch-clamp electrophysiology in voltage-clamp mode from a cell is used to apply voltage steps via the recording pipette while imaging [28]. Fluorescence changes induced by applied voltage steps can be used to generate a calibration table. An alternative method of applying voltage to cells is to apply external electric fields [28]. It requires a chamber that allows simultaneous imaging and application of electrical fields. This method can be used for high-throughput screening of VSDs to study their characteristics. The conversion table (Table 1) was generated with external electric fields, confirmed with patch-clamp technique, and allows to estimate the voltage change if ANNINE-6 or ANNINE-6plus are excited at 1020 nm or 1040 nm, and a 540 nm emission longpass filter is used for imaging.

In bulk-loaded tissue, the voltage signal is "diluted" by the fluorescence of cells not participating in a specific voltage change as cell surfaces are closer to each other (about 40–70 nm [60]) than the optical resolution limit. Additionally, fine structures like axons and dendrites are tightly intermingled. Therefore, it is only possible to determine the average membrane voltage change in bulk-loaded tissue. If an excitation wavelength of 1020 nm or 1040 nm and a 540 nm emission longpass filter are used for imaging, Table 1 allows to determine the average membrane voltage change for a detected relative fluorescence change.

We found that Table 1 can also be used to calibrate voltage signals from dendrites of internally labeled Purkinje neurons. However, it is not clear if this can be generalized to other cell types. The ratio of plasma membrane area to membrane area of organelles will be critical.

Table 1
Conversion of relative fluorescence change (%) to voltage for ANNINE-6 or ANNINE-6plus in the outer membrane leaflet (bath application of the dye) at an excitation wavelength of 1020 nm or 1040 nm using a 540 nm emission longpass filter. If the dye is internally applied, the sign of the relative fluorescence change must be inverted to read the correct voltage

Relative fluorescence change [%]	Excitation at 1020 nm: membrane potential change [mV]	Excitation at 1040 nm: membrane potential change [mV]
50	−74	−68
40	−62	−57
30	−48	−44
20	−34	−31
10	−18	−16
0	0	0
−10	19	18
−20	41	38
−30	65	60
−40	94	85
−50	127	117

4   Troubleshooting

4.1   Bleaching Is Observed

Try to increase the number of dye molecules by using higher dye concentration or a larger volume of dye solution. Reduce laser power. In general, no bleaching should be observed when excited near the red spectral edge of absorption. Check if the PMTs are still sensitive as they might have degraded.

4.2   Imaging Intensity Is Too Low

Try to increase the number of dye molecules by using a higher dye concentration or a larger volume of applied dye solution. Increase laser power. The next possibility is to reduce the excitation wavelength in 10 nm steps starting from 1040 nm. This reduces the sensitivity of the VSD but increases the absorption cross section. For Ti:sapphire lasers, it also increases the output power; therefore, the image becomes brighter. Check if the PMTs are still sensitive as they might have degraded.

4.3 No Voltage Signals Are Observed

If the intensity is too low, then see above. Check noise level of your imaging system. The noise should be shot-noise limited. To test this, record the intensity of a homogeneous fluorescent sample (for example, a dye solution) over time with different laser intensities. Calculate the mean (over time) and corresponding variance of single pixels and groups of pixels for each laser intensity setting and plot variance over mean intensity. For a shot-noise-limited imaging system, an interpolated line through the data points will cut the coordinate system close to its origin. If the line deviates significantly from the origin, it is necessary to search for the noise source. Calculate a stimulus-triggered average response to reduce noise. Noise decreases with the square root of the number or repetitions. For example, 25 averages reduce the noise to $1/5$ of the non-averaged recording. An alternative is to smooth the data spatially, temporally, or spatio-temporally.

5   Outlook

Despite the progress made, several problems remain to be solved for two-photon voltage imaging. Labeling single neurons in vivo is a challenging technique and requires practice and optimization. Labeling of groups of more than approximately five neurons is currently not possible with synthetic dyes. Cell-type-specific labeling is typically the domain of genetically encoded indicators. However, there are also attempts to combine genetic targeting with synthetic VSDs [61–63]. The basic idea is to express an enzyme in a specific cell type. The synthetic VSD is made water-soluble by adding the target group of the enzyme. If the now hydrophilic VSD is exposed to cells expressing the enzyme, the hydrophilic group will be cleaved, the dye becomes hydrophobic, and labels the enzyme-expressing cell. An alternative would be to use the Halo Tag to bind tagged VSD molecules to the membrane [64]. If such processes work under in vivo conditions remains to be tested. It would open up two-photon voltage imaging to a wide range of applications and would make the experiments – hopefully – significantly easier.

Acknowledgments

The authors thank Espen Hartveit, Steven D. Aird, Neil Dalphin, Ray X. Lee, Mohamed M. Eltabbal, Alisher Duspayev, Claudia Cecchetto, and Leonidas Georgiou for valuable feedback on the manuscript and the Okinawa Institute of Science and Technology Graduate University for internal funding.

References

1. Cohen LB, Salzberg BM (1978) Optical measurement of membrane potential. Rev Physiol Bioch P 83:35–88

2. Cohen LB, Salzberg BM, Davila HV, Ross WN, Landowne D, Waggoner AS, Wang CH (1974) Changes in axon fluorescence during activity – molecular probes of membrane potential. J Membr Biol 19(1–2):1–36. https://doi.org/10.1007/Bf01869968

3. Fluhler E, Burnham VG, Loew LM (1985) Spectra, membrane-binding, and potentiometric responses of new charge shift probes. Biochem 24(21):5749–5755. https://doi.org/10.1021/bi00342a010

4. Grinvald A, Hildesheim R, Farber IC, Anglister L (1982) Improved fluorescent probes for the measurement of rapid changes in membrane potential. Biophys J 39(3):301–308

5. Hübener G, Lambacher A, Fromherz P (2003) Anellated hemicyanine dyes with large symmetrical solvatochromism of absorption and fluorescence. J Phys Chem B 107(31):7896–7902. https://doi.org/10.1021/jp0345809

6. Kuhn B, Fromherz P (2003) Anellated hemicyanine dyes in a neuron membrane: molecular Stark effect and optical voltage recording. J Phys Chem B 107(31):7903–7913. https://doi.org/10.1021/jp0345811

7. Loew LM, Bonneville GW, Surow J (1978) Charge shift optical probes of membrane potential – theory. Biochem 17(19):4065–4071. https://doi.org/10.1021/bi00612a030

8. Loew LM, Simpson LL (1981) Charge-shift probes of membrane potential – a probable electrochromic mechanism for para-aminostyrylpyridinium probes on a hemispherical lipid bilayer. Biophys J 34(3):353–365

9. Grinvald A, Frostig RD, Lieke E, Hildesheim R (1988) Optical imaging of neuronal activity. Physiol Rev 68(4):1285–1366

10. Grinvald A, Hildesheim R (2004) VSDI: a new era in functional imaging of cortical dynamics. Nat Rev Neurosci 5(11):874–885. https://doi.org/10.1038/nrn1536

11. Antic S, Zecevic D (1995) Optical signals from neurons with internally applied voltage-sensitive dyes. J Neurosci 15(2):1392–1405

12. Palmer LM, Stuart GJ (2006) Site of action potential initiation in layer 5 pyramidal neurons. J Neurosci 26(6):1854–1863. https://doi.org/10.1523/Jneurosci.4812-05.2006

13. Meyer E, Müller CO, Fromherz P (1997) Cable properties of dendrites in hippocampal neurons of the rat mapped by a voltage-sensitive dye. Eur J Neurosci 9(4):778–785. https://doi.org/10.1111/j.1460-9568.1997.tb01426.x

14. Fromherz P, Müller CA (1994) Cable properties of a straight neurite of a leech neuron probed by a voltage-sensitive dye. Proc Natl Acad Sci U S A 91(10):4604–4608. https://doi.org/10.1073/pnas.91.10.4604

15. Antic S, Major G, Zecevic D (1999) Fast optical recordings of membrane potential changes from dendrites of pyramidal neurons. J Neurophysiol 82(3):1615–1621. https://doi.org/10.1152/jn.1999.82.3.1615

16. Canepari M, Zecevic D (2010) Membrane potential imaging in the nervous system: methods and applications. Springer, New York

17. Peterka DS, Takahashi H, Yuste R (2011) Imaging voltage in neurons. Neuron 69(1):9–21. https://doi.org/10.1016/j.neuron.2010.12.010

18. Knöpfel T (2012) Genetically encoded optical indicators for the analysis of neuronal circuits. Nat Rev Neurosci 13(10):687–700. https://doi.org/10.1038/nrn3293

19. Antic SD, Empson RM, Knöpfel T (2016) Voltage imaging to understand connections and functions of neuronal circuits. J Neurophysiol 116(1):135–152. https://doi.org/10.1152/jn.00226.2016

20. Ephardt H, Fromherz P (1993) Fluorescence of amphiphilic hemicyanine dyes without free double-bonds. J Phys Chem 97(17):4540–4547. https://doi.org/10.1021/j100119a048

21. Fromherz P (1995) Monopole-dipole model for symmetrical solvatochromism of hemicyanine dyes. J Phys Chem 99(18):7188–7192. https://doi.org/10.1021/j100018a061

22. Fromherz P, Heilemann A (1992) Twisted internal charge-transfer in (aminophenyl)pyridinium. J Phys Chem 96(17):6864–6866. https://doi.org/10.1021/j100196a004

23. Fromherz P, Hübener G, Kuhn B, Hinner MJ (2008) ANNINE-6plus, a voltage-sensitive dye with good solubility, strong membrane binding and high sensitivity. Eur Biophys J Biophy 37(4):509–514. https://doi.org/10.1007/s00249-007-0210-y

24. Lambacher A, Fromherz P (2001) Orientation of hemicyanine dye in lipid membrane measured by fluorescence interferometry on a silicon chip. J Phys Chem B 105(2):343–346. https://doi.org/10.1021/jp002843i

25. Röcker C, Heilemann A, Fromherz P (1996) Time-resolved fluorescence of a hemicyanine

dye: dynamics of rotamerism and resolution. J Phys Chem 100(30):12172–12177. https://doi.org/10.1021/jp960095k

26. Ephardt H, Fromherz P (1991) Anillnopyridinium – solvent-dependent fluorescence by intramolecular charge-transfer. J Phys Chem 95(18):6792–6797. https://doi.org/10.1021/j100171a011

27. Kuhn B and Roome CJ (2019) Primer to voltage imaging with ANNINE dyes and two-photon microscopy. Front. Cell. Neurosci. 13:321. https://doi.org/10.3389/fncel.2019.00321

28. Kuhn B, Fromherz P, Denk W (2004) High sensitivity of Stark-shift voltage-sensing dyes by one- or two-photon excitation near the red spectral edge. Biophys J 87(1):631–639. https://doi.org/10.1529/biophysj.104.040477

29. Acker CD, Hoyos E, Loew LM (2016) EPSPs measured in proximal dendritic spines of cortical pyramidal neurons. eNeuro 3(2). https://doi.org/10.1523/ENEURO.0050-15.2016

30. Kuhn B, Denk W, Bruno RM (2008) In vivo two-photon voltage-sensitive dye imaging reveals top-down control of cortical layers 1 and 2 during wakefulness. Proc Natl Acad Sci U S A 105(21):7588–7593. https://doi.org/10.1073/pnas.0802462105

31. Acker CD, Yan P, Loew LM (2011) Single-voxel recording of voltage transients in dendritic spines. Biophys J 101(2):L11–L13. https://doi.org/10.1016/j.bpj.2011.06.021

32. Yan P, Acker CD, Zhou WL, Lee P, Bollensdorff C, Negrean A, Lotti J, Sacconi L, Antic SD, Kohl P, Mansvelder HD, Pavone FS, Loew LM (2012) Palette of fluorinated voltage-sensitive hemicyanine dyes. Proc Natl Acad Sci U S A 109(50):20443–20448. https://doi.org/10.1073/pnas.1214850109

33. Fisher JA, Barchi JR, Welle CG, Kim GH, Kosterin P, Obaid AL, Yodh AG, Contreras D, Salzberg BM (2008) Two-photon excitation of potentiometric probes enables optical recording of action potentials from mammalian nerve terminals in situ. J Neurophysiol 99(3):1545–1553. https://doi.org/10.1152/jn.00929.2007

34. Millard AC, Campagnola PJ, Mohler W, Lewis A, Loew LM (2003) Second harmonic imaging microscopy. Methods Enzymol 361:47–69

35. Helmchen F, Denk W (2005) Deep tissue two-photon microscopy. Nat Methods 2(12):932–940. https://doi.org/10.1038/nmeth818

36. Lu R, Sun W, Liang Y, Kerlin A, Bierfeld J, Seelig JD, Wilson DE, Scholl B, Mohar B, Tanimoto M, Koyama M, Fitzpatrick D, Orger MB, Ji N (2017) Video-rate volumetric functional imaging of the brain at synaptic resolution. Nat Neurosci 20(4):620–628. https://doi.org/10.1038/nn.4516

37. Roome CJ, Kuhn B (2018) Simultaneous dendritic voltage and calcium imaging and somatic recording from Purkinje neurons in awake mice. Nat Commun 9(1):3388. https://doi.org/10.1038/s41467-018-05900-3

38. Nadella KMNS, Ros H, Baragli C, Griffiths VA, Konstantinou G, Koimtzis T, Evans GJ, Kirkby PA, Silver RA (2016) Random-access scanning microscopy for 3D imaging in awake behaving animals. Nat Methods 13(12):1001–1004. https://doi.org/10.1038/Nmeth.4033

39. Reddy GD, Kelleher K, Fink R, Saggau P (2008) Three-dimensional random access multiphoton microscopy for functional imaging of neuronal activity. Nat Neurosci 11(6):713–720. https://doi.org/10.1038/nn.2116

40. Katona G, Szalay G, Maak P, Kaszas A, Veress M, Hillier D, Chiovini B, Vizi ES, Roska B, Rozsa B (2012) Fast two-photon in vivo imaging with three-dimensional random-access scanning in large tissue volumes. Nat Methods 9(2):201–208. https://doi.org/10.1038/nmeth.1851

41. Pages S, Cote D, De Koninck P (2011) Optophysiological approach to resolve neuronal action potentials with high spatial and temporal resolution in cultured neurons. Front Cell Neurosci 5:20. https://doi.org/10.3389/fncel.2011.00020

42. Heo C, Park H, Kim YT, Baeg E, Kim YH, Kim SG, Suh M (2016) A soft, transparent, freely accessible cranial window for chronic imaging and electrophysiology. Sci Rep 6:27818. https://doi.org/10.1038/srep27818

43. Roome CJ, Kuhn B (2014) Chronic cranial window with access port for repeated cellular manipulations, drug application, and electrophysiology. Front Cell Neurosci 8:379. https://doi.org/10.3389/fncel.2014.00379

44. Vranesic I, Iijima T, Ichikawa M, Matsumoto G, Knöpfel T (1994) Signal transmission in the parallel fiber-Purkinje cell system visualized by high-resolution imaging. Proc Natl Acad Sci U S A 91(26):13014–13017

45. Petersen CC, Grinvald A, Sakmann B (2003) Spatiotemporal dynamics of sensory responses in layer 2/3 of rat barrel cortex measured in vivo by voltage-sensitive dye imaging combined with whole-cell voltage recordings and neuron reconstructions. J Neurosci 23(4):1298–1309

46. Braitenberg V, Schüz A (1998) Cortex: statistics and geometry of neuronal connectivity, 2nd edn. Springer, Berlin/New York

47. Djurisic M, Antic S, Chen WR, Zecevic D (2004) Voltage imaging from dendrites of mitral cells: EPSP attenuation and spike trigger

zones. J Neurosci 24(30):6703–6714. https://doi.org/10.1523/JNEUROSCI.0307-04.2004

48. Kitamura K, Häusser M (2011) Dendritic calcium signaling triggered by spontaneous and sensory-evoked climbing fiber input to cerebellar Purkinje cells in vivo. J Neurosci 31(30): 10847–10858. https://doi.org/10.1523/JNEUROSCI.2525-10.2011

49. Holtmaat A, Bonhoeffer T, Chow DK, Chuckowree J, De Paola V, Hofer SB, Hübener M, Keck T, Knott G, Lee WC, Mostany R, Mrsic-Flogel TD, Nedivi E, Portera-Cailliau C, Svoboda K, Trachtenberg JT, Wilbrecht L (2009) Long-term, high-resolution imaging in the mouse neocortex through a chronic cranial window. Nat Prot 4(8):1128–1144. https://doi.org/10.1038/nprot.2009.89

50. Mostany R, Portera-Cailliau C (2008) A craniotomy surgery procedure for chronic brain imaging. J Vis Exp (12):e680. https://doi.org/10.3791/680

51. Grinvald A, Lieke E, Frostig RD, Gilbert CD, Wiesel TN (1986) Functional architecture of cortex revealed by optical imaging of intrinsic signals. Nature 324(6095):361–364. https://doi.org/10.1038/324361a0

52. Bonhoeffer T, Grinvald A (1996) Optical imaging based on intrinsic signals: the methodology. In: Toga AW, Mazziotta JC (eds) Brain mapping: the methods, 1st edn. Academic Press, San Diego, pp 55–97

53. Theer P, Kuhn B, Keusters D, Denk W (2005) Two-photon microscopy and imaging. In: Meyers RA (ed) Encyclopedia of molecular biology and molecular medicine, vol 15. 2nd edn. VCH, Weinheim/New York, pp 61–87

54. Pologruto TA, Sabatini BL, Svoboda K (2003) ScanImage: flexible software for operating laser scanning microscopes. Biomed Eng Online 2:13. https://doi.org/10.1186/1475-925X-2-13

55. Schneider CA, Rasband WS, Eliceiri KW (2012) NIH image to ImageJ: 25 years of image analysis. Nat Methods 9(7):671–675

56. Schindelin J, Arganda-Carreras I, Frise E, Kaynig V, Longair M, Pietzsch T, Preibisch S, Rueden C, Saalfeld S, Schmid B, Tinevez JY, White DJ,

Hartenstein V, Eliceiri K, Tomancak P, Cardona A (2012) Fiji: an open-source platform for biological-image analysis. Nat Methods 9(7):676–682. https://doi.org/10.1038/nmeth.2019

57. Vecellio M, Schwaller B, Meyer M, Hunziker W, Celio MR (2000) Alterations in Purkinje cell spines of calbindin D-28 k and parvalbumin knock-out mice. Eur J Neurosci 12(3):945–954

58. Kuhn B, Ozden I, Lampi Y, Hasan MT, Wang SS (2012) An amplified promoter system for targeted expression of calcium indicator proteins in the cerebellar cortex. Front Neural Circuits 6:49. https://doi.org/10.3389/fncir.2012.00049

59. Zimmermann T (2005) Spectral imaging and linear unmixing in light microscopy. Adv Biochem Eng Biotechnol 95:245–265. https://doi.org/10.1007/b102216

60. Thorne RG, Nicholson C (2006) In vivo diffusion analysis with quantum dots and dextrans predicts the width of brain extracellular space. Proc Natl Acad Sci U S A 103(14):5567–5572. https://doi.org/10.1073/pnas.0509425103

61. Hinner MJ, Hübener G, Fromherz P (2004) Enzyme-induced staining of biomembranes with voltage-sensitive fluorescent dyes. J Phys Chem B 108(7):2445–2453. https://doi.org/10.1021/jp036811h

62. Hinner MJ, Hübener G, Fromherz P (2006) Genetic targeting of individual cells with a voltage-sensitive dye through enzymatic activation of membrane binding. Chembiochem 7(3):495–505. https://doi.org/10.1002/cbic.200500395

63. Ng DN, Fromherz P (2011) Genetic targeting of a voltage-sensitive dye by enzymatic activation of phosphonooxymethyl-ammonium derivative. ACS Chem Biol 6(5):444–451. https://doi.org/10.1021/cb100312d

64. Los GV, Encell LP, McDougall MG, Hartzell DD, Karassina N, Zimprich C, Wood MG, Learish R, Ohana RF, Urh M, Simpson D, Mendez J, Zimmerman K, Otto P, Vidugiris G, Zhu J, Darzins A, Klaubert DH, Bulleit RF, Wood KV (2008) HaloTag: a novel protein labeling technology for cell imaging and protein analysis. ACS Chem Biol 3(6):373–382. https://doi.org/10.1021/cb800025k

High-Speed, Random-Access Multiphoton Microscopy for Monitoring Synaptic and Neuronal Activity in 3D in Behaving Animals

Mate Marosi, Gergely Szalay, Gergely Katona, and Balázs Rózsa

Abstract

Understanding neural computation requires methods for three-dimensional readout of neural activity simultaneously on somatic and dendritic scales. Random-access point scanning offers measurement speed increased by several orders of magnitude compared to classical raster scanning methods; however, it is highly sensitive to sample movements during behavior, when fluorescence information could be completely lost due to dislocation of the point of interest. Volume scanning, on the other hand, is insensitive to even relatively large movements, but the scanning speed can be as low as 0.1–1 Hz for full 3D volumes, which is insufficient for recording fast Ca^{2+} activity. The 3D Drift acousto-optical (AO) scanning method can extend each scanning point of random-access AO point scanning to small 3D line, surface, or volume elements for fast recording of complex neuronal activity in vitro and in vivo, allowing free selection of any portion of the volume measured that can help finding the optimal trade-off between point scanning and full-volume scans. The video-like information, of the volume or surface elements measured, allows offline motion correction even during behavior and brain movement.

In this chapter, fast 3D measurement of over 150 dendritic spines with 3D lines, over 100 somata with squares and cubes, or multiple spiny dendritic segments with surface and volume elements in awake and behaving animals will be demonstrated.

Key words Two-photon microscopy, Acousto-optical scanning, Calcium imaging, In vivo, 3D functional imaging, AO 3D Drift scanning

1 Introduction

Scientists have always desired to study and better understand the structure and function of various organs including the brain, for that they need to "look inside" the tissue. Anatomical methods give excellent opportunity to study fine morphological details of various tissues but give no information about the function of these assemblies. Electrophysiological recording, on the other hand, allows studying neural activity from individual neurons or even larger networks of neurons in vivo with excellent temporal resolution;

Espen Hartveit (ed.), *Multiphoton Microscopy*, Neuromethods, vol. 148,
https://doi.org/10.1007/978-1-4939-9702-2_14, © Springer Science+Business Media, LLC, part of Springer Nature 2019

however, it provides no information of the anatomical location or connections of these cells and cannot investigate subcellular compartments except for some very specially located cells.

Light microscopy has an important role in biological research since it allows to simultaneously investigate neuronal activity and morphology from living tissue with high spatial resolution, resolving not only cells and their firing rates, but also subcellular compartments and subthreshold activity. However, the optical resolution is limited by the wavelength of the emitted light (according to Abbe's rule) and does not rival that of electron microscopy. On the other hand, the scope of electron microscopy is limited when observing living specimens [1]. Other vital imaging technologies, such as MRT (magnetic resonance tomography), PET (positron emission tomography), or X-ray imaging, can neither resolve subcellular structures nor provide the exquisite molecular selectivity that would allow single molecules to be detected with billions of others in the background [2, 3].

Fluorescence microscopy occupies a unique place in biological microscopy [4]. Fluorescent objects can be selectively excited and visualized, even in living systems [5]. Two-photon excitation microscopy [6] is a fluorescence imaging technique that allows imaging of living tissue [6, 7] up to a depth of about 1 mm [8–10] and studying large neural populations with single-cell resolution in vivo [11–14].

In the last decades, scientists have modified fluorescent proteins not only to visualize cell morphology and morphological changes but also to monitor changes in intracellular calcium, pH, protein–protein interactions, and the function of single enzymes [15]. In particular, two-photon excitation microscopy excels at high-resolution imaging in intact thick tissues such as brain slices, embryos, whole organs, and live animals (intra-vital imaging) [16].

1.1 3D Random-Access Scanning

Neuronal diversity, layer specificity of information processing, area-wise specialization of neural mechanisms, internally generated patterns, and dynamic network properties all show that understanding neural networks and computation requires fast read-out of information flow and processing. Moreover, this fast recording is required not only from a single plane or point, but also at the level of large neuronal populations situated in large 3D volumes [17–19]. In addition, coding and computation within neuronal networks are generated not only by the somatic integration domains but also by highly nonlinear dendritic integration centers that, in most cases, remain hidden from somatic recordings [20–23]. The ability to simultaneously read out neural activity at both the population and subcellular levels is essential. Therefore, novel methods are crucial to simultaneously record activity patterns of somatic, dendritic, dendritic spine, and axonal assemblies of neurons with high spatial and temporal resolution in large scanning volumes in the brains of awake, behaving animals.

Several new optical methods have recently been developed for the fast readout of neuronal network activity in 3D. For example, it is possible to record 3D structures using spatial light modulators [24], liquid lenses [25], acousto-optical (AO) deflectors [18, 26, 27], temporal multiplexing [28], axicon or planar illumination-based imaging [29], fast z-scanning based on an axially moving mirror [30], piezo actuators [31, 32], simultaneous multi-view light sheet microscopy [33], two-photon light sheet microscopy [34], optical fibers addressed by AO deflectors [35], light field microscopy [36], phase-locked ultrasound lens [37], or phase mask combined with holographic illumination [38]. Although all of these are 3D technologies, only multi-photon excitation makes it possible to use all the scattered photons emitted from a single focal point for deep brain imaging [8]. Among the available 3D scanning solutions for multi-photon microscopy, only 3D acousto-optical (AO) scanning is capable of performing 3D random-access point scanning [18, 27, 39].

The phrase "acousto-optics" refers to the field of optics that studies the interaction between sound and light waves. In the field of fluorescent imaging, we use acousto-optical devices to diffract laser beams through ultrasonic gratings. The acousto-optical effect is based on a periodic change of the refractive index in a high-refractive index medium (usually tellurium dioxide, TeO_2), which is the result of the sound wave-induced pressure fluctuations in the crystal. This grating diffracts the light beam just like a normal optical grating, but here the gradient can be rapidly adjusted by changing the driving signal. Acousto-optical deflectors (AODs) control the optical beam spatially and use ultrasonic waves to diffract the laser beam depending on the acoustic frequency. If we introduce a sine wave at the piezoelectric driver, it will generate an optical deflection in the acousto-optic medium according to the frequency of the ultrasonic wave (Fig. 1a).

Besides deflection, AODs can also be used for fast focal plane shifting [18, 27, 39, 40]. If the frequency of the acoustic wave that fills up the crystal is changing linearly in time (i.e., the frequency is chirped), then at a given point in the crystal, the optical gradient increases as a function of time, so different portions of the optical beam are deflected in different directions (Fig. 1b). Thus, a focusing or, alternatively, a defocusing effect occurs, depending on the frequency slope (sweep rate) of the chirped acoustic wave. The focal length of an acousto-optical lens (F) can be calculated as in Fig. 1b, or as detailed in Katona et al. [18]. 3D point-by-point scanning is achieved by using two groups of x and y deflectors. During focusing, the second x and y deflector's driver function is supplemented with the same piezoelectric driver that elicits radio frequency (RF) sound waves, but with a linearly increasing (chirped) frequency. The amount of virtual focusing, i.e., the distance between the nominal focal plane of the objective and the actual focal distance, is dependent on the chirp of the acoustic

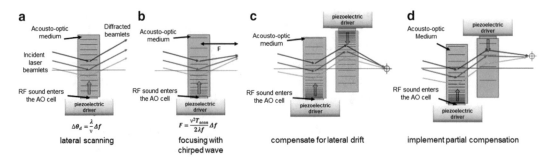

Fig. 1 Schematic of operating principle of acousto-optical deflectors (AODs). (**a**) AODs spatially control the optical beam. A piezoelectric driver elicits radio frequency (RF) sound waves due to the externally applied sinusoidal voltage. Sound enters and traverses through the diffracting (TeO$_2$) medium (acousto-optic medium) while interacting with light throughout the aperture. Light is diffracted on the sound wave's refraction index changes, as on a steady optical grating, providing diffracted light beams whose angle is dependent on the sound wave's frequency. $\Delta\Theta_d$ = optical deflection in the acousto-optic medium, λ = the optical wavelength of the beam, ν is the velocity of the acoustic wave, f = sine wave frequency. (**b**) Acousto-optical focusing. AO deflector arrangement similar to (**a**), but here the sound frequency is changing, resulting in a varying grating size along the propagation axis. As a consequence, different parts of the laser beam are diffracted with varying angles, creating a focus point whose position can be changed by the parameters of the sound applied to the deflector. F denotes focal distance. F = focal length of an acousto-optical lens, λ = the optical wavelength of the beam, ν = velocity of the acoustic wave, Δf = change of the sound frequency, T_{scan} = the modulation rate of the sound frequency. (**c**) Fully compensated lateral drift of the focal spot by a counter-propagating acoustic wave with a linearly increasing frequency. (**d**) Implemented partial compensation with AODs with a counter-propagating acousto-optical wave

frequency. To maintain a stable focal distance, the frequency gradient should be preserved in the crystal. The frequency should therefore be continuously increased (or decreased) at the piezoelectric driver to preserve the focal distance. This will result in a lateral drift of the focal point, which can be easily compensated by introducing a second acousto-optical deflector with a counter-propagating acousto-optical wave into the optical pathway (Fig. 1c). In summary, the first AO pair is used for z focusing, whereas lateral scanning is performed entirely by the second pair (details in Sect. 2.1).

The point scanning method yields high pointing stability, but requires relatively long switching times because it is necessary to fill the large AO deflector apertures with sound each time when a new point is addressed in 3D. In other words, the system can address a different location in 3D, when the mechanical waves, denoting the current location, leave the AO crystal. The time needed for this can be calculated as the ratio of the crystal size and the speed of sound in the crystal.

1.2 Scanning Speed of AO 3D Scanning

AO scanning provides an increased measurement speed and signal collection efficiency by several orders of magnitude in comparison to classical raster scanning. This is because the pre-selected regions of interest (ROIs) can be precisely and rapidly targeted without wasting measurement time on unnecessary regions. More quantitatively, if we compare the relative gain in measurement speed (ν_{gain}) and the improvement in the signal-to-noise ratio (SNR$_{gain}$)

for 3D AO scanning compared to traditional raster scanning of the same sample volume, we can say that the product $v_{gain} \times (SNR_{gain})^2$ is equivalent to the ratio of the total image volume (V_{total}) to the volume covered by the pre-selected scanning points (V_i):

$$\left(SNR_{gain}\right)^2 \times v_{gain} = \frac{V_{total}}{\sum_{i=1}^{N} V_i}, \tag{1}$$

This ratio can be up to 10^6 for very sparse labeling where we measure only a small set of points ($300 \times 0.5 \times 0.5 \times 2.5$ μm [size of the point spread function (PSF)] $\approx 2 \times 10^2$ μm^3) from the whole volume accessible by a large field-of-view objective ($550 \times 550 \times 6$ 50 μm $\approx 2 \times 10^8$ μm^3) (for additional details, see [18]).

1.3 Measurements During Motion

Despite the evident benefits of 3D random-access point scanning, the method can face a major technical limitation: fluorescence data are lost or contaminated with large-amplitude movement artifacts during in vivo recordings. This is because the actual locations of the recorded ROIs are continuously changing during in vivo measurements due to tissue movement caused by heartbeats, blood flow in nearby vessels, respiration, and locomotion of the awake animal on a treadmill or on a floating ball or by other movements, such as licking or whisking [41]. In addition, the amplitudes of motion-induced transients can even be larger than the ones induced by one or a few action potentials detected by genetically encoded calcium indicators (GECIs) [42]. Therefore, it is difficult to separate post hoc the genuine fluorescence changes associated with neural activity from the artifacts caused by brain movement.

Using novel 3D Drift AO microscopy, instead of keeping the same scanning position, the excitation spot can be drifted quickly in any direction in 3D space while continuously recording fluorescence data with no limitation on sampling rate (sampling rate is limited only by properties of the photomultipliers (PMT) during drifts). In this way, the pre-selected, individual scanning points can be extended to 3D line, surface, or volume elements to cover not only the pre-selected ROIs but also the neighboring background areas or volume elements, while utilizing an increased data-sampling rate, by limiting the scanning only to the areas with high information content. Using this scanning method, the error of the SNR caused by motion artifacts can be decreased up to one order of magnitude. This reduction of error enables fast functional measurement of neuronal networks at the level of tiny neuronal processes, such as dendritic spines, even in head-restrained, behaving animals, with a z-scanning range of more than 650 μm.

1.4 Principles of 3D Drift AO Microscope and Scanning Methods

Compared to the previously described, the new continuous trajectory scanning method [18] allows fast lateral scanning in 2D. 3D trajectory scans, however, still need to be interrupted by time-consuming jumps when moving along the z-axis. In other words, scanning along the z-axis still suffers from the same limitation as during point-by-point scanning. Our aim was to generalize the previous methods by deriving a one-to-one relationship between the focal spot coordinates and speed and the chirp parameters of the four AO deflectors to allow fast scanning drifts with the focal spot, not only in the horizontal plane but also along any straight line in 3D space by drifting (3D Drift AO scanning), starting at any point in the scanning volume. To realize this, a nonlinear chirp with parabolic frequency profiles had to be used. The partial drift compensation realized with these parabolic frequency profiles allows the directed and continuous movement of the focal spot in arbitrary directions and with arbitrary velocities, determined by the temporal shape of the chirped acoustic signals (Fig. 1d). During these fast 3D drifts of the focal spot, the fluorescence collection is uninterrupted, lifting the pixel dwell-time limitation of the previously used point scanning method. Maximum scanning speed is still limited by physical constraints such as the AO cycle or minimum pixel dwell time, with the latter limited both by the sampling rate of the electronics and the biological sample, since we need a certain amount of time to acquire enough fluorescent information. In contrast to traditional point-by-point scanning (Fig. 2), here, the z-scanning and the x, y scanning units of the microscope can drift the focal spot in 3D in an arbitrary direction and with an arbitrary speed (Fig. 2).

1.4.1 Theory Behind the 3D Drift AO Scanning

To define an adequate driving signal, we derived a one-to-one relationship between the temporal trajectory of the focal spot and the chirp parameters of the four AO deflectors used to move the focal spot along any 3D line, starting at any point in the scanning volume. We need the following three groups of equations: (1) the simplified matrix equation of the AO microscope, (2) the basic equation of the AO deflectors, and (3) the temporally nonlinear chirps for the acoustic frequencies (f) in the deflectors deflecting in the x-z (and y-z) plane

$$f_i(x,t) = f_i(0,0) + \left(b_{xi} \times \left(t - \frac{D}{2 \times v_a} - \frac{x}{v_a} \right) + c_{xi} \right) \times \left(t - \frac{D}{2 \times v_a} - \frac{x}{v_a} \right) \quad (2)$$

(where $i = 1$ or $i = 2$ indicates the first and second x deflector, D is the diameter of the AO deflector, and v_a is the propagation speed of the acoustic wave within the deflector). When we calculate and then combine the expression $z_x(t)$ with $x_0(t)$, the similarly calculated $z_y(t)$, and $y_0(t)$, and add all the required initial positions (x_0, y_0, z_0) and speed parameter values (v_{x0}, v_{y0}, $v_{zx0} = v_{zy0}$) of the focal spot, we can determine all the parameters required to calculate the nonlinear chirps according to Eq. 2 in the four AO deflectors ($\Delta f_{0x} = f_{1x}(0,0) - f_{2x}(0,0)$, b_{x1}, b_{x2}, c_{x1}, c_{x2} and $\Delta f_{0y} = f_{1y}(0,0) - f_{2y}(0,0)$,

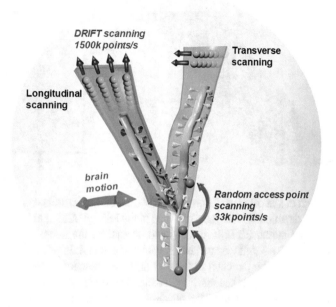

Fig. 2 Schematic of point-by-point scanning and continuous trajectory scanning measurements performed with 3D Drift AO microscopy. Schematic of the in vivo measurements performed with 3D Drift AO microscopy. The z scanning and x, y scanning units of the microscope can drift the focal spot in 3D. Therefore, individual scanning points can be extended to small surfaces or even volume elements by using longitudinal or transverse scanning methods. (Reproduced (modified) from Szalay et al. [43])

b_{y1}, b_{y2}, c_{y1}, c_{y2}). The calculations and AO driver signal generation are detailed in [43].

1.4.2 The Point Spread Function (PSF) During 3D Drift AO Scanning

During 3D Drift AO scanning, we can not only scan individual points, but can also scan at high speed (up to 10 μm/μs) along any segments of any 3D lines situated in any location in the entire scanning volume with minimal loss of resolution (Fig. 3). In this way, fluorescence information can be continuously collected when scanning the entire 3D line in the same short period of time (≈ 20 μs) as is required for single-point scanning in the point-by-point scanning mode. The data acquisition rate is limited only by the maximal sampling rate of the PMT units. Therefore, we can generate folded surface (or volume) elements, fit them to long, tortuous dendrite segments and branch points in an orientation which minimizes fluorescence loss during brain motion, and image them at high speed.

Using 3D Drift AO microscopy can extend the pre-selected individual scanning points to small 3D lines, surfaces, or volume elements to cover not only the pre-selected ROIs but also the neighboring background areas or volume elements while utilizing an increased data-sampling rate.

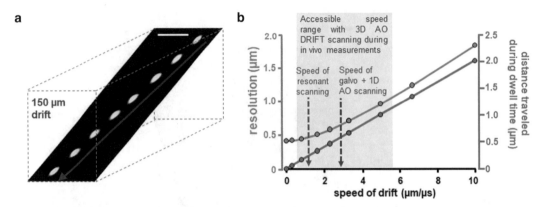

Fig. 3 Effect of fast 3D AO drifts on the size of the PSF. 3D AO drifts were generated with up to 10 μm/μs speed in different directions to analyze elongation of the PSF. (**a**) Maximum intensity projection image of a 6 μm fluorescent bead during 3D Drift AO scanning. Images of the fluorescent beads at different time points were overlaid. The 3D AO drift was generated with a speed of 4.55 μm/μs. (**b**) Full width at half maximum (FWHM) of the PSF along the direction of the fast drift as a function of the speed of 3D drifts (gray line). Note the increasing PSF at higher speeds. FWHM along the perpendicular direction did not change. This blurring effect of the PSF can be explained by the movement within the pixel dwell time (blue line). The red arrow indicates the maximal scanning speed of the typical resonant-galvo system. The green arrow indicates the scanning speed of a typical 2D system with one integrated AO deflector. (Reproduced (modified) from Szalay et al. [43])

2   Materials

2.1 Microscope Design

Figure 4 shows the schematic of the 3D-AO microscope. Briefly, the 105 fs long laser pulses are delivered by a Ti:Sapphire laser at an 80 MHz repetition rate (Mai Tai, Spectra-Physics). A Faraday isolator (BB9-5I, Electro-Optics Technology) eliminates coherent back reflections. Motorized mirrors (*m1*, *m2*, *m9*, and *m10*) (PF03-03-P01 and BB1-E03, Thorlabs) stabilize the position of the beam on the surface of the quadrant detectors in the two beam stabilization units according to Fig. 4 (PDQ80A and TPA101, Thorlabs). All the electronics are mounted in a control hub (TCH002, Thorlabs) and programmed to perform closed-loop beam alignment at high speed (≈300 Hz).

Temporal dispersion compensation is adjusted with a motorized four-prism compressor (*P1–P4*) (Fig. 4). Compared to the previous realization from our laboratory [18], a retro reflector (*m7* and *m8*) is used to make the two laser beams, crossing the four-prism sequence, parallel (BBSQ05-E03, Thorlabs) (Fig. 4). In contrast, using only a single mirror, this combination preserves the Gaussian beam shape. The beam is then expanded by the *L1* and *L2* lenses (Fig. 4). The z coordinates of the desired locations are targeted by two AO deflectors (*AO1* and *AO2*) (Fig. 4) optimized for diffraction efficiency and bandwidth. These generate two cylindrical lenses (*AO z-focusing* unit) (Fig. 4). A separate two-dimensional AO scanner unit (*AO3* and *AO4*, *2D-AO scanning*

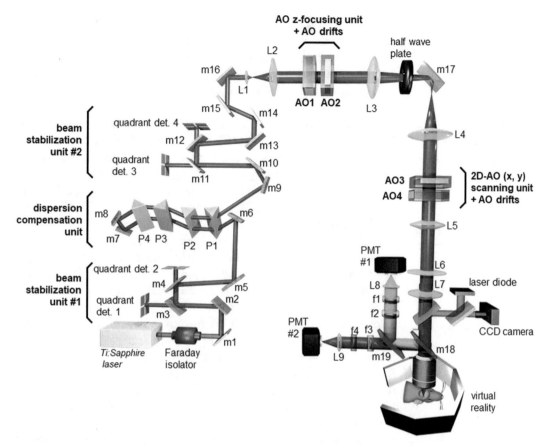

Fig. 4 Design and characterization of the two-photon AO microscope setup. Simplified optical path block diagram of the 3D Drift AO microscope and four-prism compressor unit for in vivo imaging on behaving animals. Red/green: light path from the laser to the PMTs; PMT: photomultiplier tubes; m1–17: broadband dielectric mirrors; m18–19: dichroic mirrors P1–4: prisms; L1–9: lenses; AO1–4: acousto-optical deflectors. (Modified from Szalay et al. [43])

unit) targets the x and y coordinates (Fig. 4). Chirped sine waves (Fig. 1b–d) is sent through the *AO3* and *AO4* deflectors to generate lateral scanning drifts and partially compensates the lateral drifts generated by the *AO1* and *AO2* deflectors. The joint work of the deflectors of the *AO z-focusing* and the *2D-AO scanning* units generates the drifts along 3D lines. A half-wave plate between the *2D-AO scanning* and *AO z-focusing* units ensures optimal incident polarization direction for maximum diffraction efficiency in the AO3 and AO4 deflectors. Angular dispersion compensation is provided by a spherical field lens (*L6*) in the second telecentric lens system (*L5* and *L7*) (Fig. 4). The back-reflected fluorescent light is separated from the IR excitation light by a dichroic mirror (*m18*) and then split into two parts ("green" and "red" channels) by a dichroic mirror (*m19*) (Fig. 4). In both channels, the remaining IR light is filtered out (*f2* and *f3*), and then the light is collected by lenses *L8* and *L9* into the large aperture photomultipliers (*PMT1* and *PMT2*) following band-pass filtering (*f1* and *f4*) (Fig. 4). The PMT cathode material is GaAsP (H10770PA-40, Hamamatsu). We use a water-immersion 20× objective (XLUMPlanFL 20×/NA 1.0, Olympus) or a 25× Nikon objective (CFI75 Apochromat 25× WMP, NA 1.1) for spine imaging.

2.2 Animal Preparation, Surgical Procedures

In vivo recordings from the primary visual cortex are performed on C57Bl/6J, Thy1-Cre, vip/IRES-Cre mice (P60–120). Animals are allowed free access to food and water and are maintained in temperature-, humidity-, and light-controlled conditions. Transgenic lines are developed and bred at the Medical Gene-technological Unit (OGR) of the Institute of Experimental Medicine of the Hungarian Academy of Sciences.

Chronic glass-covered cranial windows are implanted according to the following procedure. Briefly, the cranial window preparation requires the removal of a piece of skull. For this surgical procedure, the mice need to be anesthetized. Either isoflurane (1.5%, in carbogen – 95% O_2 and 5% CO_2 via a nose cone) or an intraperitoneal injection of a mixture of midazolam, fentanyl, and medetomidine (5 mg, 0.05 mg and 0.5 mg/kg body weight, respectively) is used. The animal is then placed in a stereotaxic frame (David Kopf Instruments, USA) and on a feedback-controlled heating pad (TMP-5b, Supertech, Hungary). A 3 mm craniotomy is made with a pneumatic dental drill. The center of the craniotomy is placed over the right hemisphere, 2.5 mm lateral to the midline and 3 mm caudal to bregma above the primary visual cortex (V1). A sterile double-cover glass coverslip custom made from two cover glasses (Electron Microscopy Sciences), glued together with UV-curving optical glue (Norland Optical Adhesive), is gently laid over the dura mater and glued to the skull with cyanoacrylate glue ("super glue"), as described previously [44, 45]. Dental acrylic (Super-Bond C&B) is then applied across the exposed skull surface. A custom-made titanium head holder bar is embedded in the dental acrylic to secure the mouse to the stage for subsequent imaging. Postoperative care consists of daily intraperitoneal injection of Carprofen (0.5 mg/ml, 500 μl) for 5 days and subcutaneous Ringer lactate injection (0.1–0.15 ml) in case of dehydration. Virus injection is carried out immediately following the opening of the skull. A glass micropipette (tip diameter ≈10 μm) used for the injections is backfilled with 0.5 μl vector solution (≈6 × 10^{13} virus particles/ml). The injection is done slowly (at a rate of 20 nl/s for first 50 nl, followed by 2 nl/s for the remaining quantity) into the cortex (Micro4 MicroSyringe Pump controller and Nanoliter2010 Injector, WPI, USA), at a depth of 450–200 μm under the pia mater. For population imaging, we use AAV9.Syn.GCaMP6f.WPRE.SV40 or AAV9.Syn.Flex.GCaMP6f.WPRE.SV40 (in the case of Thy-1-Cre animals); both viruses are purchased from Penn Vector Core (PA, USA). For sparse labeling, we inject a 1:1 mixture of AAV9.Syn.Flex.GCaMP6f.WPRE.SV40 and AAV1.hSyn.Cre.WPRE.hGH diluted 10,000 times. With this vector combination, most cells will contain flexed GCaMP6f in high concentration, while Cre protein will be expressed only in a very small proportion of the cells. A single Cre molecule is sufficient for activating multiple flexed proteins; this way, the cells containing Cre vector will express GCaMP in a very high number, while all other cells will express none. With only very few cells labeled, dendritic signal is much more visible. Before the imaging sessions, mice are kept head-restrained in the dark under the 3D microscope for at least 1 hour to accommodate them to the setup.

For two-photon measurement and behavior experiments, mice are fixed in a single-dimension VR system (Femtonics Gramophone, Femtonics, Hungary), allowing a head-restrained mouse to respond to visual or other stimuli by controlling its speed of motion on a rotating disk. The Gramophone system can either be used for high-accuracy velocity recording in conjunction with a two-photon microscope, or as a control interface for behavioral training in a virtual linear maze.

2.3 Visual Stimulation

An LCD monitor (1280 × 768 pixel resolution) is placed 20 cm from the eye contralateral to the hemisphere being imaged (covering ~100° × 70° of the visual field). The monitor is used to display visual stimuli generated by a program written in MATLAB using the "Psychtoolbox" add-on package (http://psychtoolbox.org/). To prevent stray light from entering the objective, a black cover can be used. This light shield consists of a black silicone cylinder that can be slid onto the objective and fits perfectly to the head bar. The gap between the cranial window and the front lens of the objective is filled with ultrasound gel (Rextra Ltd., Hungary). Each trial of the visual stimulation is started by showing a black background with a single white bar appearing at the edge of the screen. After 14 s, a non-moving grating appears on the screen for 1 s, then the grating moves in a direction orthogonal to its orientation for 5 s (drifting speed 1 cycle per 1 s), is stopped for 1 s, and then disappears, leaving a black screen for an additional 1 s period. In our testing, we have used trials with eight different grating directions and an angular interval of 45°.

2.4 Data Processing

Most of the analyses, including video rearrangement, motion correction, calculation of a running average, and $\Delta F/F$ calculation, are performed with the built-in analysis tools in the acquisition software (MES, Femtonics Ltd., Hungary). Raw fluorescence data (F) recorded along surface elements in 3D are spatially normalized, and then projected onto a 2D plot by applying the equation: $\Delta F/F = (F\,(d_L, d_{tr1}, t) - F_0(d_L, d_{tr1}))/F_0(d_L, d_{tr1})$, where t denotes time, and d_L and d_{tr1} indicate the longitudinal and the transverse distance along the ribbon, respectively. For 3D projection, 3D rendering, and 3D hyperstack analysis, the open-source software ImageJ is used.

2.5 Motion Correction in 3D

Data resulting from the 3D ribbon scanning, multi-layer/multi-frame scanning, and chessboard scanning methods are stored in a 3D array as time series of 2D frames. The 2D frames are sectioned to smaller segments matching the AO drifts to form the basic unit of our motion correction method. We select the frame with the highest average intensity in the time series as a reference frame. Then we calculate the cross-correlation between each frame and bar and the corresponding bars of the reference frame to yield a set of displacement vectors in the data space. The displacement vector for each frame and for each bar is transformed to the Cartesian coordinate system of the sample, with knowledge of the scanning orientation for each bar. Noise bias is avoided by calculating the displacement vector of a frame as the median of the motion vectors of its bars. This common displacement vector of a single frame is transformed back to the data space. The resulting displacement vector for each bar in every frame is then used to shift the data of the bars using linear interpolation for subpixel precision. Gaps are filled with data from neighboring bars, whenever possible.

3   AO 3D Drift Scanning

The different surface and volume elements can be organized into different scanning patterns. Here we provide step-by-step demonstrations of six scanning methods (ribbon; snake; multi-3D line scanning; chessboard; multi-layer/multi-frame; snake scanning; and multi-cube scanning) which are the most useful during in vivo measurements (Fig. 5b).

3.1 3D Ribbon Scanning to Compensate for In Vivo Motion Artifacts

To achieve 3D ribbon scanning (Figs. 2, 5b.ii, and 6), the first step is to select guiding points along any cellular structure based on the z-stack taken at the beginning of the experiment (Fig. 5a). The second step is to fit a 3D trajectory to these guiding points using piecewise cubic Hermite interpolation. Two main strategies can be used to form ribbons along the selected 3D trajectory: we can generate drifts either parallel to the trajectory (longitudinal drifts), or orthogonal to the trajectory (transverse drifts) (Figs. 2 and 6). In both cases, the orientation shall maximize how parallel these surface elements lie to the plane of brain motion or to the nominal focal plane of the objective. The basic idea behind the latter is that the point spread function is elongated along the z-axis: fluorescence measurements are therefore less sensitive to motion along the z-axis.

To demonstrate 3D ribbon scanning, a small number of pyramidal neurons in layer 2/3 of the mouse primary visual cortex (V1) were labeled with a Ca^{2+} sensor – GCaMP6, using an Adeno-

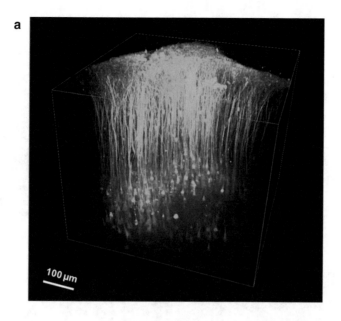

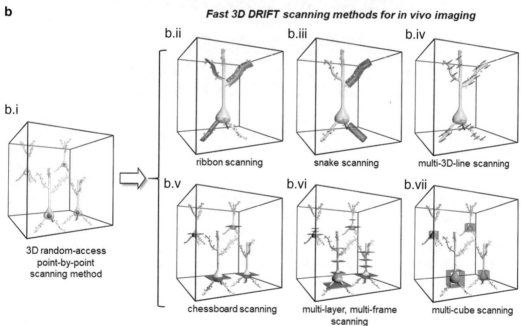

Fig. 5 Summary of fast 3D Drift AO scanning methods developed for in vivo imaging in behaving mice. (**a**) 650 × 650 × 600 μm block of the cortex labeled with GCaMP6f. Example *z*-stack is taken for guiding points selection. (**b**) Schematics comparing the currently used 3D random-access point-by-point scanning method (on the left) with the six novel 3D Drift AO scanning methods (on the right); elements illustrate the ROIs selected for measurements (**b.i**: 3D random-access point-by-point scanning method; **b.ii**: ribbon scanning; **b.iii**: snake scanning; **b.iv**: multi-3D-line scanning; **b.v**: chessboard scanning; **b.vi**: multi-layer, multi-frame scanning; **b.vii**: multi-cube scanning). (Adapted from Szalay et al. [43])

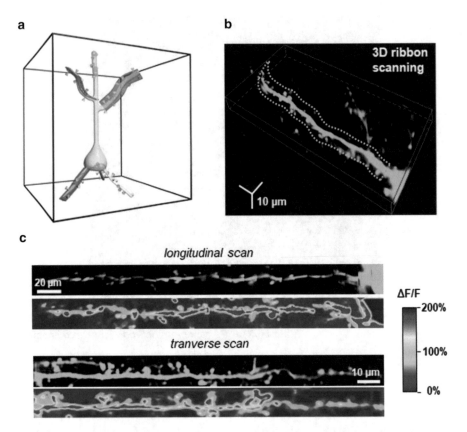

Fig. 6 Ribbon scanning method. (**a**) Schematic depiction of the ribbon scanning methods. (**b**) 3D image of a dendritic segment of a selected GCaMP6f-labeled pyramidal neuron (V1 layer2/3). A 3D ribbon (blue dashed lines) was selected for fast 3D Drift AO scanning within the red cuboid. (**c**) Fast 3D ribbon scanning was performed along the blue ribbon indicated in (**a**) using either the longitudinal (top) – 139.3 Hz – or the transversal (bottom), 70.1 Hz, scanning. Raw fluorescence data were measured along the selected 3D ribbon and were projected into 2D along the longitudinal and transversal axes of the ribbon following elimination of motion artifacts. Average Ca^{2+} responses along the ribbon during spontaneous activity are color-coded. $\Delta F/F$ calculation was performed with the built-in analysis tools in the acquisition software (MES, Femtonics Ltd., Budapest, Hungary). Raw fluorescence data (**c**) recorded along ribbons in 3D were spatially normalized and then projected onto a 2D plot by applying the formula for $\Delta F/F$; see the main text for more details. (Modified from Szalay et al. [43])

associated virus vector for delivery. Then, according to the z-stack taken in advance (Fig. 5a), guiding points were selected and fitted to the 3D trajectory which covered a spiny dendritic segment of a labeled pyramidal cell (Fig. 5b).

Transverse drifts were used to scan along the selected 3D ribbons to measure the selected 140 μm dendritic segment and spines at 70.1 Hz temporal resolution (Fig. 5c). Next, using longitudinal drifts allows a much faster measurement (between 139.3 and 417.9 Hz temporal resolution) of the same dendritic segment because fewer (but longer) 3D lines are required to cover the same ROI. In the next step, 3D recorded data were projected into 2D as a function of perpendicular and transverse distances along the surface of the ribbon. Note that in this way, the dendritic segment is straightened to a frame (Fig. 5c) to record its activity in 2D movies. This projection also allows the use of an adapted version of previous methods developed for motion artifact elimination in 2D scanning [41, 46].

3.1.1 Recording Dendritic Segments with Multiple 3D Ribbon Scanning

It has been reported that for many neurons, synaptic integration occurs not only at the axon initial segment, but also within the dendritic tree [23, 47–49]. Here, dendritic segments form nonlinear computational subunits which also interact with each other, for example, through local regenerative activities generated by nonlinear voltage-gated ion channels [21, 50, 51]. However, in many cases, the direct results of local dendritic computational events remain hidden in somatic recordings [23, 49, 52]. Therefore, to understand computation in neuronal networks, we also need novel methods for the simultaneous measurement from multiple dendritic segments (and spines). However, in vivo recording over large z-scanning ranges has remained an unsolved problem because the brain movement generated by heartbeat, breathing, or physical motion prevents the 3D measurement of these fine structures. Therefore, the novel 3D ribbon scanning methods could be a solution to simultaneously record the activity of multiple dendritic segments (Fig. 6) spanning larger z-ranges.

As in the 3D measurement of single dendritic segments, a reference z-stack needs to be taken in advance (Fig. 5a). The next step is to select guiding points in 3D along multiple dendritic segments and fit 3D trajectories. Finally, the 3D ribbons are extended to each of the selected dendritic segments (Figs. 6 and 7a, b). As shown in Fig. 2b, the surface of the ribbons must be adjusted such that they are parallel to the average motion vector of the brain, to minimize the effect of motion artifacts. In the next example described here, 12 dendritic segments were selected from a GCaMP6f-labeled V1 pyramidal neuron for fast 3D ribbon scanning (Fig. 7a, b). The 3D data recorded along each ribbon were 2D projected as a function of distance perpendicular to the trajectory and along the trajectory of the given ribbon. Then, these 2D projections of the dendritic segments were ordered as a function of their length and placed next to each other for better visualization. Note that, in this way, all the dendritic segments are straightened and visualized in parallel. In this way, we are able to transform and visualize 3D functional data as a standard calcium imaging video displayed in real time. The 2D projection used here allows fast motion artifact elimination and simplifies data storage, data visualization, and manual ROI selection (Fig. 7c).

Since each ribbon can be oriented differently in 3D space, the local coordinate system of measurements varies as a function of distance along a given ribbon, and between ribbons covering

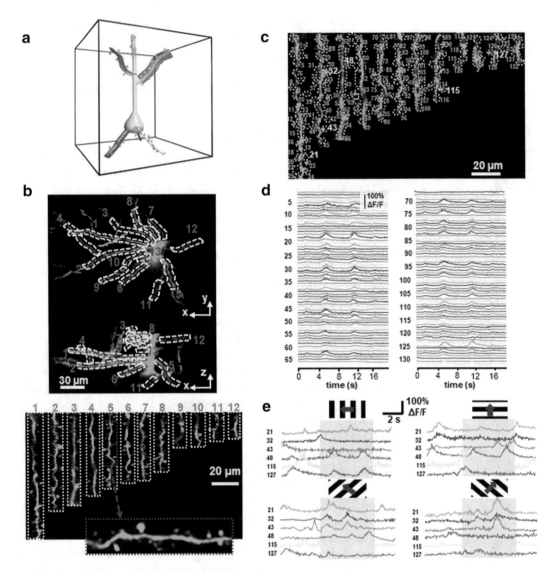

Fig. 7 Imaging of multiple spiny dendritic segments with 3D ribbon scanning in behaving animals. (**a**) Schematic of the dendritic ribbon measurement. (**b**) Maximum intensity projection in the x-y and x-z planes of a GCaMP6f-labeled layer 2/3 pyramidal neuron. Fluorescence was recorded simultaneously along the 12 dendritic regions measured. (**b**, bottom) Fluorescence data were projected into a 2D image as a function of the distance along the longitudinal and transverse directions of each ribbon. This transformation allowed the simultaneous recording, motion artifact elimination, and visualization of the activity of the 12 selected dendritic regions as a 2D movie. The image is a single frame from the movie recorded at 18.4 Hz. Red inset: enlarged view of dendritic spines shows the preserved two-photon resolution. (**c**) Numbers indicate 132 ROIs: dendritic segments and spines selected from the recording in (**b**). (**d**) Spontaneous Ca²⁺ transients derived from the 132 numbered regions indicated in (**c**). (**e**) Ca²⁺ transients from the six exemplified dendritic spines (white numbers on **c**). Transients were induced by moving grating stimulation. Time of moving grating stimulation in four example directions is indicated with a gray shading. (Modified from Szalay et al. [43])

different dendritic segments. The brain motion generates artifacts with different relative directions in each ribbon, so the 2D movement correction methods [46] cannot be directly used for the flattened 2D movie generated from ribbons. To resolve this issue, a motion compensation method is needed that takes into account not only the displacement calculated from the recorded videos, but also the position of a given segment in the original 3D space [43].

After motion correction, the activity from over 130 spines and dendritic regions could be analyzed in this experiment and revealed spontaneous and visual stimulation-induced activities (Fig. 7d and e, respectively).

3.2 Multi-layer, Multi-frame Imaging of Neuronal Networks and Chessboard Scanning

To avoid the loss of somatic fluorescence information during movement, each scanning point is extended to small squares or rectangles (Fig. 5b.v and b.vi). Two main strategies can be used to optimize the orientation of these recording areas for motion correction: (i) the squares can be set to be parallel to the direction of motion, or (ii) the squares can be set to be parallel to the nominal focal plane of the objective. This second strategy will be demonstrated in the following (Figs. 8 and 9). Similarly to 3D ribbon scanning, a 2D projection of the 3D data during multi-layer, multi-frame recording can be generated, even during image acquisition, by simply arranging all the squares, and hence each soma, into a "chessboard" pattern for better visualization and movie recording (Fig. 8a–c). As described for the 3D ribbon scanning, the average brain displacement vector (as a function of time) has to be calculated, and subsequently subtracted from all frames to correct motion artifacts. Finally, sub-regions from each 2D projection are selected for calculating the corresponding Ca^{2+} transients (Fig. 8d), similarly to the case of ribbon scanning. Chessboard scanning combines the advantage of low phototoxicity of low-power temporal oversampling [39] with the 3D scanning capability of AO microscopy. The AO microscope allows simultaneous imaging along multiple small frames placed in arbitrary locations in the scanning 3D space with speeds exceeding that of resonant scanning.

Naturally, the multi-layer, multi-frame scanning method is not limited to cell bodies. With this scanning method, many neurons with their dendritic (or axonal) arbor can be simultaneously imaged in several planes (Fig. 9a). To demonstrate this, four layers of neuropil – containing a GCaMP6f-labeled cell body and dendritic processes of a cortical pyramidal cell – were selected (Fig. 9b) and the activity in all layers was recorded simultaneously at a temporal resolution of ~100 Hz (Fig. 9c, d). Following motion artifact elimination, the maximal relative fluorescence changes were overlaid on the background fluorescence images (Fig. 9c). To show an alternative quantitative analysis, Ca^{2+} transients from some selected somatic and dendritic ROIs can also be calculated (Fig. 9d).

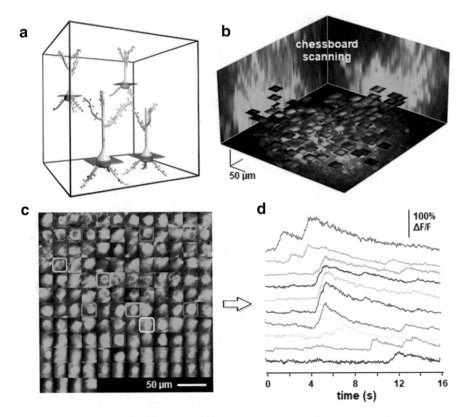

Fig. 8 Chessboard scanning of neuronal networks in behaving animals. (**a**) Schematic of chessboard scanning. Red squares represent the recorded areas around the somata. (**b**) Neurons from a mouse V1 region were labeled with GCaMP6f sensor. Neuronal somata and surrounding background areas (small horizontal frames) were selected according to a z-stack taken at the beginning of the measurement. Scale bars 50 μm. (**c**) Selected frames (from **b**) are "transformed" into a 2D "chessboard." Therefore, the activity can be recorded as a 2D movie. The image is a single frame from the video recording of 136 somata. (**d**) Representative somatic Ca²⁺ responses derived from the color-coded regions in (**c**) following motion artifact compensation. (Modified from Szalay et al. [43])

3.3 Volume Scanning with Multi-cube and Snake Scanning

Even though the brain moves along all three spatial dimensions during imaging, fluorescence information can still be preserved, and motion artifacts can also be effectively eliminated by scanning at reduced dimensions, along surface elements, in 3D. However, under some circumstances (e.g., in larger animal, certain surgeries, behavioral protocols, etc.), the amplitude of motion can be larger than the z dimension of the PSF and the missing third scanning dimension cannot be adequately compensated for. To sufficiently preserve fluorescence information even in these cases, the missing scanning z dimension may be regained by extending the surface elements to volume elements. To demonstrate this in two examples, we describe how 3D ribbons can be extended to folded cuboids – "snake scanning" (Fig. 5b.iii and 10a, b) and multi-frames to multi-cuboids – "multi-cube scanning" (Figs. 5b.vii and 11).

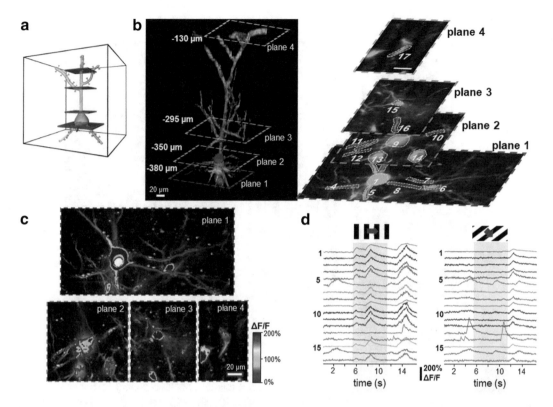

Fig. 9 Multi-layer imaging of a layer 2/3 pyramidal neuron and its dendritic arbors. (**a**) Schematic and 3D view of a layer 2/3 neuron labeled with the GCaMP6f sensor. Rectangles indicate four simultaneously imaged layers (plane 1–4). Numbers in μm indicate distances from the pia mater. Somata and neuronal processes of the three other labeled V1 neurons situated in the same scanning volume were removed from the z-stack for clarity. (**b**) Average baseline fluorescence in the four simultaneously measured layers. (**c**) The averaged baseline fluorescence images from (**b**) shown in gray (same color dashed lines for same planes) and overlaid with the color-coded relative Ca²⁺ changes ($\Delta F/F$). (**d**) Representative Ca²⁺ transients were derived from the numbered sub-regions shown in (**b**) following motion artifact elimination. Responses were induced by moving grating stimulation in two different directions at the temporal intervals indicated with gray shading. (Modified from Szalay et al. [43])

3.3.1 Snake Scanning

A spiny apical dendritic segment of a GCaMP6f-labeled layer 2/3 neuron was selected from a sparsely labeled region of a mouse primary visual cortex (V1) for snake scanning (Fig. 10b). Using the z-stack taken at the beginning, guiding points were selected, interpolated to a 3D trajectory, and used to generate a 3D ribbon which covered the dendritic segment. Next, this ribbon was extended to a volume, and 3D snake scanning from the selected folded cuboid was performed (Fig. 10). Dendritic Ca²⁺ responses were induced by visual stimulation with a moving grating and were projected into 2D, as a function of distance along the dendrite and distances along one of the perpendicular directions (Fig. 10c). Finally, maximum intensity projections of the data were generated along the second (and orthogonal) perpendicular axis to show average

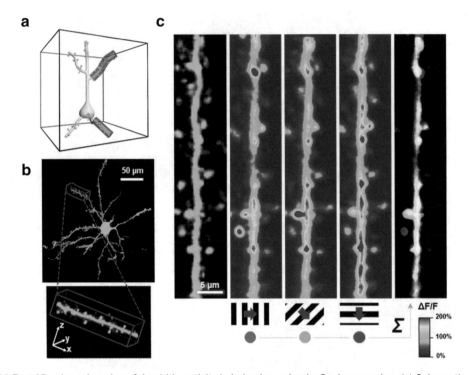

Fig. 10 Fast 3D volume imaging of dendritic activity in behaving animals: Snake scanning. (**a**) Schematic of the 3D measurement. 3D ribbons selected for 3D scanning can be extended to 3D volume elements (3D "snakes") to fully preserve fluorescence information during movement. (**b**) z-projection of a pyramidal neuron in the V1 region sparsely labeled with GCaMP6f. (**c**) Fast snake scanning was performed at ~10 Hz in the selected dendritic region shown in (**b**). Fluorescence data were projected as a function of the distance along the longitudinal and the transverse directions, then data were projected at maximum intensity along the second transverse direction. Left, average of 30 frames from the recorded video. Middle, peak Ca^{2+} responses following visual stimulation. Responses were induced by moving grating stimulation in three different directions. Right, relative fluorescence responses for each of the three different stimuli were transformed to either red, green, or blue, and summed: the result is shown as an RGB image. (Modified from Szalay et al. [43])

responses for three different stimulations (separately and together) following motion correction (Fig. 10c). Accordingly, with this volume scanning method, we can simultaneously record and distinguish between transients from neighboring spines, even when they are located in hidden and overlapping positions. Arguably, these key elements of the technique developed here are necessary to precisely understand dendritic computation.

3.3.2 Multi-cube Imaging

To demonstrate this second volume-scanning method, measurement points were extended from frames to small 3D cubes (Fig. 11a, b) around the cell bodies of GCaMP6-labeled cortical pyramidal cells. In this way, all somatic fluorescence points are preserved in case of motion. In the following example, simultaneous measurements of 10 somata were performed using relatively large cubes (each cube was between 46 × 14 × 15 voxels; 1

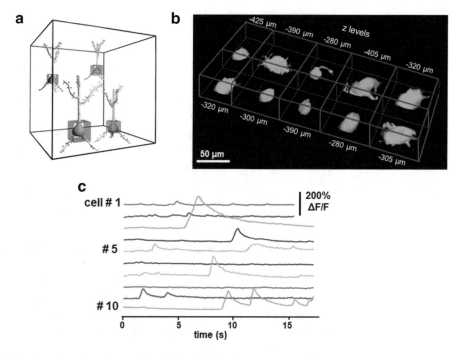

Fig. 11 3D volume imaging of somatic activity: Multi-cube scanning. (**a**) Schematic of the 3D multi-cube measurements. (**b**) Ten representative scanned cubes with individual neuronal somata for simultaneous 3D volume imaging (with their distance from the pial surface). (**c**) Volume imaging was performed simultaneously in the ten cubes (shown in panel **b**) with 8.2 Hz temporal resolution in GCaMP6-labeled pyramidal cells. Finally, following 3D motion correction, data (spontaneous Ca²⁺ transients) are shown as relative fluorescence changes ($\Delta F/F$). (Modified from Szalay et al. [43])

voxel = 1.5 μm × 3 μm × 4 μm) with a temporal resolution of ~8 Hz. This spatial and temporal resolution made it possible to resolve the sub-cellular Ca²⁺ dynamics. Similarly to multi-frame recordings, ROIs can be ordered next to each other for visualization (Fig. 11b). Spontaneous Ca²⁺ transients were derived from each cube using small sub-volumes (Fig. 11c). As described above, the net displacement vector was calculated to eliminate motion artifacts and sub-volume positions were corrected at each time point during calculation of Ca²⁺ transients. The use of volume scanning reduces the amplitude of motion artifacts in Ca²⁺ transients by ~19-fold during large-amplitude movements in awake and behaving mice [43].

3D volume scanning methods (snake scanning and multi-cube scanning) can be effectively used for 3D measurement of neuronal networks and spiny dendritic segments (distributed over the whole scanning volume); moreover, these methods are capable of completely eliminating even large-amplitude motion artifacts.

3.4 Multi-3D Line Scanning

In the previous part of this chapter, 1-dimensional scanning points were extended to 2D or 3D objects. In this section, we extend scanning points along only a single dimension to perform measurements at a considerably higher speed (Fig. 12). In many experimental conditions, sample movement is small or negligible, and brain motion can be approximated as movement along a single 3D trajectory (Fig. 12b). In this case, instead of multiple surface or volume elements, we can extend each 3D random-access point to multiple short 3D lines (Fig. 12a). As previously, points from the z-stack were selected. Next, the average trajectory of brain motion was calculated (Fig. 12b). In the third step, short lines were generated with 3D Drift AO scanning (each pre-selected point is the mid-point of the lines and the orientation of the lines was set to be parallel to the average trajectory of motion; Fig. 12a–c). In our example, the activity of 169 spines (Fig. 12c and e) were detected. If we switch back from the multi-3D line-scanning mode to the classical 3D point-by-point scanning mode, oscillations induced by heartbeat, breathing, and physical motion appear immediately in the recorded Ca^{2+} transients (Fig. 12f).

These data show that, in cases where the amplitude of the movement is small and can be approximated by a single 3D line or curve, we can effectively use multi-3D-line scanning to rapidly record over 160 dendritic spines in behaving animals.

4 Summary

With the optical developments detailed above (such as material dispersion compensation, motorized prism compressor, application of the angular dispersion compensation lens, optimization of AO deflectors for longer wavelength, increased transmittance efficiency, and stabilization of the laser beam's position), the optical resolution and transmittance efficiency of the system have been improved. With these improvements, imaging at longer wavelengths became possible that is required for imaging of genetically encoded indicators. Also, the field of view was extended to at least 500 μm × 500 μm × 650 μm when imaging with, e.g., the GCaMP6 sensor from behaving animals. This allows the measurement of hundreds of cells simultaneously, while the elimination of angular and material dispersion enables reliable somatic measurements to be made throughout the entire volume.

There are several ways to extend single scanning points to surface and volume elements with 3D Drift AO scanning as the combinations of different 3D lines. In this chapter, we have introduced six novel scanning methods for in vivo measurements in large scanning volumes: multi-3D line scanning, 3D ribbon scanning, snake scanning, multi-layer/multi-frame imaging, chessboard scanning, and multi-cube scanning (see Fig. 3). Each of these methods is optimal for a specific neurobiological goal. For example, the first three are optimal for dendritic measurements, while the last two are ideal for somatic recordings. On the other hand,

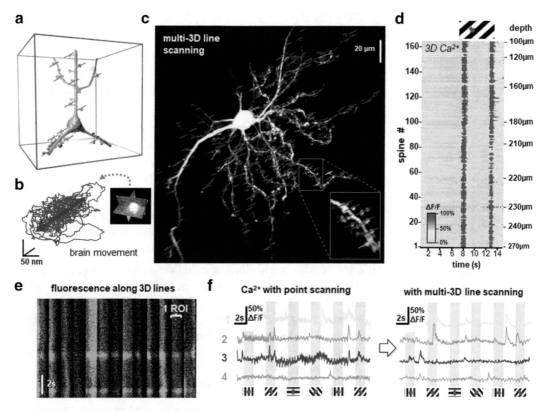

Fig. 12 Multi-3D line scanning of spine assemblies in behaving animals. (**a**) Schematic of the measurement. Each scanning line is associated with one spine. (**b**) Amplitude of brain motion (blue) was recorded by the fast 3D motion-detection method. Average motion direction is shown in red. The scale bars are 50 nm. (**c**) z-projection of a layer 2/3 pyramidal cell, labeled with GCaMP6f. Red lines indicate the scanning lines running through 164 pre-selected spines. All scanning lines were set to be parallel to the average motion shown in (**b**) (**d**) Corresponding 3D Ca^{2+} responses. (**e**) Exemplified individual raw Ca^{2+} transients recorded along 14 spines. Note the movement artifacts in the raw fluorescence. (**f**) Exemplified Ca^{2+} transients measured using point scanning (left) and multi-3D line scanning (right). Note the improvement in the SNR when multi-3D line scanning was used. Transients were induced with moving grating stimulation. (Modified from Szalay et al. [43])

the surface scanning methods are optimized for speed while the methods based on volume scanning are optimal when compensation for large amplitude movements is important. All of these scanning methods allow motion artifact correction on a fine spatial scale and, hence, the in vivo measurement of fine structures (e.g., dendrites, dendritic spines) in behaving animals. Therefore, we can preserve the fluorescence information from the 3D pre-selected ROIs even in the brain of behaving animals, while at the same time maintaining sampling rates between 10 Hz and 1 kHz that are necessary to resolve neural activity at the individual ROIs.

4.1 Benefits of the New 3D Scanning Methods in Neuroscience

In this chapter, we have demonstrated several further technical advances over previous 3D methods:

(i) They enable a scanning volume with GECIs which is more than two orders of magnitude larger than previous realizations, with preserved spatial resolution.

(ii) They offer a method of fast 3D scanning in any direction, with an arbitrary velocity, without any limitation in sampling rate during drifts.

(iii) They are free of mechanical and electrical inertias, which makes it possible to flexibly select surface and volume elements matching multiple somatic and dendritic locations, thereby effectively focusing measurement time to the preselected ROIs.

(iv) To preserve high spatial resolution, they compensate fast motion artifacts in 3D during 3D surface scanning and volume imaging even in behaving animals.

(v) They provide good SNR and enable generalization of the low-power temporal oversampling to reduce phototoxicity. When raster scanning of the entire cubature is replaced with 3D Drift AO scanning, the product $v_{gain} \times (SNR_{gain})^2$ is proportional to the ratio of the overall cubature versus the volume covered by the scanned ROIs, and this ratio can be larger than 10^6. Therefore, the flexibility in selection of ROIs results in an increase in speed and/or SNR of several orders of magnitude.

The limits of our understanding of neural processes now lie at the fast dendritic and neuronal activity patterns occurring in intact brain tissue (in 3D), and how they are integrated over larger network volumes. Until now, these aspects of neural circuit function have not been measured in awake, behaving animals. Our new 3D scanning methods, with preserved high spatial and temporal resolution, provide the missing tool for these activity measurements.

Among other advantages, it will be possible to use these methods to investigate spike-timing-dependent plasticity and the underlying mechanisms during in vivo conditions [52–55], the origin of dendritic regenerative activities [45, 47, 49, 50], the propagation of dendritic spikes [18, 27, 50], receptive field structures [12], dendritic computation between multiple spiny and aspiny dendritic segments [22, 50], spatiotemporal clustering of different input assemblies [18, 56], associative learning [57], multisensory integration [58], the spatial and temporal structure of the activity of spine, dendritic, and somatic assemblies [59–61], and function and interaction of sparsely distributed neuronal populations, such as parvalbumin-, somatostatin-, and vasoactive intestinal polypeptide-expressing neurons [62, 63].

These 3D scanning methods may also provide the key to understanding synchronization processes mediated by neuronal circuitry locally and on a larger scale: these are thought to be important in the integrative functions of the nervous system [64] or in different diseases [65]. Importantly, these complex functional questions can be addressed using these methods at the cellular and sub-cellular level, and simultaneously at multiple spiny (or aspiny) dendritic segments, and at the neuronal network level in behaving animals.

These technical achievements enabled the realization of the following fast 3D measurements and analysis methods in awake, behaving animals:

(i) Simultaneous functional recording of over 150 spines

(ii) Fast parallel imaging of activity in over 12 dendritic segments

(iii) Precise separation of fast signals in space and time from individual spines (and dendrites) from the recorded volume

(iv) Simultaneous imaging of large parts of the dendritic arbor and neuronal networks in a z-scanning range of over 650 μm

(v) Imaging a large neuronal network: over 100 neurons at subcellular resolution in a scanning volume of up to 500 μm × 500 μm × 650 μm. The SNR is more than an order of magnitude larger than for 3D random-access point scanning

(vi) More than 10× better single action potential resolution during motion in neuronal network measurements

4.2 Future Perspectives

(i) Deep brain imaging: with the advances both in available laser sources and acousto-optical deflectors, as well as with the appearance of novel red-shifted dyes, deeper scanning in the tissue will be possible. There will be no additional limitations of the depth of scanning compared to those already present in any other, conventional 2D system. This will hopefully enable imaging through the entire cortical thickness of a mouse brain.

(ii) Increased field of view: the combination of objectives that were developed for mesoscope, and 3D Drift AO scanning methods and new detection light paths may enable imaging from a 5 × 5 mm field of view with similar optical resolution and drift scan rate as with normal objectives.

(iii) Resolve complex cell networks and different cell types in multi-layer/3D neuronal ensembles in combination with novel red-shifted Ca^{2+} indicators and two-photon optogenetical tools.

(iv) Combining imaging with photo-stimulation: the AO scanhead can be combined with a second 3D scanning system or with a 2D galvanometric scanner to facilitate simultaneous 3D imaging and photo-stimulation at multiple points in the 3D volume. This may become a useful tool to map and understand neuronal networks in the functioning brain.

Acknowledgments

The authors thank Linda Judák for the in vivo recordings and Alexa Bojdán for contributing to the figures.

Conflict of Interest

Dr. Gergely Katona and Dr. Balázs Rózsa are founders of Femtonics Ltd. Balázs Rózsa is a member of its scientific advisory board.

References

1. Danilatos GD (1991) Review and outline of environmental SEM at present. J Microsc 162:391–402. https://doi.org/10.1111/j.1365-2818.1991.tb03149.x

2. Kerr JND, Denk W (2008) Imaging *in vivo*: watching the brain in action. Nat Rev Neurosci 9:195

3. Kherlopian AR, Song T, Duan Q et al (2008) A review of imaging techniques for systems biology. BMC Syst Biol 2:74

4. Svoboda K, Yasuda R (2006) Principles of two-photon excitation microscopy and its applications to neuroscience. Neuron 50:823–839

5. Lichtman JW, Magrassi L, Purves D (1987) Visualization of neuromuscular junctions over periods of several months in living mice. J Neurosci 7:1215–1222

6. Denk W, Strickler JH, Webb WW (1990) Two-photon laser scanning fluorescence microscopy. Science 248:73–76. https://doi.org/10.1126/science.2321027

7. Benninger RKP, Piston DW (2013) Two-photon excitation microscopy for the study of living cells and tissues. Curr Protoc Cell Biol Chapter 4:Unit 4.11.1-24. https://doi.org/10.1002/0471143030.cb0411s59

8. Helmchen F, Denk W (2005) Deep tissue two-photon microscopy. Nat Methods 2:932–940

9. Kobat D, Horton NG, Xu C (2011) *In vivo* two-photon microscopy to 1.6-mm depth in mouse cortex. J Biomed Opt 16:106014. https://doi.org/10.1117/1.3646209

10. Theer P, Hasan MT, Denk W (2003) Two-photon imaging to a depth of 1000 μm in living brains by use of a Ti:Al2O3 regenerative amplifier. Opt Lett 28:1022–1024. https://doi.org/10.1364/OL.28.001022

11. Stosiek C, Garaschuk O, Holthoff K, Konnerth A (2003) *In vivo* two-photon calcium imaging of neuronal networks. Proc Natl Acad Sci U S A 100:7319–7324. https://doi.org/10.1073/pnas.1232232100

12. Ohki K, Chung S, Ch YH et al (2005) Functional imaging with cellular resolution reveals precise micro- architecture in visual cortex. Nature 433:597–603. https://doi.org/10.1055/s-2004-818954

13. Peron SP, Freeman J, Iyer V et al (2015) A cellular resolution map of barrel cortex activity during tactile behavior. Neuron 86:783–799. https://doi.org/10.1016/j.neuron.2015.03.027

14. Dombeck DA, Khabbaz AN, Collman F et al (2007) Imaging large-scale neural activity with cellular resolution in awake, mobile mice. Neuron 56:43–57. https://doi.org/10.1016/j.neuron.2007.08.003

15. Miyawaki A (2011) Proteins on the move: insights gained from fluorescent protein technologies. Nat Rev Mol Cell Biol 12:656–668

16. Mostany R, Miquelajauregui A, Shtrahman M, Portera-Cailliau C (2014) Two-photon excitation microscopy and its applications in neuroscience. Methods Mol Biol 1251:25–42. https://doi.org/10.1007/978-1-4939-2080-8_2

17. Scanziani M, Häusser M (2009) Electrophysiology in the age of light. Nature 461:930–939

18. Katona G, Szalay G, Maák P et al (2012) Fast two-photon *in vivo* imaging with three-dimensional random-access scanning in large tissue volumes. Nat Methods 9:201–208. https://doi.org/10.1038/nmeth.1851

19. Khodagholy D, Gelinas JN, Thesen T et al (2015) NeuroGrid: recording action potentials from the surface of the brain. Nat Neurosci 18:310–315. https://doi.org/10.1038/nn.3905

20. Poirazi P, Brannon T, Mel BW (2003) Pyramidal neuron as two-layer neural net-

work. Neuron 37:989–999. https://doi.org/10.1016/S0896-6273(03)00149-1

21. Polsky A, Mel BW, Schiller J (2004) Computational subunits in thin dendrites of pyramidal cells. Nat Neurosci 7:621–627. https://doi.org/10.1038/nn1253

22. Losonczy A, Magee JC (2006) Integrative properties of radial oblique dendrites in hippocampal CA1 pyramidal neurons. Neuron 50:291–307. https://doi.org/10.1016/j.neuron.2006.03.016

23. Johnston D, Narayanan R (2008) Active dendrites: colorful wings of the mysterious butterflies. Trends Neurosci 31:309–316

24. Nikolenko V (2008) SLM microscopy: scanless two-photon imaging and photostimulation using spatial light modulators. Front Neural Circuits 2. https://doi.org/10.3389/neuro.04.005.2008

25. Grewe BF, Voigt FF, van 't Hoff M, Helmchen F (2011) Fast two-layer two-photon imaging of neuronal cell populations using an electrically tunable lens. Biomed Opt Express 2:2035. https://doi.org/10.1364/BOE.2.002035

26. Reddy GD, Saggau P (2005) Fast three-dimensional laser scanning scheme using acousto-optic deflectors. J Biomed Opt 10:064038. https://doi.org/10.1117/1.2141504

27. Fernández-Alfonso T, Nadella KMNS, Iacaruso MF et al (2014) Monitoring synaptic and neuronal activity in 3D with synthetic and genetic indicators using a compact acousto-optic lens two-photon microscope. J Neurosci Meth 222:69–81. https://doi.org/10.1016/j.jneumeth.2013.10.021

28. Cheng A, Gonçalves JT, Golshani P et al (2011) Simultaneous two-photon calcium imaging at different depths with spatiotemporal multiplexing. Nat Methods 8:139–142. https://doi.org/10.1038/nmeth.1552

29. Holekamp TF, Turaga D, Holy TE (2008) Fast three-dimensional fluorescence imaging of activity in neural populations by objective-coupled planar illumination microscopy. Neuron 57:661–672. https://doi.org/10.1016/j.neuron.2008.01.011

30. Botcherby EJ, Smith CW, Kohl MM et al (2012) Aberration-free three-dimensional multiphoton imaging of neuronal activity at kHz rates. Proc Natl Acad Sci USA 109:2919–2924. https://doi.org/10.1073/pnas.1111662109

31. Göbel W, Kampa BM, Helmchen F (2007) Imaging cellular network dynamics in three dimensions using fast 3D laser scanning. Nat Methods 4:73–79. https://doi.org/10.1038/nmeth989

32. Katona G, Kaszás A, Turi GF et al (2011) Roller Coaster Scanning reveals spontaneous triggering of dendritic spikes in CA1 interneurons. Proc Natl Acad Sci U S A 108:2148–2153. https://doi.org/10.1073/pnas.1009270108

33. Tomer R, Khairy K, Amat F, Keller PJ (2012) Quantitative high-speed imaging of entire developing embryos with simultaneous multiview light-sheet microscopy. Nat Methods 9:755–763

34. Wolf S, Supatto W, Debrégeas G et al (2015) Whole-brain functional imaging with two-photon light-sheet microscopy. Nat Methods 12:379–380

35. Rozsa B, Katona G, Vizi ES et al (2007) Random access three-dimensional two-photon microscopy. Appl Optics 46:1860–1865. https://doi.org/10.1364/AO.46.001860

36. Prevedel R, Yoon YG, Hoffmann M et al (2014) Simultaneous whole-animal 3D imaging of neuronal activity using light-field microscopy. Nat Methods 11:727–730. https://doi.org/10.1038/nmeth.2964

37. Kong L, Tang J, Little JP et al (2015) Continuous volumetric imaging via an optical phase-locked ultrasound lens. Nat Methods 12:759–762. https://doi.org/10.1038/nmeth.3476

38. Quirin S, Jackson J, Peterka DS, Yuste R (2014) Simultaneous imaging of neural activity in three dimensions. Front Neural Circuits 8. https://doi.org/10.3389/fncir.2014.00029

39. Cotton RJ, Froudarakis E, Storer P et al (2013) Three-dimensional mapping of microcircuit correlation structure. Front Neural Circuits 7. https://doi.org/10.3389/fncir.2013.00151

40. Kaplan A, Friedman N, Davidson N (2001) Acousto-optic lens with very fast focus scanning. Opt Lett 26:1078–1080. https://doi.org/10.1364/OL.26.001078

41. Greenberg DS, Kerr JND (2009) Automated correction of fast motion artifacts for two-photon imaging of awake animals. J Neurosci Meth 176:1–15. https://doi.org/10.1016/j.jneumeth.2008.08.020

42. Chen T-W, Wardill TJ, Sun Y et al (2013) Ultrasensitive fluorescent proteins for imaging neuronal activity. Nature 499:295–300. https://doi.org/10.1038/nature12354

43. Szalay G, Judák L, Katona G et al (2016) Fast 3D imaging of spine, dendritic, and neuronal assemblies in behaving animals. Neuron 92:723–738. https://doi.org/10.1016/j.neuron.2016.10.002

44. Goldey GJ, Roumis DK, Glickfeld LL et al (2014) Removable cranial windows for long-term imaging in awake mice. Nat Prot

9:2515–2538. https://doi.org/10.1038/nprot.2014.165

45. Mostany R, Portera-Cailliau C (2008) A craniotomy surgery procedure for chronic brain imaging. J Vis Exp. https://doi.org/10.3791/680

46. Kaifosh P, Zaremba JD, Danielson NB, Losonczy A (2014) SIMA: Python software for analysis of dynamic fluorescence imaging data. Front Neuroinform 8. https://doi.org/10.3389/fninf.2014.00080

47. Schiller J, Major G, Koester HJ, Schiller Y (2000) NMDA spikes in basal dendrites of cortical pyramidal neurons. Nature 404:285–289. https://doi.org/10.1038/35005094

48. Magee JC, Johnston D (2005) Plasticity of dendritic excitability. Curr Opin Neurobiol 15:334–342. https://doi.org/10.1016/j.conb.2005.05.013

49. Larkum ME, Nevian T, Sandier M et al (2009) Synaptic integration in tuft dendrites of layer 5 pyramidal neurons: a new unifying principle. Science 325:756–760. https://doi.org/10.1126/science.1171958

50. Chiovini B, Turi GF, Katona G et al (2014) Dendritic spikes induce ripples in parvalbumin interneurons during hippocampal sharp waves. Neuron 82:908–924. https://doi.org/10.1016/j.neuron.2014.04.004

51. Tran-Van-Minh A, Cazé RD, Abrahamsson T et al (2015) Contribution of sublinear and supralinear dendritic integration to neuronal computations. Front Cell Neurosci 9. https://doi.org/10.3389/fncel.2015.00067

52. Araya R (2014) Input transformation by dendritic spines of pyramidal neurons. Front Neuroanat 8. https://doi.org/10.3389/fnana.2014.00141

53. Bloodgood BL, Sabatini BL (2007) Ca(2+) signaling in dendritic spines. Curr Opin Neurobiol 17:345–351. doi: S0959-4388(07)00057-8 [pii]\r10.1016/j.conb.2007.04.003

54. Harvey CD, Yasuda R, Zhong H, Svoboda K (2008) The spread of Ras activity triggered by activation of a single dendritic spine. Science 321:136–140. https://doi.org/10.1126/science.1159675

55. Sjostrom PJ, Rancz EA, Roth A, Hausser M (2008) Dendritic excitability and synaptic plasticity. Physiol Rev 88:769–840. https://doi.org/10.1152/physrev.00016.2007

56. Larkum ME, Nevian T (2008) Synaptic clustering by dendritic signalling mechanisms. Curr Opin Neurobiol 18:321–331

57. Kastellakis G, Cai DJ, Mednick SC et al (2015) Synaptic clustering within dendrites: an emerging theory of memory formation. Prog Neurobiol 126:19–35

58. Olcese U, Iurilli G, Medini P (2013) Cellular and synaptic architecture of multisensory integration in the mouse neocortex. Neuron 79:579–593. https://doi.org/10.1016/j.neuron.2013.06.010

59. Ikegaya Y, Aaron G, Cossart R et al (2004) Synfire chains and cortical songs: temporal modules of cortical activity. Science 304:559–564. https://doi.org/10.1126/science.1093173

60. Takahashi N, Kitamura K, Matsuo N et al (2012) Locally synchronized synaptic inputs. Science 335:353–356. https://doi.org/10.1126/science.1210362

61. Villette V, Malvache A, Tressard T et al (2015) Internally recurring hippocampal sequences as a population template of spatiotemporal information. Neuron 88:357–366. https://doi.org/10.1016/j.neuron.2015.09.052

62. Klausberger T, Somogyi P (2008) Neuronal diversity and temporal dynamics: the unity of hippocampal circuit operations. Science 321:53–57

63. Kepecs A, Fishell G (2014) Interneuron cell types are fit to function. Nature 505:318–326

64. Womelsdorf T, Valiante TA, Sahin NT et al (2014) Dynamic circuit motifs underlying rhythmic gain control, gating and integration. Nat Neurosci 17:1031–1039

65. Engel J, Thompson PM, Stern JM et al (2013) Connectomics and epilepsy. Curr Opin Neurol 26:186–194

INDEX

Espen Hartveit (ed.), *Multiphoton Microscopy*, Neuromethods, vol. 148,
https://doi.org/10.1007/978-1-4939-9702-2, © Springer Science+Business Media, LLC, part of Springer Nature 2019

Printed in the United States
By Bookmasters